GAME THEORY AND CLIMATE CHANGE

PARKASH CHANDER

GAME THEORY AND CLIMATE CHANGE

COLUMBIA UNIVERSITY PRESS
NEW YORK

Columbia University Press
Publishers Since 1893
New York   Chichester, West Sussex
cup.columbia.edu
Copyright © 2018 Columbia University Press

Library of Congress Cataloging-in-Publication Data
Names: Chander, Parkash, 1947- author.
Title: Game theory and climate change / Parkash Chander.
Description: New York : Columbia University Press, [2018] |
Includes bibliographical references and index.
Identifiers: LCCN 2017035937 (print) | LCCN 2017039006 (ebook) |
ISBN 9780231184649 (cloth : alk. paper) | ISBN 9780231545594 (e-book)
Subjects: LCSH: Game theory. | Climatic changes—Economic aspects.
Classification: LCC HB144 (ebook) | LCC HB144 .C435 2018 (print) |
DDC 363.738/74015193—dc23

Columbia University Press books are printed on permanent
and durable acid-free paper.
Printed in the United States of America

Cover design and art: Lisa Hamm

*For Bayla and Anura, with the hope that this work
will lead to more effective climate change policies
and less climate change as they grow up*

CONTENTS

Preface xi

1. PURPOSE AND SCOPE 1

1.1. The Multidisciplinary Nature of Climate Change 2

2. THE BASIC FRAMEWORK 17

2.1. Environment as an Economic Good 18
2.2. Transfer Functions 21
2.3. The Basic Model 26

3. RATIONALE FOR COOPERATION 43

3.1. Equilibrium Concepts 44
3.2. Efficiency and Cooperation 49
3.3. The Rationale 54

4. THE CORE OF A STRATEGIC GAME 67

4.1. Strategic Games 69
4.2. Coalitional Functions 77
4.3. Partition Functions 88
4.4. Noncooperative Foundations of the γ-Core 93

5. ENVIRONMENTAL GAMES 103

5.1. The Environmental Game 104
5.2. The γ-Coalitional Function of the Environmental Game 107
5.3. Nonemptiness of the γ-Core 112
5.4. Some Specific γ-Core Imputations 115
5.5. The γ-Core and the Coase Theorem 124
5.6. The Seasonal Haze in Southeast Asia 131

6. COALITION FORMATION GAMES 135

6.1. Rationale for the γ-Core 137
6.2. Extensions to n-Player Games 148
6.3. Other Approaches to Coalition Formation 155
6.4. Concluding Remarks 161

7. DYNAMIC ENVIRONMENTAL GAMES 163

7.1. The Dynamic Model 165
7.2. The Dynamic Environmental Game 174
7.3. The Nash Equilibria of the Dynamic Environmental Game 175
7.4. The Subgame-Perfect Nash Equilibrium 182
7.5. The Subgame-Perfect Cooperative Agreements 190

8. LIMITS TO CLIMATE CHANGE 207

8.1. The Model 211
8.2. Pollution Rights and Climate Change 218
8.3. A Road Map for Stabilizing the Climate 223

9. THE JOURNEY FROM KYOTO TO PARIS 233

9.1. Main Features of the Kyoto Protocol 235
9.2. A Model for Interpreting the Kyoto Protocol 236
9.3. The Necessity of Transfers 241
9.4. The Paris Agreement and the Road Ahead 257
9.5. The Developing Countries and the Paris Agreement 268

10. INTERNATIONAL TRADE AND CLIMATE CHANGE 273

10.1. A Model with Two Consumption Goods 276
10.2. International Trade in Goods 277
10.3. International Trade in Both Goods and Emissions 284
10.4. Environmental Agreements and International Trade in Goods 287

CONCLUSION 291

References 299
Author Index 313
Subject Index 317

PREFACE

C limate change is an international environmental issue that has gained importance and public attention in recent years. From a theoretical perspective, tackling this issue is equivalent to solving the problem of providing a global public good that has some special characteristics. I address this issue in the book and pursue two complementary objectives. On the one hand, I propose a unified conceptual framework within which to think about climate change and other similar international issues, and, on the other hand, I present this framework in terms that lead to policies and applications.

The text is quite focused and by no means encyclopedic. It presents the material in a unified analytical framework, combining the economic, the ecological, and the game-theoretic components of the problem and thereby permitting the reader to see their respective roles in a focused but comprehensive study. As is well known, the market mechanism, in the context of international environmental issues, has to be replaced by some other allocation mechanism—negotiations and voluntary agreements, for example. Thus, the book introduces many new concepts and results that have not been reported previously.

As to the target readership, the text is meant for both academic researchers and practitioners and policymakers. It attempts to offer to the researchers a rigorous and complete exposition of what game theory can bring to this field. It offers an analytical framework for understanding why the Paris Agreement on climate change may succeed whereas the Kyoto Protocol failed. It begins with an introduction written for someone with little knowledge of game theory and leads to the state of the art in advanced game theory designed for an expert in game theory. It is a steep road to traverse, but the book, I believe, does it ably. For practitioners and policymakers, the book includes two chapters that specifically focus on policy.

The material in the book can be used for a full one-semester course for graduate students specializing in environmental economics and economics of climate change and/or in applied game theory. To be specific, chapters 2, 3, 5, and 9 can be useful for teaching environmental economics at an undergraduate level, and chapters 4–8 can be used for PhD-level courses. Thus, some readers may want to skip chapters 4–8 while others may want to skip chapters 2 and 3. Chapters 4–7 can also be used to supplement a public economics course at the upper undergraduate level. I used them as part of a fourth-year undergraduate course in public economics at the National University of Singapore (NUS). Chapters 9 and 11 focus on policy and are written in such a way that they can be read even with minimal knowledge of the preceding chapters in the book. Chapter 10 is also concerned with an important policy question in that it examines whether combining negotiations on climate change and free trade can lead to greater world welfare.

This work has been supported over the years by the Jindal School of Government and Public Policy at O.P. Jindal Global University and by the Department of Economics at NUS. I am also grateful to the Center for Operations Research and Econometrics (CORE) and the Beijer Institute for Ecological Economics for hosting me on numerous occasions during the early stages of this work. I appreciate the cheerful and patient encouragement of my colleagues Parimal Bag, Sudarshan Ramaswamy, Scott Barrett, Luc Leruth, Aditya Goenka, Euston Quah, and Myrna Wooders. I am also thankful to students in the public economics course at NUS who asked many critical questions, and especially Chen Li Ling, who also cheerfully helped me in preparing a number of diagrams for the book.

Last but not least, I wish to express my greatest thanks to my wife, Alka, for the part she took in motivating me toward completion of this work.

GAME THEORY AND CLIMATE CHANGE

1

PURPOSE AND SCOPE

Serious problems require serious policy responses, and no problem is more threatening to the human future than climate change, or global warming, caused by emissions of greenhouse gases (GHGs). Our understanding of the physical science of climate change has been growing steadily,[1] but there has been no meaningful progress toward fixing the economic and strategic forces that are causing it, except perhaps the Paris Agreement adopted by 196 countries in 2015 at Paris, France. This is often attributed to a failure of political will, but, as will be made clear in this book, it is the result of a particularly challenging economic and strategic setup. Accordingly, the book proposes game-theoretic solutions to the problem (which are efficient and immune to strategic behavior by sovereign nations) in the hope that it will lead to a better understanding of the economic and strategic forces behind climate change in general and of the Paris Agreement and its predecessor—the Kyoto Protocol—in particular.[2] It discusses why and how the Paris Agreement may succeed in controlling climate change whereas the Kyoto Protocol failed. However, applications of the theory developed in this book are not restricted to climate change

---

1. For a thorough account of the scientific evidence on the state of the problem, the reader is referred to the assessment reports issued by the Intergovernmental Panel on Climate Change (referenced in the bibliography as IPCC 1990, 1995, 2001, 2007, 2013). The negotiations themselves have been taking place under a body that was created by the General Assembly of the United Nations in 1992 and established in Geneva as the United Nations Framework Convention on Climate Change (UNFCCC).

2. See Barrett (2003) for a comprehensive study of a variety of actual international agreements.

alone. It is applicable also to other similar problems of global environmental externalities such as ozone layer depletion, acid rain, and sea and ocean pollution, to name a few.

1.1. THE MULTIDISCIPLINARY NATURE OF CLIMATE CHANGE

Climate change, which is caused mainly by emissions of GHGs, is a global environmental externality. It is an externality because it is a form of interaction between agents that takes place outside the market for exchange of goods and services, it is global because the externality extends beyond boundaries of any kind, and it is environmental because it affects people through the environment in which they live. Each one of these three characteristics of climate change puts it outside the scope of standard social sciences, especially standard economics. As a result, some basic concepts have to be modified and some new ones have to be invented to enable these sciences to cope with the issues involved.

1.1.1. The Three Disciplines

The choice of the title of this book puts game theory in the forefront because this discipline provides key intellectual tools to handle the problem. But more generally, three different disciplines are involved: ecological science, economics, and game theory. This section summarizes what the book has to offer regarding each one of them.

Ecological Science

Because the book deals with an ecological phenomenon, the reason for the involvement of this discipline should be obvious. Without pretense of contributing much new to this discipline, the book borrows from it whatever ideas are useful or relevant for the analysis. In particular, the main idea that is imported from this discipline is the concept of a *transfer function*, which is a tool whereby ecological science describes the effect of exogenous interventions on the measurable characteristics of the environment. How this tool operates is illustrated by an example of air pollution in chapter 2. In the rest of the book, only a simple linear form of the transfer function is used so that the complexity in this respect does not divert our attention away from the more important complexity of strategic interactions among the agents, which is the main focus.

Another idea that is borrowed from ecological science is the notion of a *stock* externality as distinct from a *flow* externality. Economics started to take note of this distinction only in the late 1980s when it could no longer ignore the ecological phenomena such as buildup of acidic depositions in soil due to acid rains or of accumulation of GHGs in the atmosphere due to their continual emissions on Earth. Before that time, externalities were treated just in terms of more or less bucolic examples of externalities, such as the celebrated bees and orchard fable and the locomotives that emit sparks and set fire to farmers' fields.

The treatment of GHGs as a stock externality in the economic analysis forces introduction of time in the formulation of the climate change problem. One implication of this is the emitters and recipients of the externality may be separated in time in such a way that they cannot meet or interact (e.g., they may belong to different non-overlapping generations),[3] making it impossible for them to engage in Coasian bargaining. Involvement of the government of each country, which can act on behalf of its future generations, is thus essential if climate change is indeed to be treated as a stock externality. Furthermore, ecological science also teaches us that phenomena such as climate change extend over a very long period. This has deep consequences for the choice of a discount rate that is appropriate for payoffs occurring only in a very distant future. Keeping these facts in view, chapters 7 and 8 treat climate change as a stock externality in contrast to its treatment as a flow externality in chapters 5 and 6.

At a more general level, introduction of economic reasoning in environmental matters definitely gives ecological science an anthropocentric perspective. That is so when economic analysis, whether positive or normative, is introduced in models in which human behavior affects a natural phenomenon. Indeed, a central concept in positive economics is that of an equilibrium, which is typically the outcome of utility or payoff

---

3. If elapse of time is substituted for river flow and distance, this is analogous to upstream pollution that affects people downstream.

maximization by agents representing the humans. This puts the natural phenomenon described by ecology as an object that is subject to human behavior and gives it essentially a human perspective. This is a fortiori so with normative economics, in which human behavior with respect to nature is advocated on the basis of criteria chosen and formulated by humans or agents representing them.

Moving one step higher in this philosophical discussion makes one realize that the anthropocentric view is the source, and the only source, of attributing value to environment.[4] Indeed, the notion of value itself is void if not supported by humans. As a natural corollary, the notion of pollution rights may be interpreted as an export from economics to ecological science. Because natural resources have value, they must have prices that are meaningful in a market economy. Moreover, prices mean exchange, and exchange presupposes property rights. In this way, the notion of pollution rights appears to be an export from economics to ecological science.

The theory developed in this book eventually leads to a fairly fundamental rethinking of the notion of value of environment, which is a matter of debate between "environmental economics" and "ecological economics."[5] Chapter 8 revisits this issue. It proposes a value that is free from the typical notion of scarcity in economics and from the exclusively physical thermodynamic theory of energy relied on in ecological economics. It is derived from a dynamic model in which each country is assumed to have developed so much that the capacity to produce consumption goods, whose production results in GHGs emissions and climate change, is no longer a binding constraint. Instead, the environment is the only limiting factor, and all countries have equal opportunity to exploit it. The countries differ, if at all, only in terms of how they are affected by their own emissions

---

4. This fact has been also recognized in the famous Brundtland Report (UN 1987). To quote, "The environment does not exist as a sphere separate from human actions, ambitions, and needs, and attempts to defend it in isolation from human concerns have given the very word 'environment' a connotation of naivety in some political circles." Similarly, Karp (2016) argues that an anthropocentric view, putting humans at the center of the narrative, can lead to more effective remedies, though the environment may have intrinsic value apart from any effect, however indirect, it has on current or future welfare.

5. For a summary presentation, see, for example, Kolstad (2000), especially chapters 1 and 3. Common and Stagl (2005) offer a comprehensive exposition.

and climate change. Consideration and analysis of such an idealized world model leads to a reconciliation of the two different views regarding the value of environment, as then it is optimal to minimize the polluting energy content of delivered goods and services. The analysis also leads to a proposal that can serve as a reference in the negotiations on climate change. Roughly speaking, the proposal means that a country affected relatively more by its own emissions and climate change should have fewer rights to emit GHGs that cause local pollution and climate change.[6] Such a proposal is free from any normative considerations such as equal per capita emission rights or grandfathering of current national emissions, but, as discussed in chapter 9, it may be considered unfair by some countries, especially by the low-lying island states and least developed small countries, which are affected by climate change but contribute little to it.

Economics

As to economics, known to be the science of resource allocation, the first extensions and novelties, introduced in chapter 2, are at the basic level of making precise whether and how the environment can be treated as an economic good. Connections are established between the standard notions of externalities, private goods, and public goods. In fact, all three are involved, and disentangling their respective roles in the climate change problem is a necessary starting point.

With this in mind, the basic framework for this work is presented in chapter 3 in terms of the two main classical strands of economic theory; namely, general equilibrium, a positive one, and welfare economics, a normative one. In both cases, the fact that the agents are sovereign nation-states is kept in mind, as this is the single most important fact underlying the climate change problem that distinguishes it from a standard externality problem. The proposed framework is also in the spirit of public economics, which recognizes the role of government in a market economy when widespread externalities are involved.

---

6. Combustion of fossil fuels often causes local pollution (e.g., smog) as well as climate change. A country is said to be more seriously affected by its own emissions and climate change if marginal damages from its own emissions and climate change are uniformly higher.

In 1960, Ronald Coase introduced a very powerful idea of great importance, and this idea has had arguably the single most important influence on policy for the problem of externalities in the past five decades. He argued that in the absence of any transaction costs, assignment of arbitrary but well-defined pollution rights and bargaining among the parties involved can internalize any externality and ensure efficiency. This was later dubbed "the Coase theorem." However, Coase took for granted the existence of a government or authority that can allocate and enforce pollution rights between the parties involved. He did not foresee the global problem of climate change where there is no such authority at the supranational level. Thus, the Coase theorem as such cannot be applied to the externality problem of climate change.[7] Furthermore, Coase did not consider stock externalities, which as noted above can make Coasian bargaining impossible among the current and future affected parties.

Chapters 5 and 8 propose suitable generalizations of the Coase theorem that are applicable in such instances of externalities that were originally ignored by Coase (1960). Chapter 5 treats greenhouse gases as a flow externality and shows that in the absence of a supranational authority, the nation-states will themselves assign rights to each other that are self-enforcing in the sense that none of the nation-states will have incentive to violate them. Furthermore, as the original Coase theorem predicts, such self-enforcing rights can lead to efficiency if they can be traded on a competitive market. However, unlike the assertion of the original Coase theorem, not every arbitrary assignment of rights can lead to efficiency, as every assignment of rights is not self-enforcing. Similarly, in the dynamic model in chapter 8, which, unlike chapter 5, treats GHGs as a stock externality, the steady-state Nash equilibrium emissions are interpreted as self-enforcing pollution rights that can lead to efficiency.

For the same reason for which the Coase theorem does not apply to climate change, the Pigouvian argument (Pigou 1920) that appropriate taxes

---

7. This limitation of the Coase theorem is not restricted to climate change alone. First, there are many other similar global externalities (e.g., depletion of the ozone layer). Second, in many parts of the world, as in the Middle Ages, authorities who could allocate and enforce pollution rights are missing. One may wonder whether efficiency can be achieved at all in such instances.

and subsidies can internalize externalities cannot be applied either. Thus, in the absence of a supranational authority, only voluntary negotiations and agreements among the sovereign countries can ensure efficiency, if at all. What should be those negotiations and agreements, and what should they bear on? That is broadly the theme pursued in this book.

Climate change has the characteristics of a global public good (or rather a bad). However, it does not fit the conventional notion of a public good. That is because reduction of greenhouse gas emissions by a country can be more than offset by another country, which may respond by increasing its own emissions. By contrast, in the conventional model of a public good, contributions to the public good by an agent cannot be neutralized by other agents. An agent can *at most* not contribute anything to the public good; it cannot reduce the public good provision made by others. Thus, the incentive to free ride in the case of climate change is much stronger than in a conventional public good model.[8] Dealing with it requires concepts and solutions that are more appropriate for tackling climate change, which, as noted above, provides stronger incentives to free ride. These are discussed later in the subsection on game theory.

Disentangling the private and public good aspects of climate change is an important step in understanding the virtues of the "cap-and-trade" mechanism, which was proposed in the Kyoto Protocol and has been retained in the Paris Agreement. In summary, the argument presented in chapters 5 and 9 is as follows: Decisions pertaining to "caps" are, in the aggregate, decisions on the provision of a public good, which explains and justifies that they be decided and agreed upon by the governments of the nation-states when negotiating an aggregate emission target. By contrast, decisions to "trade" in emissions pertain to private goods, namely, the pollution permits, and this explains and justifies that they be left to markets. The economic analysis in chapters 5 and 9 shows that these two types of decision-making processes are complementary to each other and together form a mechanism that can lead to a solution of the climate change problem.

---

8. See, for example, Groves and Ledyard (1977) and Chander (1993) for solutions to the problem of incentives in a conventional public good model.

Beside negotiations on climate change, countries are also currently engaged in negotiations on trade liberalization through the World Trade Organization (WTO). Though the negotiations on these two international problems have been conducted independently of each other, it is often claimed that increased free trade in goods will aggravate the climate change problem. That is because the scale of economic activities then will be higher and production of pollution-intensive goods will shift from countries with stricter pollution controls to those with lax ones. Chapter 10 considers this issue and explores a model with trade in consumption goods and pollution permits to determine whether merging negotiations on climate change and free trade could have led to a "better" agreement on climate change and improved world welfare. It derives sufficient conditions under which merging the negotiations would not have led to additional gains in welfare.[9] It also shows that the claim that more free international trade in goods and services will aggravate climate change and shift the production of the "carbon-intensive" goods from countries with strict environmental regulations toward those with lax ones is questionable.

Game Theory

Then, whither game theory? Games are essentially mathematical objects, to which it has become apparently fashionable in economics to have recourse.[10] The use of game theory in this book, however, finds its roots less in fashion than in realizing that economics per se does not offer conceptual tools that are rich enough for dealing with the two most essential aspects of climate change; namely, (1) the absence of a supranational authority that can enforce its policy decisions on the nation-states, and (2) the externality has public good characteristics—though it does not fit the *classic* notion of a public good.

---

9. Indeed, the Paris Agreement on climate change was concluded on December 12, 2015, without waiting for the outcome of the WTO Ministerial Conference, which started shortly thereafter on December 16, 2015.

10. My basic references on game theory are Osborne and Rubinstein (1994), Myerson (1991), and Mas-Colell, Whinston, and Green (1995).

Thus, the decision-making processes involved in tackling climate change and other global environmental externalities are neither of the same nature as those operating in designing the usual domestic macroeconomic policies (where an institution empowered by the nation-state makes decisions and can use its authority to implement them) nor are they akin to international trade negotiations on tariffs or opening of markets (where only private goods are involved). What is required is the optimal control of a global externality with public good characteristics. Thus, incentive mechanisms for optimal provision of a public good—Clarke (1971), Groves and Ledyard (1977), and others—may seem appropriate for tackling the problem. But these mechanisms begin by assuming that the power to design and implement the mechanism has been handed over to an independent authority, presumably with the agreement of all agents. It does not consider that the agents may not have incentives to hand over such power to an independent authority. Thus, the incentive mechanisms for public goods, though extremely important and of an independent interest, are not consistent with voluntary negotiations among sovereign countries.

Newer concepts are needed to understand the nature of interactions that may take place between sovereign countries, each pursuing its own interests, and the kind of outcomes those interactions may lead to. Classical economics does not provide enough of them,[11] whereas game theory offers many ways to describe and understand a wide variety of interactions among independent agents. This book makes no encyclopedic review of game theory. Instead, it innovates as well as selectively uses those existing concepts that are appropriate for the game formulations of climate change studied in this book. They are dealt with in full detail with the aim of showing how deeply and rigorously they can solve the problem, and those that are not directly relevant are left aside deliberately and without apology.

---

11. That is due to several reasons, but mainly because resource scarcity is at the root of the discipline (there is no economic, i.e., resource allocation, problem if there is no scarcity), whereas the object of game theory is of wider scope, namely, all forms of interactive decision making, though resource scarcity may also be a concern of game theory.

Games can be analyzed in two ways: noncooperative analysis of games focuses on the strategies that each player would undertake to maximize his own payoff and the subsequent equilibrium that would be reached when all players do so, while cooperative analysis of games focuses on how incentives can be designed such that the players will adopt the strategies to achieve the outcome with the maximum total payoff.

In the case of the climate change problem, there is definite room for applying cooperative game analysis, as the sum of benefits (i.e., damage prevented) from controlling climate change outweighs the total cost of controlling climate change. However, though the total benefits exceed the total cost of controlling climate change, the benefits and costs may not be spread out evenly among the countries in that the costs may be higher than the benefits for some countries unless the countries are identical. Because in reality the countries are indeed not identical, not every country would be willing to engage in controlling climate change.

To incentivize every country to engage in controlling climate change, a solution offered by cooperative game theory is that of side payments. If the premise that the sum of benefits from controlling climate change exceeds the costs of controlling climate change is indeed correct, then it might be possible to distribute the social surplus that will be generated such that after the side payments, every country is better off. Designing such side payments is a trying task, and this is where the book becomes fairly technical.

Both types of game analysis are involved in this work. A general strategic game provides a natural primitive framework for analyzing an interactive decision problem such as climate change, in which decisions of the agents (nation-states) affect each other. Accordingly, a central concept that is used throughout the book is that of the core of a strategic game, which, as will be shown, can be interpreted both as a cooperative and a noncooperative game theory solution concept. More specifically, the standard approach in classical cooperative game theory is to convert a strategic game into a coalitional game and analyze the core of the game so defined. The core of a strategic game in this book, called the γ-core, is defined in the same way except that the conversion is not standard. More specifically, the *worth* of a coalition is defined as equal to its Nash equilibrium payoff in an induced game in which the coalition acts as one single player and all other players act individually as in a Nash equilibrium of the original game. In other words, the worth of a coalition is the payoff that the coalition can obtain without cooperation *from* and *among* the rest of players.

Chapter 4 presents the γ-core and related concepts in the primitive framework of a *general* strategic game, and these concepts are then applied to different game formulations of climate change in the rest of the book. More specifically, chapter 5 introduces a strategic game formulation of climate change, called the environmental game, which corresponds to the short-run model that treats greenhouse gases as a flow externality. Chapter 7 introduces a dynamic game formulation of climate change in discrete time, called the dynamic environmental game. This dynamic game consists of a sequence of linked strategic games. Chapter 8 introduces a dynamic game formulation of climate change in continuous time. Chapter 10 introduces a trade and climate change model and interprets and analyzes it as a game.

The motivation for the γ-core concept comes from the fact that the classical α- and β-cores are not suitable solution concepts for the environmental game. They result in too large a set of outcomes; in fact, the entire set of efficient outcomes. This is because under these core concepts, the players outside a deviating coalition can more than offset the pollution abatement carried out by the coalition. Thus, there is no minimum payoff that a coalition can ensure for itself by reducing its own pollution. In contrast, the γ-core concept is based on a more plausible assumption concerning the behavior of the players outside a deviating coalition and results in a smaller set of outcomes.

More specifically, the γ-coalitional function is such that $w^{\alpha}(S) \leq w^{\beta}(S) \leq w^{\gamma}(S)$ for all $S \subset N$, where w^{α}, w^{β}, and w^{γ} are the coalitional functions derived from a general strategic game under the behavioral assumptions implicit in the α-, β-, and γ-core concepts, respectively. Thus, the γ-core is a stronger concept than the conventional α- and β-cores. Furthermore, for the class of strategic games that admit a *unique* strong Nash equilibrium, the γ-core consists of the *unique* imputation in which the payoffs of the players are equal to their strong Nash equilibrium payoffs. However, the γ-core may be nonempty even if the game admits no strong Nash equilibrium. Thus, for a class of strategic games, the γ-core is a weaker concept than that of a strong Nash equilibrium.

Chapter 4 also proposes, but does not pursue, an alternative γ-core concept. More specifically, the alternative γ-core is defined analogously except that the players in deviating coalitions are not capable of writing binding agreements, as in the concept of a coalition-proof Nash equilibrium. This alternative concept is based entirely on the noncooperative approach to game theory. But it is a weaker concept in the sense that the payoffs that deviating coalitions can achieve if they cannot write binding agreements are generally lower. For this reason, the alternative formulation of the concept is introduced and contrasted, but not applied or pursued further.

A growing branch of literature seeks to unify cooperative and noncooperative approaches to game theory through underpinning cooperative game-theoretic solutions with noncooperative equilibria, the "Nash program" for cooperative games, initiated by Nash (1953).[12] In the same vein, I show that the γ-core payoff vectors (a cooperative solution concept) of a general strategic game can be supported as equilibrium payoff vectors of a noncooperative game, and the grand coalition is the unique equilibrium outcome if the γ-core is nonempty and the game is "partially" superadditive. In this way, the book also contributes to the Nash program by rationalizing the γ-core as an equilibrium outcome of a noncooperative game. In fact, the book makes significant progress in reconciling cooperative and noncooperative game theory, and, therefore, it might be of interest to noncooperative game theorists as well—even though its contribution may seem to be primarily in the realm of cooperative game theory. To be specific, the book includes applications and contextual characterizations of the Nash equilibrium, the strong Nash equilibrium, the coalition-proof Nash equilibrium, the subgame-perfect Nash equilibrium, the closed-loop and open-loop Nash equilibria, and subgame-perfect stationary strategy equilibria of an infinitely repeated game. It also introduces a newer concept called a subgame-perfect agreement that combines

---

12. Analogous to the micro-foundations of macroeconomics, which aim at bridging the gap between the two branches of economic theory, the Nash program seeks to unify the cooperative and noncooperative approaches to game theory. Numerous papers have contributed to this program, including Rubinstein (1982), Perry and Reny (1994), Pérez-Castrillo (1994), and Lehrer and Scarsini (2013), for example.

the two concepts of the γ-core (a cooperative solution concept) and the subgame-perfect Nash equilibrium (a noncooperative solution concept).

The noncooperative game in which the γ-core payoff vectors are equilibrium outcomes is intuitive and explicitly models the process by which the players may agree to form a partition when they know in advance what their payoffs will be in each partition. It consists of infinitely repeated two-stages.[13] In the first stage of the two-stages, which begins with the finest partition as the status quo, each player announces whether it wishes to stay alone or form a nontrivial partition. In the second stage of the two-stages, the players form a partition as per their announcements. The two-stages are repeated if the outcome of the second stage of the two-stages is the finest partition as in the status quo from which the game began in the first place. It is shown that if the partition function representation of the strategic game is "partially superadditive," then breaking apart into singletons upon deviation by a coalition, as the γ-core concept requires, is a subgame-perfect equilibrium strategy of the remaining players.

The cooperative game theory is largely devoted to understanding how a group of agents may share the benefit of forming the grand coalition. However, there is another more recent strand of literature that aims at understanding whether agents will have incentives to form the grand coalition at all (i.e., whether they will *actually* decide to form the grand coalition). In the presence of externalities, a player can derive benefits from the activities of a coalition without joining it (i.e., free ride). Therefore, the grand coalition may not form. Several coalition formation games have been proposed in the recent literature (see, e.g., Ray and Vohra 1997, among others). The economic applications of these games, however, have focused primarily on strategic or partition function games with *identical* players. Though closely related to the approach in this book and extremely important and of independent interest, the results from this strand of literature cannot be applied to study climate change. To be applicable, a theory must necessarily deal with heterogeneous players—developed versus developing or low-lying small island states versus large countries.

---

13. If the game is limited to a single play of the two stages, then, as shown by Ray and Vohra (1997) and Yi (1997), the grand coalition is not an equilibrium outcome.

Furthermore, coalition formation theory has not been so far extended to dynamic games. The theory developed in this book applies to both strategic and dynamic games with heterogeneous players.

Chapter 7, as mentioned earlier, treats greenhouse gases as a stock externality and formulates the problem of climate change as a dynamic game of finite or infinite time horizon, called the dynamic environmental game. An important new issue that arises in the context of the dynamic game formulation of climate change is that of subgame perfection of an agreement. To be precise, an agreement is subgame-perfect if no coalition of countries has incentive to withdraw from it in *any* subgame.[14] I prove existence and comprehensively characterize a subgame-perfect agreement. From a methodological point of view, the concept of a subgame-perfect agreement brings together two of the most important solution concepts in game theory, namely, the subgame-perfect Nash equilibrium of a noncooperative game and the core of a cooperative game—a link between the two is apparently missing in the extant literature.

Chapter 9 interprets the cap-and-trade mechanism that was proposed in the Kyoto Protocol and has been retained in the Paris Agreement as a blend of economics and game theory. In addition to what was said earlier about the disentangling of the public and private good aspects of climate change and about the extensions of the Coase theorem to global flow and stock externalities, it is shown that as far as emissions under a cap-and-trade scheme are concerned, the market equilibrium of pollution permits, when it is competitive, belongs to the core of an appropriately formulated market game. Thus, the two disciplines of economics and game theory when put together contribute, each in its own way, to the understanding of the cap-and-trade mechanism proposed in the Kyoto Protocol and retained in the Paris Agreement.

---

14. See Chander (2017a) for motivating the requirement of subgame-perfection of an agreement. Becker and Chakrabarti (1995) show that the need for subgame-perfection of an agreement can also arise in dynamic games without externalities and, thus, propose the recursive core as the set of allocations such that no coalition can improve upon its consumption stream in any subgame.

1.1.2. The Common Methodology and Climate Policy

The common language of mathematical modeling is prevalent throughout the book. The main justification for it lies in the fact that this methodology allows us to make direct connections between the three disciplines involved, which is a necessary condition if the aim is to contribute new results rather than to only describe problems and known solutions.

There can be at least two kinds of mathematical models: those used to derive theoretical results (these are abstract models) and those used for computing actual policy (these are simulation models). The book deals mostly with the former, because the current state of the field requires that an elaborate theoretical background be made available to provide a solid foundation to the highly desirable computational work for the problem of climate change. Allusions to the latter are indeed made throughout the book to serve as illustrations.

Policy considerations may be seen as the ultimate goal of a work of applied theory like this book. The book indeed approaches policy considerations to the extent that an actual policy can be enlightened by theory or a policy recommendation can be derived from theory. This is done specifically in chapters 8 and 9, which study the two international agreements on climate change: the 2015 Paris Agreement and the failed 1997 Kyoto Protocol. In these chapters, I discuss why the Paris Agreement may succeed whereas the Kyoto Protocol failed. More specifically, chapter 8 interprets the Paris Agreement in terms of a differential game model and demonstrates how the theory can be applied to assess and propose a road map for its successful implementation. It shows that development and adoption of cleaner technologies is the key to the success of the Paris Agreement and recommends that development of cleaner technologies should not be left to market forces alone. In parallel to the Paris Agreement and the International Solar Alliance of 121 countries initiated by the proactive Prime Minister Narendra Modi of India, a global agreement to fund research for development of cleaner technologies is needed. Such a fund to which governments of all countries as well as private individuals and entities may contribute can boost the efforts to develop and adopt cleaner technologies and overcome intellectual property rights barriers.

One broad conclusion concerning climate policy that follows from this work is that "direct" or "indirect" financial transfers to balance the costs and benefits of controlling climate change are a necessity and not a matter of approach or choice. In the absence of transfers, sovereign and peaceful countries, unless they are identical, cannot be induced to take actions that are necessary for controlling climate change.[15]

A conclusion is drawn in the final chapter of this book. It summarizes both the theory and the policy implications of this work. It also discusses problems that have been left open and that may be addressed in future work. Some readers may find it useful to go through the concluding chapter before proceeding to read the rest of the text.

---

15. However, many preceding studies, for reasons of tractability, get around the necessity of transfers by assuming identical countries such that the costs and benefits of controlling climate change can be balanced without transfers. In contrast, transfers are implicit in Coase's (1960) classical solution for tackling externalities, which requires direct transfers between the parties involved for reducing an externality. As shown in Ellerman and Decaux (1998) and Chander (2003), transfers were also implicit in the Kyoto Protocol via the Clean Development Mechanism and the assignment of emission quotas that could be traded on an international market. The Montreal Protocol, which has been hailed as an example of successful international cooperation, explicitly requires transfers, though not in as large amounts as those implicit in the Kyoto Protocol.

2

THE BASIC FRAMEWORK

The game-theoretic approach proposed and applied in this book cannot be developed from game theory alone. Game theory is indeed only a framework within which to understand phenomena that occur when decision makers interact.[1] What those phenomena are and what the objects of the decisions are must come from other sciences. Among them, economics and ecological science provide appropriate descriptions of the objects and the phenomena we are concerned with in this book.

This and the next chapter are devoted to laying down the basic framework within which my game-theoretic approach is developed and applied in the rest of the book. Only elements of economics and ecological science are involved at this stage, leading to construction of models that provide an integrated understanding of the realities involved, so that when game theory considerations are brought into the picture—in chapter 5—they can be applied to well-defined objects, lending themselves to decision making.

As far as the economics part is concerned, the framework I choose comes from two main branches of economic theory: general equilibrium and welfare economics.[2] In this chapter, I examine how environmental problems can be handled—if at all—using the descriptive tools with which these two theories operate.

---

1. Therefore, it would be better named "interactive decision theory" as suggested by Aumann (1987). However, Aumann does not list environmental problems among the applications of game theory, perhaps because at that time they were not considered important enough.

2. See, for example, Mas-Colell, Whinston, and Green (1995), among others.

2.1. ENVIRONMENT AS AN ECONOMIC GOOD

Environmental pollution has been recognized in the economics literature for a long time. Most often, it appears under the general heading of *externalities*, a concept originally of broader scope than that of environmental pollution.[3] Yet the currently accepted textbook definition of externalities serves my purpose fairly well[4]: externalities are interactions among economic agents that affect the consumption sets, preferences, and/or production sets of the agents and do not occur through markets and prices.

Indeed, for environmental pollution to be dealt with in the framework of general equilibrium theory, as announced above, this definition of externalities allows me to describe them in terms that are compatible with the language and concepts of this theory. Among them, the notion of "commodity" and specification of the "commodity space" are basic. Externalities were not included in the specification of the commodity space when it was first introduced in Arrow and Debreu (1954) and were explicitly excluded in Debreu's (1959) description of the commodity space. But a suggestion by Arrow (1970) corrected for that and opened the way to make externalities amenable to general equilibrium analysis. The novelty of Arrow's suggestion was in the way he proposed to treat externalities, namely, to treat them *as commodities* in the same sense as ordinary commodities were defined in 1954, and thus to simply include externalities as additional dimensions of the commodity space.[5]

---

3. I deliberately limit myself to those forms of externalities that are relevant for my purpose in this book. Notice that the definition about to be given excludes "pecuniary" externalities.

4. See, for instance, Mas-Colell, Whinston, and Green (1995). Much earlier, Baumol and Oates (1975) proposed almost the same definition. Before that, lengthy debates and controversies took place for decades around the expressions "external economies and diseconomies," starting with Sidgwick (1883) and continuing throughout the twentieth century with Marshall, Pigou, Meade, Buchanan, and finally Coase (1960). Laffont (1977b) presents an extremely general view of externalities, while Cornes and Sandler (1996) also restrict the debate to what is said in the above definition.

5. As well explained in Cornes and Sandler (1996), the purpose of this inclusion was, in Arrow's mind, to provide a logical explanation for the absence of markets for externalities. The argument was that for such "commodities" *competitive* markets cannot operate.

2.1.1. Externalities as Two-Dimensional Commodities

Upon close scrutiny, Arrow's commodity called "externality" has a special characteristic in that it is two-dimensional in nature. One dimension refers to the generation of the externality (what is generated and by whom), while the other dimension refers to the reception of it (what is received, or perceived to be received, and by whom).[6] In the vocabulary of environmental pollution, this corresponds, respectively, to *emissions* as they accompany economic activities and *ambient pollutants* as they are present in the environment and hence inflicted on those who are exposed to them. This two-dimensional view may be seen as an extension of the early tools of analysis of externalities, an extension specifically due to the environmental applications that have burgeoned over the past six decades.

The justification for the treatment of an externality as a two-dimensional economic good is not only that it has descriptive merit, but more importantly it also points to two quite distinct roles of the agents involved: on the one hand, emissions are an unintentional by-product (often a joint product) of specific actions taken by an agent or agents,[7] and, on the other hand, reception of pollutants by an agent or agents may not be due to actions taken by them but may occur passively and often unwillingly.[8]

To reinforce the justification for the distinction made above, let me mention that when I describe my models in more detail later, it will appear that emissions are of the same nature as what are called *private* goods in economic theory, whereas ambient pollution most often has characteristics of a *public* good, local or global. Indeed, an ambient pollutant may affect just one,[9] or a few, or many agents simultaneously, possibly all individuals in the world and even those who are yet to be born! Thus, externalities often have the "public good" characteristic of "nonrivalry" (a term introduced by Musgrave 1969): the fact that some agents

---

6. As illustrated by an upstream-downstream example of pollution in Kolstad (2000, chapter 9).
7. By contrast, abatement/mitigation (i.e., reducing emissions) is most often intentional on the part of an agent or agents.
8. I ignore momentarily the fact that recipients may decide to act to prevent being polluted or adapt to it. But those are actions that anticipate or follow exposure to pollutants.
9. For the case of just one recipient, Arrow (1970) speaks of "personalized" externalities.

are affected by it does not modify the amount by which other agents are affected by it. Actually, the term "public bad" would be a more appropriate description of the detrimental character of ambient pollution. "Good," however, applies well to abatement/mitigation. In this sense, cleanup activities that reduce ambient pollution are equivalent to the production of a public good.

2.1.2. Reciprocal Externalities

A further aspect of environmental externalities that is highly relevant in our context is how the agents are involved. The relation between generators and recipients of an externality may be *unilateral* (as in the case of an upstream-downstream externality), but it may also be *reciprocal*, as in the case of factories, each of which draws water from a lake and discharges impure water into the same lake—each factory requires pure water (Shapley and Shubik 1969).

2.1.3. Global Externalities

The size of the geographic area where an externality is exerted is another aspect to be considered, from two points of view: physical and institutional. While physical considerations determine the size of the area affected by an ambient pollutant, institutional factors come into play when the area is structured by boundaries of, say, municipalities, provinces, countries, or regions: the externality then becomes "transboundary" or "transnational." *Local* externalities are of the same nature as local public goods, known as "club" goods since Buchanan (1965) introduced them; at the other extreme, *global* externalities are those pollution phenomena that are international and often categorized as problems of "global public goods or bads."[10] Climate change is perhaps the most important and difficult to deal with of all global externalities.[11]

---

10. See, for example, Kaul, Grunberg, and Stern (1999).

11. Some other examples are depletion of the ozone layer and the spread of AIDS or other such contagious diseases. A more recent unfortunate addition is international terrorism.

2.1.4. Stock Externalities

In the Arrow-Debreu commodity space, the quantities of commodities are defined per unit of time, and—if the analysis ought to consider them at different points of time—they are dated and treated as different commodities. This specification also fits the case where externalities are considered over time: emissions as well as ambient pollutants can be expressed in quantities per unit of time as well as dated and, thus, can be subjected to intertemporal analyses.

However, one additional element may have to be taken into account; namely, the property that some ambient pollutants can accumulate, such as carbon dioxide (CO_2), for instance. Here, quantities of the ambient pollutant are linked to each other over time because of accumulation, subject to some possible decay. Externalities where accumulation does not occur (e.g., noise or smoke) are called "flow" externalities while those where it does are dubbed "stock" externalities.

On the recipients' side and seen from the ambient pollution point of view, another property of stock externalities is that their duration may extend over long periods of time such that many successive generations of economic agents are affected. Such intergenerational externalities may raise social tensions if some future recipients, who are contemporaries of the current emitters, consider the emitters to be "responsible" for the damage the recipients will incur in the future. This is an important issue concerning climate change.

2.2. TRANSFER FUNCTIONS

In my characterization of an externality as a two-dimensional economic good, one important point was left unspecified that should answer the following question: How are the physical magnitudes that describe emissions and ambient pollution related? At first sight, this connection may seem evident because these magnitudes are sometimes identical, as in the case of noise at close quarters. But most often, the relationship is quite complex and surprisingly neglected in the economics literature: in the case of water pollution, what is measured at the dumping stage, say in a river, is of course quite different from what is received downstream by a user of the river water for drinking and recreational purposes. Most conspicuously in the climate change problem, there is a long way to go between the emissions of greenhouse gases and the ensuing changes in the world temperature. Any serious development of environmental economics must devote considerable attention to this connection, a subject matter that belongs to a scientific discipline of its own; namely, ecology.

2.2.1. Externalities and Transfer Functions

Ecology may be defined as the science of the relationship of living organisms to their environment.[12] A central concept there is that of an *ecosystem*, which describes any system in which there is interaction between living organisms and their environment. Ecological science studies how ecosystems are structured, operate, and evolve over time according to the laws of physics, chemistry, biology, and other natural sciences.

Ecological science also considers the effects of external interventions on the ecosystem, which includes interventions by humans. When these effects are formalized mathematically, they can be summarized in the form of what is generally called a *transfer function*. The arguments of a transfer function are variables that measure the size of the interventions (or emissions), while the values of the dependent variables describe and measure the states in which the ecosystem may find itself as a result of the interventions. In its most general form, an ecological *transfer function* for a *single* pollutant is a vector-valued function

$$z = F(e), \tag{2.1}$$

where z is an n-dimensional vector describing the state of the ecosystem at some point in time, with as many components as the number of locations chosen by the analyst. Specifically, the components of the n-dimensional vector z denote the quantities of the ambient pollutant at chosen locations.

---

12. This expands slightly the definition that was proposed by the German biologist Ernst Haeckel in 1866. Oddly enough for our context, the first textbook on the subject was written by the Danish botanist named Eugenius Warming!

Similarly, e in the argument of the transfer function F is an m-dimensional vector, and its components denote the quantities of emissions from all sources. The components of both z and e are expressed in physical units per unit of time. The sources of emissions and their number, m, may or may not be the same as the locations and their number, n, chosen for measuring the quantities of the ambient pollutant.

A particularly relevant case of a transfer function is a linear function

$$z = Ae, \qquad (2.2)$$

where A, called the transport matrix, is a nonnegative $n \times m$ matrix whose typical element a_{ij} denotes the amount of ambient pollutant at location i that results from one unit of emissions at source j.[13]

The linear case can serve to illustrate how transfer functions can describe various types of externalities some of which were discussed earlier. Indeed, alternative structures of the matrix A imply different types of the externalities. For instance, if the structure of matrix A is triangular with all elements in the triangle equal to one and all the others equal to zero, the externality, though additive, is not reciprocal and of an "upstream-downstream" type: the larger the number of zeros in the row corresponding to a location, the more upstream the location. Similarly, if the elements in each column of the matrix A add up to one, then the externality is neither additive nor reciprocal and lacks public good characteristics as it is not non-rival. To contrast reciprocal externalities with this type of externalities, table 2.1 illustrates a simplified version of a matrix that describes how the acid rains occur across various parts of northern Europe as a result of sulfur dioxide (SO_2) emissions.[14] The geographic area covered by this matrix consists of seven regions (one of which is actually a country, which is immaterial at this stage of discussion); namely, northern, central, and southern regions in Finland (NF, CF, and SF, respectively); the Kola (Ko) Peninsula, Karelia (Ka),

---

13. See Kolstad (2000) for an interesting application of a linear transfer function.
14. The acid rains in Europe were first modeled and studied by Mäler (1989) and Alcamo, Shaw, and Hordijk (1990).

TABLE 2.1 An Estimated SO_2 Transport Matrix for Seven Regions of Northern Europe

		EMITTING REGION (j)						
		NF	CF	SF	Ko	Ka	SP	E
		1	2	3	4	5	6	7
RECEIVING REGION (i)	NF 1	0.200	0.017	0.010	0.045	0.012	0.000	0.000
	CF 2	0.000	0.300	0.062	0.011	0.047	0.036	0.029
	SF 3	0.000	0.017	0.227	0.003	0.000	0.027	0.038
	Ko 4	0.000	0.017	0.000	0.286	0.023	0.009	0.000
	Ka 5	0.000	0.033	0.031	0.017	0.318	0.045	0.019
	SP 6	0.000	0.017	0.031	0.003	0.012	0.268	0.058
	E 7	0.000	0.000	0.031	0.000	0.000	0.018	0.221

Note: Northern, central, and southern regions in Finland (NF, CF, and SF, respectively); the Kola (Ko) Peninsula, Karelia (Ka), and St. Petersburg (SP) regions in Russia; and Estonia (E).

Source: Tuovinen, Kangas, and Nordlund (1990).

and St. Petersburg (SP) regions in Russia, and Estonia (E). The numbers 1–7 in the rows of the 7×7 matrix correspond to the regions where SO_2 emissions take place, and the numbers 1–7 in the columns correspond to the regions where depositions occur. Each element a_{ij} of the matrix is the fraction of emissions in region j that ends up as deposition in region i. These elements of the matrix reveal how the emissions of SO_2 are deposited across the area. It is seen that a major fraction of every region's emissions is deposited in the region itself, as indicated by the diagonal elements of the matrix compared with the off-diagonal elements. The values in each column of the matrix do not add up to one because the balance of the emissions is deposited outside the area under consideration. The externality is neither additive nor reciprocal, because emissions of northern Finland are not deposited in any of the seven regions, and only fractions of emissions of four other regions are deposited in northern Finland.

Among all possible structures of the transfer matrix, the one that is relevant for my purpose is the matrix A in which $m = n$ and all elements are equal to one. Thus, the transfer function is *additive* and corresponds to a reciprocal externality with public good characteristics. In this case, all components of vector z are equal, and the transfer function can be written compactly, with some notational inconsistency, as $z = \sum_{i=1}^{n} e_i$, where z now denotes a scalar.

2.2.2. Transfer Functions for Stock Externalities

Regarding the time aspect of transfer functions, an important distinction ought to be made between "stock" and "flow" of a pollutant, which is in fact at the root of the notion of stock externalities mentioned in the preceding section. The distinction is made here more precise in terms of the physical nature of what the variables z represent. Stock models are those in which the ambient pollutant variables z represent the amounts that accumulate at the chosen locations: the flow of emissions that takes place at various sources at each point of time determines not only the additional amounts of the ambient pollutant accumulated at the chosen locations but also the total stocks of accumulated ambient pollutant for the next point of time. The transfer function then reads

$$z_{t+1} = F(z_t, e_t), t = 1, 2, \ldots \tag{2.3}$$

By contrast, flow models describe situations in which the ambient pollutant variables z in each time period depend only on the current levels of emissions e with their previous quantities playing no role. The transfer function then reads

$$z_t = F(e_t), t = 1, 2, \ldots \tag{2.4}$$

where the length of each period t is such that at the end of it, the emissions have disappeared. Such disappearance is sometimes called the "assimilative capacity" of the environment. A particularly relevant case of the stock model for a single ambient pollutant is the linear form

$$z_{t+1} = (1-\delta)z_t + \beta \sum_{i=1}^{m} e_{it}, t = 1, 2, \ldots \tag{2.5}$$

where $0 < \delta \leq 1$ is the natural rate at which the stock of the ambient pollutant decays or is assimilated by nature, $0 < \beta \leq 1$ is a conversion factor of total emissions into stock, and e_{it} is the amount of emissions from source i

at time t. It may be noted that if $\delta = \beta = 1$, the stock model becomes the additive flow model discussed earlier. Thus, the flow models are technically special cases of the stock models.

2.3. THE BASIC MODEL

In this section, I present the basic flow model that I use in chapters 3, 5, and 6. The stock model is presented in chapters 7 and 8. I make use of the simplest forms of both the ecological transfer function and the economic model compatible with the general equilibrium theory. I abstract from the time component introduced in the previous subsection and limit myself to a single flow pollutant.[15] This results in a static model in which all variables are defined per unit of time, say a year.

2.3.1. Components of the Basic Flow Model

The world economy consists of $n > 1$ countries. We treat each country as a primitive economic agent and denote the set of countries by $N = \{1, \ldots, n\}$. To focus later on the strategic aspects of climate change, I consider, on purpose, a highly aggregated model that has been previously studied by Chander and Tulkens (1992, 1997).

Commodities

There are three categories of commodities: (1) a composite consumption good,[16] whose quantities are denoted by $x_i \geq 0$ if they are consumed by country i and by $y_i \geq 0$ if they are produced by country i; (2) an emitted pollutant, the quantities of which are denoted by $e_i \geq 0$ when emitted in country i; and (3) an ambient pollutant whose quantities are denoted by $z \geq 0$.

---

15. Thus, I also assume that the capital stock, the other production factors, and the technology remain fixed during the period.

16. Reducing the number of consumption goods to one may hurt our sense of realism. However, introducing several consumption goods is not essential at this stage where I only wish to present a very basic economic-ecological model. In chapter 10, I introduce a model with two consumption goods.

Preferences

Each country's preferences—that is, the aggregate preferences of the country's inhabitants—are represented by a utility function $u_i(x_i, z)$. Throughout the book, the utility functions are assumed to be differentiable, increasing in x_i, decreasing in z, and satisfying the following assumptions.

Assumption 2.1

The utility function u_i is of the form $u_i(x_i, z) = x_i - v_i(z)$; that is, the utility function is additively separable and quasi-linear.

The function $v_i(z)$ will be often referred to as the "disutility" or the "damage" function as it represents the loss in utility from ambient pollution z due to its detrimental character.

Assumption 2.2

The disutility function is strictly increasing and convex; that is, $v_i'(z) > 0$ and $v_i''(z) \geq 0$ for all $z \geq 0$.

Figure 2.1 presents typical indifference curves in the (x_i, z) space. Their unusual upward rising shape comes from the assumption that the utility is increasing in the composite consumption good amount x_i but decreasing in the amount of ambient pollution z. The quasi-linearity assumption is responsible for their congruent form; that is, for each given z, the partial derivative $\partial u_i(x_i, z) / \partial z \equiv v_i'(z)$ is constant for all values of x_i.

This last feature, introduced for analytical convenience, is not innocuous: it amounts to assuming that the intensity of each country's preference for the environment is independent of its level of the consumption good. However, the assumption is justified because changes in the amount of the consumption good due to climate change are not large. It is also worth noting that Assumption 2.2 does not rule out linear damage functions. This is on purpose as it is believed that, unlike many other pollutants, damage from climate change is linear over a wide range outside which we may never have to operate.

Thus, linear damage functions will be paid special attention throughout the book. Assumption 2.2 rules out $v_i'(z) = 0$ for all $z \geq 0$; that is,

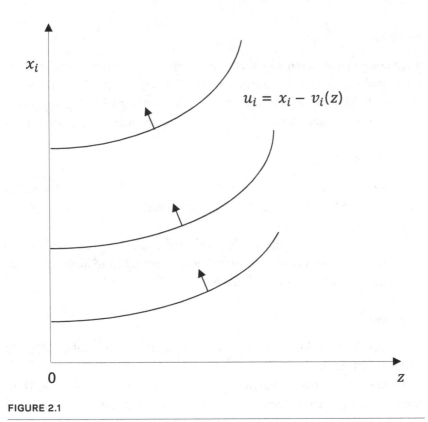

FIGURE 2.1

Typical indifference curves corresponding to a quasi-linear utility function

country i is not a pollutee. Because allowing this does not provide additional insights, I continue with the stronger assumption. To derive suitable interpretations of the concepts involved, however, my illustrative
examples will sometimes allow that a country may not be a polluter; that
is, $v'_i(z)=0$ for all $z \geq 0$.

But, as a number of scientific studies show, damage from climate
change may rise steeply once the greenhouse gases (GHG) stock crosses
a certain threshold. Figure 2.2 illustrates a damage function that is linear
over a wide range until the GHG stock hits the threshold.

One of the difficulties that arise in the international negotiations to
tackle climate change is that the threshold may be different for different
countries. For example, the threshold may be reached much sooner in the

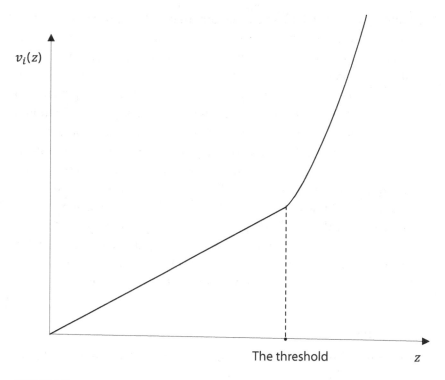

FIGURE 2.2

A damage function that is linear over a wide range until the GHG stock hits the threshold

case of low-lying small island states that would be the first ones to face the impact of climate change due to rise in sea levels. An important issue left unresolved by the recent Paris Agreement on climate change, and studied in chapters 8 and 9, is whether global warming should be limited to 1.5°C or 2.0°C. While the low-lying island states would be better off if it is limited to 1.5°C, some other countries are reluctant to agree because the cost of limiting climate change to 1.5°C is very high compared to limiting it to 2.0°C.

Production and Emissions

In each country i, production activities are represented by an aggregate production function $y_i = g_i(e_i)$, where $y_i \geq 0$ is the total output produced in country i (say, its GDP), and e_i denotes the total quantity of a polluting input (fossil fuels) used to produce y_i. Other inputs are not explicitly mentioned in the production function and are subsumed in the function g_i.[17] While this modeling of production does not explicitly mention any environmentally relevant variable, I may exploit the fact that all production entails some pollution as a joint by-product. Thus, I interpret the amount e_i of the polluting input as also the amount of emissions generated and reduction of the amount e_i of the polluting input as pollution abatement. In other words, I assume that using one unit of fossil fuels generates one unit of GHGs. This is obviously a simplification because in reality, fuel types differ in their carbon content: natural gas emits less GHGs than oil, which in turn emits less than coal. But an appropriate aggregation procedure can be applied to define a unit of fossil fuels that generates one unit of GHGs. Throughout the book, I assume that the production functions are defined and differentiable for all $e_i \geq 0$ and satisfy the following assumption.

Assumption 2.3

The production function $g_i(e_i)$ is increasing and strictly concave; that is, $g_i'(e_i) > 0$ and $g_i''(e_i) < 0$ for all $e_i \geq 0$.

The derivative $g_i'(e_i)$, if taken to the left, is the marginal cost of abatement/mitigation, expressed in units of foregone output. Alternatively, if taken to the right, it measures the marginal gain in output or marginal product entailed by increased emissions. Figure 2.3 presents a typical production function satisfying these assumptions. It is seen that the derivative $g_i'(e_i)$ is decreasing with e_i; that is, the marginal cost of abatement is decreasing with e_i. Thus, the assumption of strict concavity of the production functions is consistent with the fact (e.g., Nordhaus and Boyer 2000) that marginal abatement costs are highly nonlinear.

The Transfer Function and Externalities

Most climate models assume that the greenhouse gases have the same effect on the earth irrespective of the location of the source from which

---

17. In contrast, Chander and Khan (2001) consider a model, to be presented later, with two consumption goods and two variable inputs.

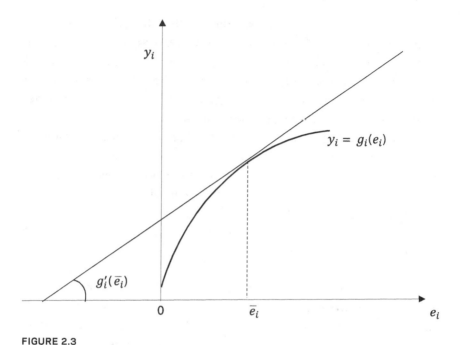

FIGURE 2.3

A typical aggregate production function and marginal abatement cost

they are emitted. In fact, they all rise through the atmosphere and spread uniformly in the stratosphere, forming a thick gaseous sphere around the earth. This leads to global warming and rise in temperatures or climate change on the earth, which causes damage. These scientific facts suggest that the transfer function in the case of greenhouse gases, as a first approximation, is simply

$$z = \sum_{i \in N} e_i, \tag{2.6}$$

where z stands for the resulting ambient pollution (i.e., the GHGs stock), which causes climate change. Because all inhabitants on the earth live under the same stratosphere, the quantity of ambient pollution (i.e., the GHGs stock) is the same for everyone, and, therefore, no subscript is attached to z. Thus, climate change is a case of an additive and reciprocal externality: each country is a polluter as well as a pollutee.

The transfer function formalizes the fact that the externalities generated by the GHG emissions of the various countries are additive: on the one hand, they are generated by production activities within *each* country, hence denoted as emissions $e_i, i=1,\ldots, n$, and, on the other hand, they are received in the same amount z by *all* countries, as formally indicated by the absence of a subscript i attached to the variable z.

Notice that z, the amount of ambient pollution, does not appear as an argument in the production function $g_i(e_i)$, although it is a common fact that production too can be affected by ambient pollution, just as consumption. Some production activities might even benefit from it; for instance, agriculture in some parts of the world may benefit from global warming. Ignoring this is again an assumption made for expositional convenience. Introducing ambient pollutant z as an additional variable in each production function g_i—with appropriate specification of the derivatives with respect to it—would render my model more realistic, but it would not add much to the analysis and results that follow. Yet it will be good to remember, at times, that besides consumers, producers too can be considered among those affected by ambient pollution, irrespective of whether or not they are emitters. I now make two additional assumptions regarding the production functions.

Assumption 2.4

There exists an $e^0 > 0$ such that (i) $g_i'(e^0) < v_i'(e^0)$ and (ii) $g_i'(0) > \sum_{j \in N} v_j'((n-1)e^0)$ for each $i \in N$.[18]

Assumption 2.4(i) means that beyond a certain level of emissions, e^0, the marginal product, $g_i'(e^0)$, of emissions for each country i is less than the marginal damage, $v_i'(z)$, because the ambient pollution z, according to (2.6), is at least e^0 if the emissions of country i are e^0 and the marginal damage function is nondecreasing, implying $v_i'(e^0) \leq v_i'(z)$ for all z. Assumption 2.4(ii) means that at zero level of emissions, the marginal product of emissions for each country i is sufficiently high. The full meaning of these assumptions will become clear after I introduce behavioral

---

18. The same assumption is also made in Chander (2007). In contrast, Chander and Tulkens (1997) make a stronger assumption.

assumptions in chapter 3. Let me limit myself at this stage to simply stating that the assumptions ensure no country will ever emit zero amount or more than the amount e^0; that is, the emissions e_i of each country i will always be such that $0 < e_i < e^0$, $i \in N$.

To summarize this subsection, the ecological components of the model consist of the emissions vector $e = (e_1, \ldots, e_n)$, the transfer function (2.6), and the ambient pollution z, whereas the economic components of the model are the production vector $y = (y_1, \ldots, y_n)$, the private good consumption vector $x = (x_1, \ldots, x_n)$, and the utility functions $u_i(x_i, z)$. Unlike the transfer function, which is determined by ecology alone, the disutility or damage functions, $(v_1, \ldots, v_n)$, are determined both by ecology and economics. While ecological science can help determine what and how much of the environment is affected by climate change, the discipline of economics is needed to assess the monetary value of the effects. This is where ecological science and economics become inseparable for quantifying climate change. Taken together and considered as a whole, these elements constitute what I call throughout this book an *economic-ecological* system. Sometimes, for short, I speak of an "economy" or "world economy," but this should not make the reader forget that we are always carrying along the ecological component.

2.3.2. Feasible Consumption Profiles

Definition 2.1

A consumption profile $(x_1, \ldots, x_n, z)$ is feasible if there exists an emission profile $(e_1, \ldots, e_n)$ such that $\sum_{i \in N} x_i = \sum_{i \in N} g_i(e_i)$ and $z = \sum_{i \in N} e_i$.

Thus, a consumption profile is feasible if and only if there exists an emission profile such that the aggregate consumption in the economy is equal to the aggregate production, and the ambient pollution, as per the transfer function (2.6), is equal to the sum of emissions of all countries.[19] I shall also say sometimes that a feasible consumption profile $(x_1, \ldots, x_n, z)$

---

19. I could have stated the feasibility condition more generally as a weak inequality, $\sum_{i \in N} x_i \leq \sum_{i \in N} g_i(e_i)$, but because utilities are strictly monotonic in the consumption good, there is no loss of generality in stating it as an equality.

is generated by an emission profile $(e_1, \ldots, e_n)$ if $\sum_{i \in N} x_i = \sum_{i \in N} g_i(e_i)$ and $z = \sum_{i \in N} e_i$.

Notice that a consumption profile is specified at the world level and does not require each country to consume an amount that is exactly equal to what it produces. Any discrepancy (positive or negative) between a country's consumption and production is implicitly assumed to be covered by transfers of resources to or from the other countries. Unlike a conventional public good model (see, e.g., Chander 1993), the feasibility condition for a consumption profile implicitly assumes no initial endowments of the consumption good for each country. Allowing each country to have a positive initial endowment of the consumption good would not change the basic nature of the model. I thus continue to assume that the initial endowment of the consumption good of each country is zero. Some of my illustrative examples later, however, allow the countries to have positive initial endowment of the consumption good.

If institutions are introduced in this basic model,[20] the interactions that occur between consumers and producers as implied by the feasibility condition can be described in various ways, and among them the following are classical. If the interactions are considered in terms of free commodity exchanges on markets and generation of externalities, one is led to formulating a positive theory of spontaneous equilibrium of the whole system. By contrast, a normative theory of policies bearing on commodities (produced and consumed) and on emissions and ambient quantities can be constructed, inspired by some criteria or norms. This is the program of general equilibrium and welfare analysis offered by economic theory to cope with worldwide problems such as climate change. I shall follow it in detail in chapter 3.

2.3.3. Additional Interpretations of the Model

The model so far may appear to ignore a number of facts. First, it seems to assume a zero initial stock of GHGs, though there is clearly an existing

---

20. I call it "basic" because a full description of environmentally relevant human activities would go far beyond those we have dealt with so far, such as cleanup and recycling, for instance. I leave these out to avoid the details of exposition from masking the essentials.

positive stock, and climate change and the damage thereby depend on the total and not just the additions to the GHGs stock. But this is not really so. Let z_0 denote the existing stock of GHGs and δ the natural rate of decay of the stock, then damage function $v_i(z) \equiv w_i((1-\delta)z_0 + z)$, where $w_i((1-\delta)z_0 + z)$ is the damage—measured in units of the consumption good—due to the total GHGs stock in the atmosphere. Second, it is a known scientific fact that significant amounts of GHGs released because of the combustion of fossil fuels are absorbed by "carbon sinks." The main natural carbon sinks are trees, soil, and the oceans. Indeed, the Kyoto Protocol recognized the absorption of CO_2 by trees and the soil to be just as valid a means to achieve emission reduction commitments as cutting CO_2 emissions from fossil fuels. This fact too is implicit in the model as the emissions $(e_1, \ldots, e_n)$ are to be interpreted as the emissions net of the amounts absorbed by the carbon sinks in each country. Absorption by the oceans, which do not belong to any country in particular, can also be incorporated into the model by replacing (2.6) with the equation $z = \beta \sum_{i \in N} e_i$, where $1 - \beta$ is the proportion of *total* emissions absorbed by the oceans. But many scientists believe that the amount absorbed by the oceans really comes from the existing stock of CO_2 in the atmosphere and not from the current emissions. Accordingly, we take $\beta = 1$ and assume that whatever amount of the current emissions is absorbed by the oceans is taken into account by the rate of decay δ in the linear equation (2.5).

Third, combustion of fossil fuels releases GHGs, which not only cause climate change and damage globally, but also often cause damage locally. For example, burning coal not only releases CO_2 but also lowers local air quality and may lead to smog in the local area. Some large cities, including Beijing and Delhi, are choked by bad air quality because of the burning of fossil fuels in combustion engine vehicles in these cities. This local damage is actually implicit in each production function $g_i(e_i)$. More explicitly, $g_i(e_i) \equiv f_i(e_i) - h_i(e_i)$, where $h_i(e_i)$ is the damage from country i's own emissions, and $f_i(e_i)$ is the gross output of the consumption good when amount e_i of fossil fuels is used. It is natural to assume that f_i is increasing and concave and h_i is increasing and strictly convex, implying strict concavity of g_i, as assumed. Thus, each g_i is to be interpreted as the production function net of domestic damage from combustion of fossil fuels. Decomposing the production function in this way should also clarify that the model does not rule out constant returns to scale in production, as we can take $g_i(e_i) = \sigma_i e_i - h_i(e_i)$, where $\sigma_i > 0$ is a constant. Indeed, in chapter 9 I explicitly assume this form of production function.

2.3.4. A Model with Two Consumption Goods

One limitation of using an aggregated model with only one consumption good is that it does not allow analysis of the impact of international trade in goods on global pollution. Thus, in chapter 9 I introduce a model with two goods, which is due originally to Chander and Khan (2001). It consists of two primary factors of production, capital and labor, and two private goods, of which good 1 is polluting and good 2 is nonpolluting. Capital is mobile across the sectors, but labor is specific to the production of the nonpolluting good. In particular,

$$y_{i1} = \begin{cases} e_i^\alpha k_{i1}^{1-\alpha} & \text{if } e_i < ak_{i1} \\ a^\alpha k_{i1} & \text{if } e_i \geq ak_{i1}; \end{cases}$$

$$y_{i2} = \ell_i^\alpha k_{i2}^{1-\alpha};$$

$$k_i = k_{i1} + k_{i2}, i = 1, \ldots, n, \tag{2.7}$$

where y_{ij} is the output of good $j (=1, 2)$ by country i, and k_i and ℓ_i are country i's total capital stock and labor endowment, respectively. Thus, capital is a production factor that is mobile across the two sectors, while labor is specific to the production of good 2. The production function for good 1 implies that for any *given* capital stock, the marginal product of emissions falls to zero if the pollution is higher than a certain level.

The utility function of each country is assumed to be linear in good 1, so that income effects are ruled out, but log linear in good 2, so that substitution possibilities are not ruled out. Specifically, the utility of country i is $u_i(x_{i1}, x_{i2}, z) = x_{i1} + \log x_{i2} - v_i(z)$, where $z = \sum_{i \in N} e_i$, and x_{i1} x_{i2} are the consumptions of goods 1 and 2. Such a utility function makes it possible to consider international trade in goods and analyze the relationship between patterns of international trade and world pollution levels. In one sense, such a model is more general because there are two consumption goods, but in another sense it is less general because it assumes specific production functions for the two goods, as described by the functions in (2.7).

2.3.5. A Compact Form of the Basic Flow Model

In this subsection, I introduce and discuss a slightly different, and possibly more intuitive, version of the basic model that has often been used in the literature. Though it will not be used in the rest of the book, it is included here as a benchmark for the basic model introduced earlier.

Mäler (1989) introduced in the environmental economics literature an extremely simple model that is also quite convenient to use because it contains all the basic ingredients necessary to make it useful and sufficiently comprehensive for many conceivable analyses and policies. Mäler's model is described simply by the following system of $n+1$ functions:

$$c_i(e_i)+d_i(z), \quad i=1,\ldots,n, \tag{2.8}$$

$$z = F(e). \tag{2.9}$$

The function $c_i(e_i)$ is called the *abatement cost function* of country i. It denotes country i's total costs (measured in monetary units) of abating emissions by an amount $\bar{e}_i-e_i$ from some fixed reference level $\bar{e}_i>0$. It is assumed to be decreasing and convex in its argument e_i with zero value at $\bar{e}_i$; that is, $c_i(\bar{e}_i)=0$. The graph of the function is as shown in figure 2.4.

The function $d_i(z)$ is called the *damage function* of country i. It denotes country i's total costs (measured in monetary units) due to the damage caused by the ambient pollutant z. It is assumed to be an increasing and convex function. Accordingly, the graph of a damage function is as shown in figure 2.5.

The function $F(e)$ in (2.9) is the transfer function and the same as in (2.6). Mäler used a matrix for defining the transfer function in his original model, similar to that in table 2.1.

In this compact model, a *state* of the economic-ecological system is simply an $n+1$ vector $(e,z)=(e_1,\ldots,e_n,z)$, and it is feasible if it is nonnegative

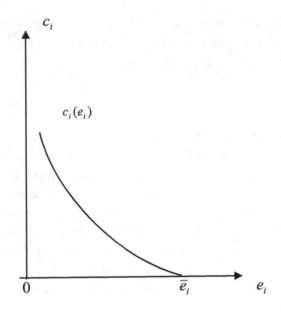

FIGURE 2.4

An abatement cost function of the compact model

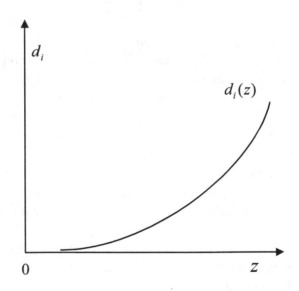

FIGURE 2.5

A damage function of the compact model

and satisfies (2.9). No other relations between the countries, other than those occurring through the externality, are accounted for. The purpose of the compact model is to characterize emissions policies, that is, values for the components of the vector e in terms of the sum total of abatement and damage costs they entail, separately for each country as in (2.8) or for all of them taken together as in

$$J = \sum_{i \in N} [c_i(e_i) + d_i(z)]. \tag{2.10}$$

The function J then simply represents the total cost of the world emissions policy $e = (e_1, \ldots, e_n)$ leading to ambient pollution $z = \sum_{i \in N} e_i$; that is, the cost of abatement for the polluters, which is mainly a private cost, *and* the damage cost for the pollutees, which is a social cost. Thus, J represents the total cost to society.

The compact model is quite intuitive and belongs to the class of partial equilibrium models. For that reason it is probably more accessible to noneconomists. However, it is not inconsistent with my model, which is more in the spirit of a general equilibrium model. In fact, the compact model is a special case of my model.

2.3.6. A Note on Adaptation and Protection

A nonnegligible aspect of human reaction to environmental degradation and especially to climate change is that the affected population may seek to adapt or take measures to protect itself against the occurring changes. This behavior has two economic characteristics. On the one hand, it is only effective if it succeeds in reducing the damage caused by environmental change; otherwise, it is useless. On the other hand, it is not free: adaptation and protection require resources or impose costs. For example, to protect myself against rain, I may have to either buy an umbrella or stay indoors until the rain stops.

It may appear that my basic model ignores the possibility of adaptation or measures for protection against climate change, parallel or in contrast to abatement, that is, emissions reduction. But that is not really so. These are implicitly taken into account by my basic model. To see this, let $a_i \geq 0$ be an aggregate measure of the physical activity devoted by country i

to adaptation and/or protection,[21] and let $h_i(a_i)$ denote the amount of resources this activity requires, expressed in units of foregone amounts of the consumption good. The damage-reducing property of the said activity can then be accounted for by introducing the variable a_i as a second argument in the disutility function v_i and writing the utility function as

$$u_i(x_i, z, a_i) = x_i - v_i(z, a_i),$$

where the disutility function v_i satisfies the following properties: $\partial v_i/\partial z \geq 0$, $\partial v_i/\partial a_i < 0$, $\partial^2 v_i/\partial a_i^2 > 0$, and $\partial^2 v_i/\partial z \partial a_i < 0$. Notice that in this formulation, adaptation and protection, in contrast to abatement, have only domestic effects in each country i: activities a_i generate no transboundary effects.[22] Therefore, they will not play a central role in the arguments that focus on international externalities. But domestically, adaptation activities compete for resources with those required for abatement. Thus, the issue arises as to the appropriate mix between the two.

Assuming utility maximization by the countries—as I shall do throughout this book—the behavior of each country i with respect to both emissions and adaptation is described by the solution of the optimization problem

$$\max_{(e_i, a_i)} u_i(x_i, z, a_i) = g_i(e_i) - h_i(a_i) - v_i(z, a_i), \qquad (2.11)$$

subject to (2.6). The first-order condition with respect to a_i reads

$$h_i'(a_i) = -\partial v_i(z, a_i)/\partial a_i, \qquad (2.12)$$

meaning that at the optimal solution, country i adapts up to the point where its marginal adaptation cost becomes equal to the value of marginal

---

21. For example, moving populations, adopting new crop patterns, and raising dikes at seashores for protection against sea-level rise.

22. Thus, from an international point of view, they are like "private" goods. However, two qualifications are in order: (1) *within* a country, adaptation and/or protection activities often have properties of a local public good (I ignore any "free-rider" problems associated with that); and (2) some forms of adaptation and/or protection may have transboundary effects, in which case the relevant variable a_i should appear in the disutility functions of many (e.g., neighboring) countries. However, such cases are of limited interest from the perspective of the climate change problem, and thus there would be little gain in expanding my model to cover them.

damage so avoided. Put simply, the benefit to the country of an additional dollar spent on adaptation at the margin must be equal to that from an additional dollar spent on abatement. The derivative on the right of (2.12) is a function of the ambient pollution z: if the other countries modify their emissions, the condition must still hold, though for a different value of a_i.

The fact that the optimality condition (2.12) holds for every value of z (i.e., in all states of the environment) leads me to reformulate the disutility function v_i in a simpler way. Consider condition (2.12) for any z. Let $a_i^*(z)$ be the amount of adaptation for which it is satisfied. The adaptation cost then is $h_i(a_i^*(z))$ and the disutility is $v_i(z, a_i^*(z))$. The objective function in (2.11) becomes a function of ambient pollution z and emissions e_i only and given by $g_i(e_i) - h_i(a_i^*(z)) - v_i(z, a_i^*(z))$, where the function $a_i^*(z)$ specifies the optimal level of adaptation for each level of ambient pollution z. Let $v_i^*(z) \equiv h_i(a_i^*(z)) + v_i(z, a_i^*(z))$. Then, the function $v_i^*(z)$ denotes the damage from ambient pollution with optimal adaptation or the "residual damage function" in the sense that it denotes the damage incurred after optimal adaptation has been made. Its derivative is equal to $\partial v_i(z, a_i)/\partial z$ because

$$\frac{dv_i^*(z)}{dz} = \frac{dh_i}{da_i^*}\frac{da_i^*}{dz} + \frac{\partial v_i}{\partial z} + \frac{\partial v_i}{\partial a_i^*}\frac{da_i^*}{dz} = \frac{\partial v_i}{\partial z}$$

after factoring out da_i^*/dz and using (2.12). As shown in figure 2.6, the function $v_i^*(z)$ is actually an envelope of a family of functions $h_i(a_i) + v_i(z, a_i)$ such that the equality $dv_i^*(z)/dz = \partial v_i(z, a_i)/dz$ holds at the tangency points. In figure 2.6 when $z = z_k$: if $a_i = a_i^*(z_2)$, country i adapts too much, and the excess damage cost is BC; if $a_i = a_i^*(z_1)$, country i adapts too little, and the excess damage cost is AC.

This shows that when optimal adaptation is made for every value of ambient pollution z, the marginal damage cost as represented by the derivative of the residual damage function $v_i^*(z)$ is net of adjustment in adaptation cost.[23] In the rest of the book, and unless indicated otherwise, I shall denote the damage function of country i with optimal adaptation simply as $v_i(z)$, dropping the superscript * to alleviate notation.

---

23. This point is of importance in the design of cost-benefit studies of investment projects in adaptation and protection (Tulkens and van Steenberghe 2009).

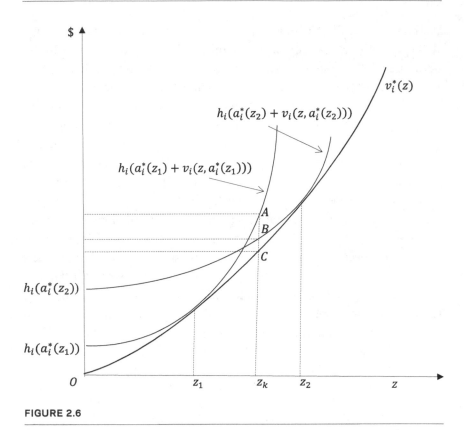

FIGURE 2.6

A damage function with optimal adaptation

I end this section with a word of caution: an adaptation policy that is optimal for a country may not be optimal for the world. That is so because for world optimality, it is not the equality between the benefits to a country of an additional dollar spent on adaptation at the margin and an additional dollar spent on abatement at the margin, but rather equality between the benefit to a country of an additional dollar spent on adaptation at the margin and the *total* benefit to *all* countries of a dollar spent on abatement at the margin. Thus, assuming a concave adaptation cost function $h_i(a_i)$, an optimal adaptation policy of a country may result in the country spending more on adaptation than is required by a world optimal adaptation policy. However, this is true only if climate change is not addressed; if climate change is brought under control, then spending on adaptation, if at all, may be insignificant.

3

RATIONALE FOR COOPERATION

For the basic model of climate change built so far, I now systematically describe the feasible emissions and consumption profiles that are of special interest.[1] I distinguish between two types of feasible emissions and consumption profiles: equilibria and optima. Equilibria are meant to describe feasible emissions and consumption profiles that occur spontaneously, depending on the behavioral assumptions made. By contrast, optima refer to feasible emissions and consumption profiles that meet some social criteria or norm, chosen a priori. Emissions and consumption profiles of this type are typical of the social sciences as opposed to the natural sciences—including ecology!—in which there is little room for normative considerations. Moreover, it is the consideration of such optima that makes the normative part of this book essentially anthropocentric, as was noted in chapter 1.

In summary, the message of this chapter is as follows. As the theory described later establishes, the spontaneous equilibria do not meet one of the most universally recognized criteria that society desires; namely, efficiency. Thus, improvements in everybody's well-being across the world are possible. This alone should justify that efforts be made to engage in collective action aimed at achieving efficiency. However, because the costs and benefits of achieving efficiency may not be evenly spread across the countries, "direct" or "indirect" transfers of the consumption good between the countries to balance the

---

1. In the case of the dynamic models in chapters 7 and 8, they are *time profiles* of emissions and consumption.

costs and benefits of achieving efficiency are necessary[2] unless the countries are identical or quite similar. This is the perspective with which I shall propose, when concluding the chapter, to interpret international environmental agreements in general and those on climate change in particular.

3.1. EQUILIBRIUM CONCEPTS

An international equilibrium is the result of a situation in which each country pursues its own goals without bothering about the impact on the well-being of other countries. At a disaggregated domestic level, the pursuit of these goals may involve exchange of goods, services, and factors of production as well as emissions and reception of externalities. However, I abstract away from all this and treat each country as a single economic agent. Accordingly, I assume for each country the usual assumption of microeconomic theory; namely, utility maximization subject to resource constraint(s). However, there are good reasons for distinguishing between two types of possible behaviors.

3.1.1. Alternative Behavioral Assumptions

In the first type of behavior, no action to control emissions is taken by any country. Emissions are determined only by the necessities of production and profit maximization, and no account whatsoever is taken by anyone of their detrimental effect. In the jargon of negotiations on climate change, this type of behavior is referred to as business as usual (BAU). Because I have assumed an aggregate production function with one variable input, this type of behavior is equivalent to each country i maximizing its consumption good output net of cost; that is, $g_i(e_i) - p_i e_i$, where p_i is the price of fossil fuels in country i. Thus, the emissions and fossil fuels use by

---

2. By this I mean transfers between countries such that the benefit to each country from reduced total world emissions after transfers is not less than the cost to each country of reducing its own emissions.

country i under BAU behavior, denoted by $\overline{\overline{e}}_i$, are characterized by the first-order conditions (FOCs)

$$g_i'(\overline{\overline{e}}_i) = p_i, i \in N, \tag{3.1}$$

which means equality between the marginal product of the input and its price. The impact of pollution resulting from production even on a country's own utility is simply ignored. Because g_i is strictly concave (i.e., g_i' is strictly decreasing), a higher fossil fuels price means lower emissions or energy use and output. Thus, under BAU behavior, the higher (lower) the fossil fuels price in a country,[3] the higher (lower) its marginal cost of abatement. That such a relationship may actually hold is illustrated by table 9.1 in chapter 9, which uses data from the mid-1990s. During that period, the prices of oil, coal, and natural gas were relatively low in the United States and so was the U.S. marginal abatement cost compared to that of the European Union or Japan. No such relationship seems to hold in the case of China, perhaps because it had a large number of public firms with a generally low concern for energy efficiency and profit maximization.

In contrast, if each country i acts so as to take into account the domestic damage caused by its own emissions, as expressed through the damage function $v_i(z)$, then that is equivalent to solving the maximization problem $max_{(x_i, z)} u_i(x_i, z) = x_i - v_i(z)$ subject to the constraints $x_i = g_i(e_i)$ and $z = \sum_{j \in N} e_j$. Because country i has no control over the production and emission decisions of other countries, it must take them as given. Choosing the level of ambient pollution z by country i is then equivalent to choosing its production and emissions, given the emissions of other countries. Hence, each country i's choice of emissions, $\overline{e}_i$, and ambient pollution, $\overline{z}$, under maximizing behavior is characterized by the FOCs

$$g_i'(\overline{e}_i) = v_i'(\overline{z}) \text{ and } \overline{z} = \sum_{j \in N} \overline{e}_j, i = 1, \ldots, n. \tag{3.2}$$

---

3. Energy prices may differ across the countries because of transportation and distribution costs and national tax policies.

In terms of behavior, this condition means that each country i chooses to push its production and emissions up to the level where any additional gain in production is just equal to the additional damage it would suffer, given the emissions of the other countries.[4] Alternatively, an interpretation in terms of abatement goes as follows: country i reduces its emissions down to the level where a further reduction would entail a cost in terms of foregone output that is just equal to the domestic damage avoided due to the reduction, given the emissions of the other countries. In either interpretation, the described behavior may require government intervention, in contrast to the BAU behavior of the previous paragraph, where decisions may be left to profit-maximizing firms.

The proviso "given the emissions of the other countries" is important. If the other countries were to change their emissions, condition (3.2) would still hold, but for different values of $\bar{e}_i$ and $\bar{z}$. In fact, the higher the other countries' emissions, $\sum_{j \in N \setminus i} \bar{e}_j$, the lower (possibly the same) country i's emissions $\bar{e}_i$. This is seen as follows. Let $s \equiv \sum_{j \in N \setminus i} \bar{e}_j$. Differentiating the first equality in (3.2) with respect to s and using the second equality, we obtain $g_i''(\bar{e}_i)d\bar{e}_i / ds = v_i''(\bar{z})(\frac{d\bar{e}_i}{ds} + 1)$. Because $g_i''(\bar{e}_i) < 0$, by strict concavity of g_i, and $v_i''(\bar{z}) \geq 0$, by convexity of v_i, this equality can be true only if $-1 < d\bar{e}_i / ds \leq 0$. Thus, this equilibrium has the property that a country with strictly convex damage functions would react to the damage it incurs from higher emissions abroad by reducing its own emissions.

Before characterizing the equilibria, let us consider an additional issue pertaining to the two types of behaviors by the countries. How do the emissions $\bar{e}_i$ under individual utility maximization compare with those under BAU behavior (i.e., $\bar{\bar{e}}_i$ compared to $\bar{e}_i$)? Clearly, equalities (3.1) and (3.2) and strict convexity of g_i imply that $\bar{\bar{e}}_i$ is greater than/equal to/less than $\bar{e}_i$ if p_i is less than/equal to/greater than $v_i'(\bar{z})$, respectively. This means that even under BAU behavior, the government of a country can

---

4. It is assumed implicitly throughout the book that each country, besides having a single utility function that represents the preferences of its population, has the ability to monitor and regulate the emissions within its geographic boundary. That this may not always be the case is illustrated in chapter 5 by the persistent problem of the Southeast Asian haze that Indonesia has been unable to tackle because of its highly decentralized governance structure and divergence of incentives at different levels of the government. There is much work to do here that is not pursued in this book.

achieve its individually optimal environmental policy by suitably adjusting the energy prices through taxes or subsidies. However, this is rarely observed to be the case. In most countries, the energy price policy is independent of the climate change policy.

3.1.2. International Equilibria

I now apply the two types of individual behaviors discussed above to understand and analyze the outcome of interactions between the countries. I concentrate first on the environmental outcomes. What the interactions may result in is described by international equilibrium concepts. We have two types of them to consider.

The Business-as-Usual Equilibrium

In the business-as-usual equilibrium, no country exercises any control whatsoever on its emissions. Thus, it refers to complete absence of environmental concerns or abatement policy.

Definition 3.1

A BAU equilibrium is a vector of energy prices $p = (p_1, \ldots, p_n)$ and an emissions profile $(\bar{\bar{e}}_1, \ldots, \bar{\bar{e}}_n)$ such that for each $i \in N$, $\bar{\bar{e}}_i = \mathrm{argmax}\,[g_i(e_i) - p_i e_i]$.

The definition essentially consists in assuming that all countries follow BAU behavior as characterized by equality (3.1). In this type of equilibrium, each country's emissions are determined by the energy prices $(p_1, \ldots, p_n)$ independently of other countries' emissions. If these prices change, the equilibrium is modified according to (3.1). Implicitly, in this equilibrium each country consumes what it produces; that is, each country i consumes $\bar{\bar{x}}_i = g_i(\bar{\bar{e}}_i)$ and the total ambient pollution is $\bar{\bar{z}} = \sum_{j \in N} \bar{\bar{e}}_j$.

The Noncooperative Equilibrium

If it is assumed that in each country the government enforces an abatement policy that is consistent with the country's utility maximization, we are led to the following equilibrium.

Definition 3.2

A noncooperative equilibrium is an emissions profile $(\overline{e}_i, \dots, \overline{e}_n)$ such that $\overline{e}_i = \text{argmax} \, [u_i(x_i, z) = x_i - v_i(z)]$ subject to $x_i = g_i(e_i)$ and $z = e_i + \sum_{j \in N\backslash i} \overline{e}_j$.

Clearly, as in the BAU equilibrium, in this equilibrium also each country consumes what it produces—that is, each country i consumes $\overline{x}_i = g_i(\overline{e}_i)$ and faces total ambient pollution $\overline{z} = \sum_{j \in N} \overline{e}_j$—but the emissions chosen by country i are not independent of emissions of the other countries.[5] In the rest of this book, I will mostly assume that the noncooperative equilibrium describes the situation that would prevail between environmentally aware countries, absent any kind of agreement or coordination between them. For this reason, I will sometimes also refer to it as the status quo. I take it to be the status quo not because it describes the prevailing situation more accurately, but because that is likely to be the prevailing situation as the countries become increasingly more aware of how climate change affects them and get ready for entering into an international agreement on climate change.[6] Thus, I will assume that each country chooses its emissions that are individually optimal for it without taking into consideration how they would affect the other countries, and the value of goods consumed in each country is exactly equal to the value of goods produced by it.[7] It may be noted that in this equilibrium, there is no reason for the emissions of different countries to be the same, except when the countries are identical. Their equilibrium marginal abatement costs need not be equal either. In chapter 5, this equilibrium will be shown to be unique and equivalent to the Nash equilibrium of the environmental game introduced there, which explains the "noncooperative" terminology used currently.

---

5. Chapter 7 presents two different versions of this equilibrium in a dynamic model in which the pollutant is treated as a stock variable.
6. See Chander and Muthukrishnan (2015) on how greater environmental awareness can indeed lead to better control of pollution.
7. This does not rule out international trade in goods but assumes implicitly that the international markets for goods are competitive, and thus the prices are the same across all countries.

3.1.3. Equilibria and the Right to Pollute

By definition, both types of equilibria imply that the countries may find themselves in a situation where each of them emits at the level implied by the first-order conditions (3.1) or (3.2) depending on whether a domestic policy for controlling emissions is in force. In either case, however, each country acts as if it has assigned to itself the right to act in this way, letting the other countries undergo the damaging effects of its emissions without asking for their consent. In fact, each country arrogates to itself an unlimited right to produce and pollute without any concern for the other countries. This situation is due to the absence of an international authority that is empowered to grant or deny this right. Thus, both types of equilibria, in fact, are the result of anarchy in the literal sense of the word. In view of Assumption 2.4 regarding the functions g_i and v_i, the emissions $\bar{e}_i$ in the noncooperative equilibrium are strictly positive for each country i.

3.2. EFFICIENCY AND COOPERATION

In contrast to the two types of international equilibria discussed above, which are supposed to describe spontaneous outcomes, this section characterizes "international optima."

3.2.1. World Efficiency

I define and characterize feasible consumption profiles that are deemed to be desirable according to the following criterion of efficiency.

Definition 3.3

A feasible consumption profile $(x_1^*, \ldots, x_n^*, z^*)$ is (Pareto) efficient if there exists no other feasible consumption profile $(x_1, \ldots, x_n, z)$ such that $u_i(x_i, z) \geq u_i(x_i^*, z^*)$ for all $i \in N$ with strict inequality for at least one i.[8]

---

8. It is customary to speak of efficiency in the sense of "Pareto" or "Pareto efficiency." But for brevity, I shall refer to Pareto efficiency simply as efficiency.

Because for each feasible consumption profile $(x_1, \ldots, x_n, z)$ by definition, there exists an emission profile $(e_1, \ldots, e_n)$ such that $\sum_{i \in N} u_i(x_i, z) = \sum_{i \in N} x_i - \sum_{i \in N} v_i(z) = \sum_{i \in N} g_i(e_i) - \sum_{i \in N} v_i(\sum_{j \in N} e_j)$, a feasible consumption profile $(x_1^*, \ldots, x_n^*, z^*)$ is efficient if and only if there exists an emission profile $(e_1^*, \ldots, e_n^*)$ such that $\sum_{i \in N} x_i^* = \sum_{i \in N} g_i(e_i^*)$, $z^* = \sum_{i \in N} e_i^*$, and $(e_1^*, \ldots, e_n^*)$ is a solution of the optimization problem $\max_{(e_1, \ldots, e_n)} \sum_{i \in N} [g_i(e_i) - v_i(\sum_{j \in N} e_j)]$. The first-order conditions for a solution of this optimization problem imply that $(e_1^*, \ldots, e_n^*)$ must be a solution of the equations

$$g_i'(e_i) = \sum_{j \in N} v_j'(z) \quad \text{and} \quad z = \sum_{j \in N} e_j, i = 1, \ldots, n. \tag{3.3}$$

These conditions say that for world efficiency to hold, no country i should push usage of its polluting input beyond the point where the marginal gain in output from it would be just equal to the sum of the damage it would impose on the world; that is, on all countries $j \in N$.[9] Alternatively, abatement should be pursued by each country up to the point where further abatement would reduce world damage by an amount just equal to the cost of the abatement for the country.

The summation in the first equality of (3.3), in contrast to (3.2), is due to the public good characteristic of the ambient pollution z: as the effect of a marginal change in the emissions of any country is felt by every country, it is only natural that the marginal damage to all countries be taken into account for achieving world efficiency.

It may be further noted that efficiency conditions (3.3) imply that a feasible consumption profile $(x_1, \ldots, x_n, z)$ is efficient only if it is generated by an emission profile $(e_1, \ldots, e_n)$ such that

$$g_i'(e_i) = g_j'(e_j) \text{ for all } i, j \in N. \tag{3.4}$$

In words, world efficiency implies equalization of marginal abatement costs across the countries—a property that is lacking in the case of the

---

9. These conditions are a version of the well-known Lindahl–Samuelson condition for efficient provision of a public good.

noncooperative equilibrium. The efficiency conditions (3.3) lead to the following characterization of efficient consumption profiles.

Proposition 3.1

Under Assumptions 2.1 2.4, all efficient consumption profiles $(x_1^*, \ldots, x_n^*, z^*)$ are generated by the same emission profile $(e_1^*, \ldots, e_n^*)$.

Proof of Proposition 3.1

Suppose contrary to the assertion that two different efficient consumption profiles $(x_1, \ldots, x_n, z)$ and $(x_1', \ldots, x_n', z')$ are generated by emission profiles $(e_1, \ldots, e_n)$ and $(e_1', \ldots, e_n')$, respectively, such that $e_i \neq e_i'$ for some i. On the one hand, if $\sum_{i \in N} e_i = z = z' = \sum_{i \in N} e_i'$, then $\sum_{j \in N} v_j'(z) = \sum_{j \in N} v_j'(z')$, and, thus, by (3.4), $g_i'(e_i) = g_i'(e_i')$ for all $i \in N$, which in turn implies, by strict concavity of the functions g_i, that $e_i = e_i'$ for all $i \in N$; and on the other hand, if $z \neq z'$, say $z < z'$ without loss of generality, then by concavity of the functions v_i, we have $\sum_{j \in N} v_j'(z) \leq \sum_{j \in N} v_j'(z')$, and, thus, by (3.3), we have $g_i'(e_i) \leq g_i'(e_i')$ for all $i \in N$, which in turn implies, by strict concavity of the functions g_i, that $e_i \geq e_i'$ for all $i \in N$. However, this contradicts that $z < z'$. ∎

The proposition implies that alternative efficient consumption profiles differ from one another *only* in terms of the amounts of the consumption good assigned to the countries and not in terms of the emissions and resulting ambient pollution. In other words, efficiency cannot be achieved unless all countries emit according to the unique efficient emission profile $(e_1^*, \ldots, e_n^*)$.[10] However, because the costs and benefits of doing so may differ across the countries, transfers of the consumption good between them to balance the costs and benefits of achieving efficiency may be necessary. Specifically, let $(e_1, \ldots, e_n)$ be the current emission profile.[11] If the countries emit instead according to the unique efficient emission

---

10. This is analogous to the formulation in Coase (1960) in which also the optimal level of the externality is unique and independent of the income distribution.
11. These could be, for example, their BAU or individually optimal international equilibrium emissions.

profile $(e_1^*, \ldots, e_n^*)$, the cost to country i of doing this is $g_i(e_i) - g_i(e_i^*)$ and the benefit is $v_i(\Sigma_{j \in N} e_j) - v_i(\Sigma_{j \in N} e_j^*)$. Though efficiency, by definition, implies $\Sigma_{i \in N} g_i(e_i^*) - \Sigma_{i \in N} v_i(\Sigma_{j \in N} e_j^*) > \Sigma_{i \in N} g_i(e_i) - \Sigma_{i \in N} v_i(\Sigma_{j \in N} e_j)$, it may well be the case, especially if the countries are sufficiently heterogeneous, that for some individual country i, we have $g_i(e_i) - g_i(e_i^*) > v_i(\Sigma_{j \in N} e_j) - v_i(\Sigma_{j \in N} e_j^*)$, that is, the cost of emitting e_i^* instead of e_i is more than the benefit to country i from the abatements of all countries. This is definitely so if, for example, $e_i^* < e_i$ and $v_i'(z) = 0$ for all $z \geq 0$. Clearly, such a country cannot be induced to emit according to the unique efficient emission profile $(e_1^*, \ldots, e_n^*)$, unless the cost $g_i(e_i) - g_i(e_i^*)$ is covered by means of transfers of the consumption good by the other countries.[12]

The above argument for the necessity of transfers to achieve world efficiency is fundamental to the approach in this book. To put it more forcefully, if the countries are sufficiently heterogeneous, then world efficiency cannot be achieved unless the solution concept or mechanism for achieving efficiency allows transfers of the consumption good between the countries.[13] In other words, the outcome of any mechanism or solution concept that rules out transfers between countries cannot be both individually rational and efficient unless the countries are identical or sufficiently similar. Because sovereign countries cannot be forced to

---

12. Indeed, transfers have been proposed in the Paris Agreement and were implicit in the Kyoto Protocol via the assignment of tradable emission quotas and the Clean Development Mechanism (see, e.g., Chander 2003). The Montreal Protocol, which has been hailed as an example of successful international cooperation, explicitly requires transfers. Transfers are also implicit in Coase's (1960) solution for externalities, according to which pollution may be reduced only in exchange for appropriate compensation.

13. Some studies get around this argument for the necessity of transfers between countries for achieving world efficiency by assuming that the countries are identical or sufficiently similar. But as the solution concepts in these studies, by definition, rule out transfers between the countries, their outcomes cannot be both individually rational and efficient if the countries are sufficiently heterogeneous. Thus, the choice between models with and without transfers is not a matter of approach or choice but is dictated by the fact that without transfers, sufficiently heterogeneous countries cannot be induced to follow emission policies— no matter how efficient the policies are from the world perspective—that cost them a lot but benefit them little.

participate in a mechanism whose outcomes are not individually rational, this essentially means that if the countries are sufficiently heterogeneous, then world efficiency cannot be achieved without appropriate transfers of the consumption good between them.[14]

3.2.2. Efficiency and Pollution Rights

Before I discuss further in section 3.3 the issues of whether and how world efficiency can be achieved, I devote the remainder of section 3.2 to studying some properties of the concept in relation to other notions that often inspire designers of an international order. In common parlance, efficiency often has an all-encompassing connotation that sometimes makes people believe that it will "solve all problems." And indeed, who is ever against it? Yet the concept is in fact pretty neutral and, thus, complementary to other notions.

Let me observe first that the concept of efficiency is totally independent of the rights notion. The notion of rights does not appear in Definition 3.3. Similarly, there is no reference to property of any kind that individuals might own: it is thus independent of the institutions that protect property. In other words, in environmental matters, achievement of efficiency is independent of the question of who "owns" the environment and how much of it is available individually—a question that I shall examine shortly. Efficiency is also a concept that does not rest on behavioral assumptions made regarding individual economic agents—in our case, the countries. It is just a well-defined feasible consumption profile for the realization of which nothing is presupposed.

3.2.3. Efficiency and Equity

Equity, or justice, or fairness (I take them as synonyms for the time being) is another concept that serves to characterize feasible consumption profiles. Here also, as in the case of rights, these terms are absent from Definition 3.3 of efficiency. Thus, the concept of efficiency is essentially neutral vis-à-vis equity.

---

14. It will be seen in chapter 7 that transfers are also necessary if climate change is formulated as a dynamic game.

Because, as Proposition 3.1 shows, every efficient consumption profile is generated by the same emission profile and, therefore, the ambient pollution and the total amount of the consumption good produced and available are the same in all efficient consumption profiles, the notions of equity that complement efficiency can be treated—in the context of my basic model—simply by choosing alternative distributions of the total amount of the consumption good produced in an efficient consumption profile: a higher allocation of the consumption good to a country—the emissions and ambient pollution remaining the same—means a higher utility for that country. Because most notions of equity when formulated in terms that are used in economic theory[15] bear precisely on relative utilities of the agents, the choice of distribution of the total amount of the consumption good in an efficient consumption profile implies a relative weighing of the utilities of the economic agents involved.

3.3. THE RATIONALE

3.3.1. Inefficiency of the Spontaneous Equilibria

I now contrast the two categories of states of the system that were identified above, namely, the spontaneous equilibria and the world efficient consumption profiles.

Proposition 3.2

Neither the BAU nor the noncooperative equilibrium is efficient.

Proof of Proposition 3.2

Each of the stated equilibria has been characterized in terms of necessary conditions that they must satisfy: conditions (3.1) or (3.2) for the two types of spontaneous equilibria and (3.3) and (3.4) for world efficiency. Because these conditions are not equivalent, the proof follows.■

---

15. A good example of this is Boadway and Bruce (1984).

This is typical of economies with externalities. Their spontaneous equilibria are inefficient. In other words, externalities generate market failures, and these are taken as a justification for government intervention. In this perspective, the BAU equilibrium is a case of double market failure regarding the environmental externality: first at the domestic level as domestic damage from domestic emissions is ignored when taking production decisions,[16] and second at the international level as the spillover effects of domestic emissions on other countries are ignored. The BAU equilibrium is thus a result of gross "negligence." In fact, we may speak not only of inefficiency, but also of anarchy.

In the case of the noncooperative equilibrium, it may be noted that there are actually two causes of inefficiency: first, the resulting ambient pollution z may be too high [violation of the first equality in (3.3)], and second, the nonequality of marginal abatement costs across countries [violation of (3.4)]. Therefore, in the noncooperative equilibrium, a gain in efficiency at the world level can be achieved even without lowering the (inefficient) level of ambient pollution z by just reallocating the national emissions between the countries so as to equate their marginal abatement costs.[17]

Because I interpret the noncooperative equilibrium as the spontaneous outcome of the interactions that occur between environmentally aware countries, the contents of Proposition 3.2 may be taken as the raison d'être for international negotiations that may aim at achieving world efficiency. Such a statement prompts two questions, however: Why negotiations? Why efficiency?

The first question is of institutional nature. If a consumption profile different from the one that prevails spontaneously is considered more desirable, then how can it be achieved? Compulsion is not an option: in the context of independent and peaceful countries, their sovereignty precludes any coercion being exerted from the outside. A common authority could conceivably compute the solution of the optimization problem, but empowering the authority to implement an efficient consumption

---

16. This is not just market failure but also government failure.

17. This property is emphasized in Kolstad (2000) as the equi-marginal cost principle and is exploited by Helm (2003) in his analysis of initial allocation of tradable emission permits.

profile would require the countries involved to renounce their sovereignty in this sphere. Thus, voluntary negotiations are the only realistic alternative, from an institutional perspective, to induce the countries to agree to a consumption profile other than that in the noncooperative equilibrium.

The second question is less straightforward to answer. Indeed, is it a universally accepted fact that efficiency is an objective of negotiations?[18] Can efficiency be considered as a natural or logical outcome, consciously or unconsciously? Momentarily, I shall take this as an assumption, leaving a discussion of its foundations for later developments, among others in the light of game theory. The notion of "ecological surplus" provides an argument in that direction.

3.3.2. The Ecological Surplus

Within the model of the economic-ecological system that I am working with, the externalities are the only source of the inefficiency just highlighted. The degree of this inefficiency can be measured by means of the notion of an "ecological surplus." Figure 3.1 illustrates this notion in the context of my basic model with just two countries, 1 and 2. Their utilities are measured along the two axes, where the origin corresponds to the theoretical case of no production and consumption in either country. In the case of the BAU equilibrium, the utilities implied by Definition 3.1 are represented by the coordinates of point B; in the case of the noncooperative equilibrium they are, as Definition 3.2 implies, represented by point A; and the set of efficient consumption profiles when translated into utilities for the two countries is depicted by the set of points on the straight line joining the points u_1^* and u_2^*.

The illustration assumes uniqueness of the two types of equilibria,[19] which permits me to identify them unambiguously by points B and A.

---

18. On the one hand, both Pigou (1920) and Coase (1960) propose solutions that aim at achieving efficiency, and, on the other hand, Barrett (2003) argues, under the heading of "broad but shallow" agreements, that in effective agreements it is often the case that efficiency is to some degree traded off against a wider participation.

19. It will be shown in chapter 5 that this is indeed so.

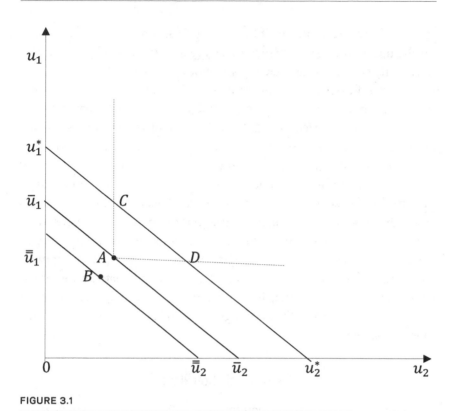

FIGURE 3.1

Equilibria, utility frontier, and ecological surplus in a two-country world

Similarly, the set of points on the straight line joining the points u_1^* and u_2^* illustrates the "utility frontier" as implied by the multiplicity of the efficient consumption profiles. Not all of these points on the line are equally likely or desirable, but selecting among them will come at a later stage of our enquiry. The utility frontier is shown to be a straight line (actually with slope equal to -1) because of my assumption that the utility functions are linear in the consumption good.[20] Briefly stated, the ecological surplus is the gain in the sum of utilities of both countries that

20. The more general textbook assumption of strictly quasi-concave utility functions implies a utility frontier that is not necessarily linear.

the world can secure for itself by moving from point B or A to a point on the utility frontier. Efficiency means realizing the ecological surplus in full. Inefficiency means leaving some of it unrealized.

But even if there is consensus on achieving efficiency, the multiplicity of efficient consumption profiles underlying figure 3.1 leaves a considerable degree of freedom as to which one of them may be chosen as an outcome of an international environmental agreement. However, a simple look at figure 3.1 immediately suggests that not all efficient consumption profiles are equally acceptable; for instance, the points outside the line segment CD are clearly not because it hurts common sense that a sovereign country will agree to an outcome that would make it worse off than what it can achieve on its own without cooperation of the other countries. Though every point on the line segment CD implies efficiency, it also amounts to choosing a distribution of the ecological surplus, and that is why negotiations among the countries may be necessary even if they all agree not to leave any ecological surplus unrealized.

3.3.3. Why Pollution Rights?

Accepting the noncooperative equilibrium as the starting point of negotiations implies that the generators of the externality have arrogated to themselves an unlimited right to pollute. It was also mentioned that the reason for this situation is the absence of any international authority empowered to grant and enforce pollution rights. Thus, if climate change is to be addressed, the self-interested countries themselves must agree to limit their rights to pollute. How is that possible?

Let me remark first that the issue is not exclusively international: If a country indeed agrees to limit its emissions, how can it in turn limit the pollution rights of economic agents such as households, firms, or public bodies such as municipalities or governmental agencies within the country? How can or should that be done? Should it assign rights to polluters or pollutees?

This last question has been debated in the literature at length. However, a simplification was proposed by Coase (1960), who argued that, whatever the initial allocation of rights either to the polluter or to the pollutee,

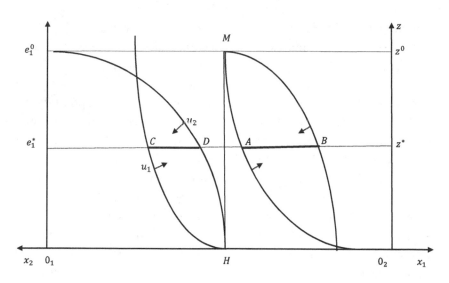

FIGURE 3.2

A one polluter–one pollutee economy and the Coase theorem

an efficient outcome will occur by letting the parties engage in mutually advantageous spontaneous bargaining. Thus, as far as efficiency is concerned, it does not matter who has the right to pollute provided the rights are clearly defined and assigned to either one of the parties. This has come to be known as the Coase theorem. It is well illustrated by the Edgeworth box–like diagram[21] in figure 3.2, which is constructed from a two-agent version of my basic model when the countries are "specialized" in the following sense: only one of them, the "polluter," is the generator of the externality and is not affected by pollution, while the other country, the "pollutee," does not generate externality but is affected by the externality generated by the polluting country.

Formally, let country 1 be the polluter with utility function $u_1(x_1, z) = x_1$ (as $v_1(z) = 0$ for all z, by assumption) and a strictly increasing concave production function $g_1(e_1), 0 \leq e \leq e_1^0$, with $g_1(0) = 0$. Let country 2 be the

21. Similar diagrams have appeared previously in Varian (1990) and Tulkens and Schoumaker (1975).

pollutee with utility function $u_2(x_2, z) = x_2 - v_2(z)$ in which $v_2(0) = 0$, and suppose country 2 produces nothing that uses fossil fuels. Let the transfer function for this one polluter–one pollutee economy be simply $z = e_1$. Let $w_1 > 0$ and $w_2 > 0$ denote the initial endowments of the consumption good of the two countries with w_1 increasing to the right of point 0_1 and w_2 to the left of point 0_2. Then, point H on the horizontal line in the box of length $w_1 + w_2$ represents the initial endowments of the consumption good of both countries, and thus $0_1 H = w_1$ and $0_2 H = w_2$.

The set of points (e_1, x_1) that satisfy the equation $x_1 + g_1(e_1) = w_1$ constitutes an indifference curve for country 1 yielding the utility level w_1; it is represented in figure 3.2 by the curve labeled u_1 passing through point H. The convexity of this curve results from the assumed strict concavity of the production function $g_1(e_1)$. Letting w_1 vary, a family of congruent indifference curves is generated, to the left of the indifference curve u_1 for lower values of w_1 and to the right of u_1 for higher values. Similarly, the set of points that satisfy the equation $x_2 - v_2(z) = w_2$ constitutes an indifference curve passing through point H and yielding utility level w_2; it is represented in figure 3.2 by the curve labeled u_2, and the convexity of this curve results from the assumed convexity of the disutility function v_2. Letting w_2 vary, a family of congruent indifference curves is generated to the left of the indifference curve u_2 for higher values of w_2 and to the right of u_2 for lower values.

This diagram illustrates Coase's reasoning about rights assignment and efficiency as follows. The line joining the points e_1^* and z^* represents the set of points of tangency of the indifference curves of the two countries, and, therefore, it represents the set of efficient allocations or the "contract curve."[22] On the one hand, if the right to exert the externality is assigned to the polluter, the individual behaviors of the polluter and the pollutee imply a noncooperative equilibrium allocation represented by point M, where the polluter emits e_1^0 and the pollutee suffers $z^0 = e_1^0$ amount of pollution.

---

22. The efficient level e_1^* of the externality is the same for any assignment of the rights in figure 3.2, as the utility functions, by assumption, are quasi-linear. With more general strictly quasi-concave utilities, the efficient level of the externality may vary because of income effects.

It is seen that in this situation, an ecological surplus can be obtained if the two parties negotiate a mutually beneficial move to any of the allocations represented by points on the line segment AB enclosed between the indifference curves passing through point M. On the other hand, if the right is assigned to the pollutee, the noncooperative equilibrium consists in the pollutee denying the externality to be exerted and the two parties consuming their respective initial endowments, a situation represented by point H. Here, too, an ecological surplus can be obtained by negotiating a move to any point on the line segment CD. Thus, in either case, existence of an ecological surplus is identified (not the same, though), and efficiency can be reached, no matter how the right is assigned. Apart from taking the decision on the assignment of rights, no state intervention is necessary for efficiency to be obtained, as mutually advantageous private bargaining among the parties would lead to efficiency.

Another major aspect of the Coase theorem, vividly illustrated by the role of points M and H in the diagram, is that the question of who should have the right is essentially a distributional one. Indeed, these points play exactly the same role as alternative initial endowments in the standard Edgeworth box illustration of an exchange economy in general equilibrium theory. And similarly, rights on the environment, whether given to polluters, pollutees, or anyone else, are like property rights on commodities: they provide wealth to their owners, either when they exert them or when they sell them.

The example behind figure 3.2 illustrates that there are commonalities between the social interactions that occur through bargaining about externalities and those that occur in the exchange of commodities. These are clear, however, only after the rights to exert the externality have been specified. As illustrated by the Edgeworth box–like diagram, there are at least two possibilities in this respect: with the rights assigned to the polluter, the bargaining is between the polluter reducing its emissions and the pollutee offering some economic good in exchange; if rights are to the pollutee, the bargaining is between the pollutee allowing the polluter to act and the polluter offering some economic good in exchange for that permission. This raises the question whether there is an objective criterion for choosing between the various allocations of rights, given that every allocation leads to an efficient outcome.

Auctioning is a conceivable method for allocating pollution rights and can be compared to auctioning land or other common property resources.[23] Because auctioning provides an opportunity to both polluters and pollutees to acquire pollution rights, the alternatives as to who among these two classes of agents should have the rights seems to disappear. But this is an illusion: because the rights in question are essentially permits to use the environment, that is, to emit polluting substances, the very notion of permits contains the fact that those who do not own them may not emit. Thus, the very idea of auctioning permits is an admission that the rights do not belong to those agents who cause the externality (i.e., the "polluters"). Those who do want to pollute must acquire the right to do so, be it directly from the prospective pollutees by means of a contract or from some public authority whose role as an intermediary is in fact to protect the pollutees' rights and properly organize the way in which they can be exerted. By contrast, the often mentioned and practiced technique of giving away rights to polluters under the name of "grandfathering" is clearly an allocation of rights to polluters.

In most civilized societies, perpetrators of damage are required to pay compensation for the damage they impose on victims. This is the most common form of the "polluters pay" principle: polluters should compensate for the discomfort and the damage they cause and pay for any cleanup required. The requirement of compensation is obviously an implication of the rights being attributed to pollutees. In the Edgeworth box–like figure 3.2, compensation for damage imposed is illustrated as follows. When exerting the externality freely, the polluter brings the system to point M. This causes the pollutee a damage that is measured in utility terms by the distance between the pollutee's two indifference curves passing through points M and H. The compensation in terms of the consumption good is thus equal to the length of the line segment DB enclosed between the pollutee's two indifference curves. Notice the subjective element in the determination of this amount, which is entirely determined by the shape of the pollutee's indifference curves; that is, its preferences.

---

23. As suggested in a note by Ellerman (2005).

It may be observed, finally, that when a natural phenomenon is at the origin of the damage—such as a natural forest fire in California or the eruption of a volcano in Iceland—nature is the generator of the externality, and the "victim pays" principle is usually accepted, albeit often amended by solidarity from others on a voluntary basis or intervention by the state. By contrast, when humans are at the origin of damage, it is expected that the perpetrators of the damage should pay compensation. The reason for this difference lies in the fairly universal notion of responsibility, which applies in the latter case while not in the former.

3.3.4. Coase and Climate Change

In continuation of my argument of why allocating pollution rights may be necessary, let us recall from chapter 1 that Coase took for granted the existence of a public authority that can assign and enforce pollution rights. He did not envisage an international problem such as climate change where there is no such authority. The Coase theorem is known to apply well to the case of what were categorized as unilateral and personalized externalities in chapter 2 and assumed in the illustrative example in figure 3.2. But in the case of climate change, as noted earlier, externalities are not only reciprocal but also have public good characteristics in that every country is both a polluter and a pollutee and affected to the same extent by climate change independently of how much others are affected by it.[24] Therefore, the Coase theorem as such is of little use for our purposes. As argued in Chander (2011) and to be discussed in chapter 5, in the absence of any authority to allocate and enforce rights, the countries themselves may decide the allocation of rights that are mutually acceptable to all countries. Thus, the problem of allocating rights is strategic in nature. It is to be tackled by means of game theory in chapter 5.

The arguments just presented bear on the acceptability of pollution rights. Their justification rests on the fact that international agreements are necessarily voluntary. Another aspect of them, also linked with their

---

24. See Kolstad (2000, chapter 6) for an illustrative example of why the Coase theorem may fail in the case of externalities with public good characteristics.

voluntary character, appears if one considers the process through which the outcome is obtained. In this perspective, and viewing this process not in its temporal dimension but rather in its logical structure, international agreements, or treaties, can be seen as contracts—bilateral or multilateral.[25] An essential component of a contract is that there is a quid pro quo, in other words an exchange between the parties that, in the case of reciprocal externalities such as climate change, could be to limit one's own pollution right in exchange for others limiting their pollution rights.

Additional complications in the assignment of pollution rights arise when climate change is treated as a stock externality. As noted in chapter 1, economics has started to make a distinction between stock and flow externalities only recently. In fact, treatment of climate change as a stock externality raises newer questions regarding the applicability of the Coase theorem and allocation of rights not just among the agents belonging to the same generation but also across those belonging to different generations. It gives the problem of allocating rights an intergenerational dimension. Beside that, the emitters and recipients of the externality may be separated in time in such a way that they cannot meet or interact (e.g., they may belong to different non-overlapping generations),[26] making it impossible for them to engage in Coasian bargaining or trade in rights. We leave these issues for chapters 7 and 8, which treat climate change as a stock externality in contrast to chapter 5, which treats it as a flow externality.

3.3.5. A Note on the Value of Environment

A question arises: Why is it justified to attach value to the environment or to some of its components, and if so, why attempt to measure that value? It is customary in environmental economics to distinguish between three possible forms of environmental values, called (1) usage value, (2) option value, and (3) existence value. Let us briefly review them before examining their relevance for climate change.

---

25. See Reuter (1995).
26. If elapse of time is substituted for river flow and distance, this is analogous to upstream pollution that affects people downstream.

Usage value is the value that is attributed to an environmental resource by those who effectively use it. In market economies, it is considered to be reflected in the market price of the resource, provided a market for it exists and is sufficiently competitive. Option value is meant to capture the value attached to the mere possibility of using the resource by people who currently make no use of it. It is of major importance in problems that involve irreversible actions. Existence value, finally, corresponds to the value attached to the existence of a resource, compared to its loss, by people who do not use it and never intend to. Option and existence values, unlike usage value, are more difficult to measure, as typically no market prices can be attached to them. For this reason, option and existence values are often ignored in actual policy/decisions concerning the environment or at best their very rough estimates are used, and usage value is the only one that is mostly used with some precision in cost-benefit analysis of environmental decisions.

What relevance do these three forms of environmental value have in the context of climate change? The impact of climatic change, typically rise in temperatures, may cause damage to resources that have usage, option, and existence value; for example, rise in sea level or extinction of some species. In most of this book, I assume that all these three values are subsumed in the disutility or the damage functions $v_i(z), i \in N$.

In the absence of a market for an environmental resource, its value is measured by estimates of what is sometimes referred to as willingness to pay or willingness to accept. However, climate change is not really a public good that is produced on purpose; it is rather to be avoided! Therefore, the expression "willingness to pay" is to be understood in the particular sense of willingness to pay to avoid the change to occur. This is illustrated in figure 3.3, where for a given country's utility function and consumption bundle $(\bar{x}, \bar{z})$, the maximum willingness to pay to avoid a change in ambient pollution by an amount LM is represented by the amount AB of the consumption good.

The alternative notion of "willingness to accept" a *detrimental change* is used mostly in the cost-benefit literature. This notion is illustrated by the amount BC for a change in the ambient pollution by an amount LM. This notion is appropriate for evaluating the benefits of adaptation decisions (i.e., the minimum acceptable compensation to be obtained to cover the

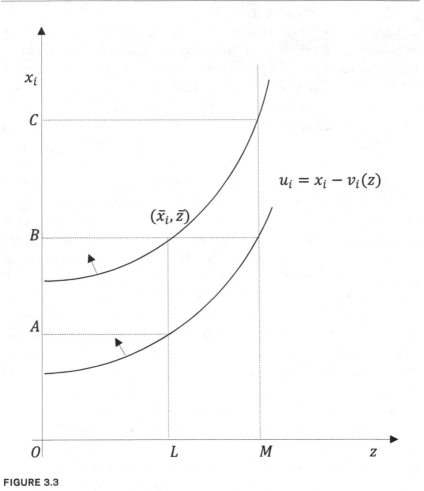

FIGURE 3.3

The marginal willingness to pay

future damage entailed by the change LM), whereas the willingness to pay to avoid the change is rather relevant for mitigation actions (i.e., emissions abatement), as this is exactly what these actions will entail. In the basic model used throughout this book, and for infinitesimal changes of the ambient pollution z, the two notions coincide and are represented simply by the derivative of the damage function, $v_i'(z)$. Thus, in what follows, no distinction needs to be made or will be made between the twin notions of willingness to pay and willingness to accept.

4

THE CORE OF A STRATEGIC GAME

T he inefficiency of the noncooperative equilibrium noted in the preceding chapter is the result of two essential features of my basic model: first, the absence of any supranational authority, and second, the absence of any markets for externalities. There are no markets for externalities because the countries do not have well-defined pollution rights, which is a prerequisite for any market to operate, and there are no well-defined pollution rights because there is no supranational authority that is empowered to assign and enforce pollution rights.[1] *Every* country arrogates to itself an unlimited right to pollute, and *no* country has the right not to be polluted. The only way in which the countries can overcome the inefficiency of the noncooperative international equilibrium, if at all, is then by negotiations and voluntary agreements among them to control climate change.[2]

What should the negotiations bear on, and what comprises a voluntary agreement? The purpose of this chapter is to define these terms in the primitive framework of a general strategic game. I choose this framework because, as will be seen in the next chapter, my basic flow model of climate change can be interpreted as a strategic game, and the dynamic game model of climate change introduced in chapter 7 also consists of a

---

1. As will be seen in the next chapter, assigning *appropriate* pollution rights that can be traded freely on a competitive market can indeed lead to an efficient outcome.
2. The current international negotiations on climate change under the aegis of the United Nations Framework Convention on Climate Change (UNFCCC) can be seen as an attempt to overcome the inefficiency of the status quo (i.e., the prevailing noncooperative equilibrium).

sequence of nested strategic games. A relevant and central concept in this framework is the core of a strategic game. As will be seen, the conventional α- and β-cores (von Neumann and Morgenstern 1953) of a strategic game are not suitable for our purpose.

This chapter motivates and introduces an alternative core concept, to be called the γ-core, for a *general* strategic game,[3] which is then applied to specific strategic games in the following chapters.[4] It will be shown that the γ-core is nicely related to the strong Nash equilibria or the coalition-proof Nash equilibria—besides the conventional α- and β-cores—of a strategic game. The next chapter introduces a specific game, called the environmental game, and establishes additional properties of the γ-core that are specific to the environmental game.

The motivation for the γ-core comes from an observation made by Mäler (1989), who pointed out that the usual α- and β-core concepts when applied to models of global pollution result in too large a set of outcomes; in fact, the entire set of efficient outcomes.[5] This is because under these core concepts, the players outside a deviating coalition can more than offset the pollution abatement carried out by the deviating coalition. Thus, there is no minimum payoff that a coalition can ensure for itself by reducing its own pollution. In contrast, a player in the standard public good model can at most not contribute anything toward the public good provision but cannot reduce the provision made by others.[6]

---

3. In contrast, Maskin (2003) and Huang and Sjöström (2006) introduce core concepts in the primitive framework of a partition function and thereby abstract away from the strategic interactions that are behind the coalitional payoffs.

4. The γ-core was introduced originally in Chander and Tulkens (1995, 1997), but in the framework of a game that does not fit the standard definition of a strategic game in the sense that the strategy set of a coalition is larger than the Cartesian product of the strategy sets of its members.

5. As pointed out by Ray and Vohra (1997), the same is also true in a model of an oligopoly. Similarly, Shapley and Shubik (1969) and Shubik (1984) note that the α-core of the "lake game" is too large.

6. This is analogous to a tale from the past according to which the churchgoers in the Middle Ages could not only not put any money in the hat circulated for collecting donations for public purposes but could even take out the money donated by others. However, this might have been allowed on purpose to render help in cases of extreme poverty.

Thus, we encounter a stronger form of "free riding" than in the standard public good model. As a result, the conventional α- and β-core concepts can no longer be meaningfully applied.

The γ-core concept is based on a more plausible assumption concerning the behavior of players outside a deviating coalition in that, unlike the α- and β-core concepts, the players outside the deviating coalition are assumed to adopt their *individually* best reply strategies and not max-min or min-max the payoff of the deviating coalition. This results in a non-cooperative equilibrium between the deviating coalition and the other players acting individually, in which the members of the deviating coalition play their best reply *joint* strategy against the *individually* best reply strategies of the remaining players. As will be shown, this implies a stronger concept in the sense that the γ-core of a strategic game is *in general* smaller than the α- and β-cores, and, as will be made clear in chapters 5 and 6, the qualifier "individually" is the key.

4.1. STRATEGIC GAMES

Because a strategic game provides a natural primitive framework for analyzing the interactive decision problem of climate change (as noted in chapter 1), we take note of some basic concepts relating to strategic games.

A *strategic game* is a triple (N, T, u) where $N = \{1, 2, \ldots, n\}$ is the *set of players*, T_i is the *set of strategies* of player i, $T = T_1 \times T_2 \times \cdots \times T_n$ is the *set of joint strategies*, u_i is the *utility* or *payoff* of player i, and $u = (u_1, u_2, \ldots, u_n)$ is the *vector of utilities* or *payoffs*.

A *strategy profile* of the players is denoted by $t = (t_1, \ldots, t_n)$ such that $t_i \in T_i$ for each $i \in N$. If the players choose strategies $(t_1, t_2, \ldots, t_n), t_i \in T_i$, then the payoff of player i is $u_i(t) \in R$.

A central feature of strategic games is that the payoff $u_i(t)$ of each player i depends not only on his own strategy t_i but also on the strategies of other players. This means each and every player must take the probable choices of other players into account in choosing his own strategy.

Throughout this book, I consider *games with transferable utility*. By this I mean that each player can increase the payoff of any other player by one unit at a cost of just one unit to itself.

A game (N, T, u) is a *game of complete information* if each player knows $N, T,$ and u. Otherwise, it is a *game of incomplete information.*[7] We will be concerned mostly with games of complete information.

4.1.1. Equilibrium Concepts in Strategic Games

Nash equilibrium is a central concept in game theory. It is based on the assumption that each player is rational in the sense that it chooses a strategy in a way that maximizes its payoff, given the strategies of other players. For the time being, I assume that the players do this independently in the sense that they do not act in concert, and no player can affect the strategy choice of any other player. Let $t_{-i} \equiv (t_1, \ldots, t_{i-1}, t_{i+1}, \ldots, t_n)$ and $(t_i, t_{-i}) \equiv (t_1, \ldots, t_{i-1}, t_i, t_{i+1}, \ldots, t_n)$.

Definition 4.1

A Nash equilibrium of the strategic game (N, T, u) is a strategy profile $\bar{t} \in T$ such that $u_i(\bar{t}) \geq u_i(t_i, \bar{t}_{-i})$ for all $t_i \in T_i$ and all $i \in N$.

Various arguments have been put forward in the literature to justify Nash equilibrium as a plausible outcome of a game. One argument is that each player assumes rationality on the part of the other players and uses that assumption to analyze the game from its standpoint. The argument relies implicitly on the assumption of complete information. That is because viewing the game from the perspective of other players in the same manner as a player views it from its own perspective is possible most naturally only under complete information. Under some circumstances, however, complete information is not needed. One such instance is when the obviously correct strategy for a player to choose does not depend on the strategy choices of the other players.

A strategy t_i of player i is *dominant* if it is as good as or better than any other strategy available to the player, no matter what strategies might be selected by the other players; that is, $u_i(\bar{t}_i, t_{-i}) \geq u_i(t_i, t_{-i})$ for all $(t_i, t_{-i}) \in T$.

---

7. To be precise, a game of incomplete information is one in which each player knows $N, T,$ and u, but not their specific realization for the other players they are playing against.

Definition 4.2

In the strategic game (N, T, u), the strategy profile $\bar{t}$ is a dominant strategy equilibrium if $\bar{t} \in T$ and for each player $i \in N$, $\bar{t}_i$ is a dominant strategy.

Clearly, a dominant strategy equilibrium is also a Nash equilibrium, but in order to choose its strategy, a player neither needs to know the strategies of the other players nor assume rationality on their part. In general, games may not have a dominant strategy equilibrium, but it can occur in some cases.

It is well known that not every strategic game has a Nash equilibrium in pure strategies. Sufficient conditions for existence of a Nash equilibrium have been investigated extensively. For the sake of completeness, I include a sufficient condition, though at the cost of exposing the reader to some advanced mathematical techniques that will not be used in the rest of the book. Readers not interested in technical details may skip the next few pages.

For each i and t_{-i}, define

$$B_i(t_{-i}) = \{t_i \in T_i : u_i(t_i, t_{-i}) \geq u_i(t_i', t_{-i}) \text{ for all } t_i' \in T_i\}.$$

The set-valued function B_i is the *best-response function* of player i, given the strategies of the other players. To show that a game has a Nash equilibrium, it is sufficient to show that there is a joint strategy $\bar{t} \in T$ such that $\bar{t}_i \in B_i(\bar{t}_{-i})$ for all $i \in N$. Let $B : T \to T$ denote the set-valued function defined as $B(t) = \times_{i \in N} B_i(t_{-i})$. Then, the required sufficient condition can be written more compactly as $\bar{t} \in B(\bar{t})$. A well-known *fixed-point theorem* gives conditions on B under which there indeed exists such a $\bar{t}$.

Lemma (Kakutani's Fixed-Point Theorem)

Let X be a compact subset of R^n and let $f : X \to X$ be a set-valued function for which (1) for each $x \in X$ the set $f(x)$ is nonempty and convex and (2) the graph of f is closed (i.e., for all sequences $\{x_n\}$ and $\{y_n\}$ such that $y_n \in f(x_n)$ for all n, $x_n \to x$, and $y_n \to y$, we have $y \in f(x)$). Then there exists an $\bar{x} \in X$ such that $\bar{x} \in f(\bar{x})$.

Proposition 4.1

The strategic game (N, T, u) has a Nash equilibrium if (1) each T_i is a compact convex set of a Euclidean space, (2) each $u_i(t)$ is defined and continuous for all $t \in T$, and (3) each $u_i(t_i, t_{-i})$ is quasi-concave with respect to $t_i \in T_i$ for all $(t_i, t_{-i}) \in T.$[8]

Proof of Proposition 4.1

For every $i \in N$, the set $B_i(t_{-i})$ is nonempty because u_i is continuous and T_i is compact. It is convex, as u_i is quasi-concave on T_i. The set-valued function B has a closed graph, as each u_i is continuous. Thus by Kakutani's theorem, B has a fixed point. As noted, every fixed point of B is a Nash equilibrium of the game. ∎

Stronger Concepts of Nash Equilibrium

The concept of Nash equilibrium assumes that each player chooses its strategy independently. Stronger concepts of Nash equilibrium obtain by relaxing this assumption and allowing that some players may form a coalition (i.e., a group of players who decide to act together as one unit, relative to the rest of the players) and choose their strategies in concert while the remaining players continue to choose their strategies individually and independently as before. Because utility is transferable, we assume that the members of a coalition choose their strategies so as to maximize the total utility of the coalition.

The first solution concept for strategic games that involves such coalitional behavior is the notion of *strong Nash equilibrium* due to Aumann (1959). A subset of players or a *coalition* is denoted by S, and the subset of players not in S is denoted by $N \backslash S$. It will be convenient to denote a *strategy of a coalition S* by $t_S \equiv (t_i)_{i \in S}$ and the set of all strategies of coalition S by the Cartesian product of the strategy sets of its members: that is, by $T_S = \underset{i \in S}{\times} T_i$. Let $t_{-S} \equiv (t_j)_{j \in N \backslash S}$ denote the strategies of the players not in S and $(t_S, t_{-S}) \equiv (t_1, \ldots, t_n)$.

---

8. That is, for every the set $\bar{t} \in T$, the set $\{t_i \in T_i : u_i(t_i, \bar{t}_{-i}) \geq u_i(\bar{t})\}$ is convex.

Definition 4.3

A strong Nash equilibrium of the strategic game (N, T, u) is a strategy profile $\bar{t} \in T$ such that $\sum_{i \in S} u_i(\bar{t}) \geq \sum_{i \in S} u_i(t_S, \bar{t}_{-S})$ for all $t_S \in T_S$ and all $S \subset N$.

In words, no coalition, taking the strategies of its complement as given, can cooperatively deviate in a way that improves its payoff. Because coalitions consisting of single players should also not be able to improve their payoffs by deviations, a strong Nash equilibrium is also a Nash equilibrium. Furthermore, because the coalition of all players should also not be able to improve its payoff, a strong Nash equilibrium implies an efficient outcome that maximizes the total payoff of all players.

As Bernheim, Peleg, and Whinston (1987) explain, a strong Nash equilibrium may fail to exist because the concept is "too strong." They note that in a strong Nash equilibrium, coalitions are allowed too much freedom (in fact, complete freedom) in choosing their joint deviations: while the whole set of players must originally be concerned with arriving at an agreement that is immune to deviation by any coalition, no deviating group of players (including the coalition of the whole) faces a similar restriction. Therefore, they propose an alternative concept of a coalition-proof Nash equilibrium, which, like the strong Nash equilibrium, also takes into account coalitional deviations but is less stringent. Indeed, they introduce the notion of self-enforceability, which requires that a deviation is valid if no proper subcoalition can reach a mutually beneficial agreement to deviate from the deviation. Likewise, any potential deviation is judged by the same criterion, and so on, the equilibrium being defined recursively. Then a coalition-proof Nash equilibrium is a strategy profile that is robust to self-enforcing deviations. Because there can be coalition-proof Nash equilibria that are not immune to all deviations, a coalition-proof Nash equilibrium may not be a strong Nash equilibrium.

To define a coalition-proof Nash equilibrium, I introduce some preliminary notation. For any strategic game (N, T, u) and any *fixed* strategy profile $\bar{t}$, define the reduced game for coalition S, given $\bar{t}$, as $(S, T_S, \bar{u}_S)$ where $\bar{u}_S = (\bar{u}_i)_{i \in S}$ and $\bar{u}_i(t_S) = u_i(t_S, \bar{t}_{-S})$ for all $t_S \in T_S$. In words, the reduced game is obtained by fixing the strategies of all players outside S and defining the payoff function of every player given this fixed strategy of the outside players. A coalition-proof Nash equilibrium is then defined recursively.

Definition 4.4

(i) In a single-player strategic game (N, T, u), $\bar{t} \in T$ is a *coalition-proof Nash equilibrium* if and only if $\bar{t}$ maximizes $u_1(t)$. (ii) Let $n > 1$ and assume that coalition-proof Nash equilibrium has been defined for games with fewer than n players. Then: (a) For any game (N, T, u) with n players, $\bar{t} \in T$ is *self-enforcing* if, for all $S \subset N, S \neq N, \bar{t}_S$ is a coalition-proof Nash equilibrium in the game $(S, T_S, \bar{u}_S)$. (b) For any game (N, T, u) with n players, $\bar{t} \in T$ is a coalition-proof Nash equilibrium if it is self-enforcing and if there does not exist another self-enforcing strategy vector $t \in T$ such that $u_i(t) > u_i(\bar{t})$ for all $i \in N$.

Clearly, a strong Nash equilibrium is a coalition-proof Nash equilibrium, and a coalition-proof Nash equilibrium is a Nash equilibrium. As noted earlier, a game may have a coalition-proof Nash equilibrium but no strong Nash equilibrium. Indeed, as will be seen, the environmental game introduced in chapter 5 admits a coalition-proof Nash equilibrium but no strong Nash equilibrium. I now introduce an alternative equivalent definition of a strong Nash equilibrium by using the notion of an induced game.

Given a strategic game $\Gamma = (N, T, u)$ and a coalition $S \subset N$, the *induced strategic game* $\Gamma^S = (N^S, T^S, u^S)$ is defined as follows:

- The player set is $N^S = \{S, (j)_{j \in N \backslash S}\}$; that is, coalition S and all $j \in N \backslash S$ are the players (thus the game has $n-s+1$ players).[9]
- The set of strategy profiles is $T^S = T_S \times_{j \in N \backslash S} T_j$, where $T_S = \times_{i \in S} T_i$ is the strategy set of player S, and T_j is the strategy set of player $j \in N \backslash S$.
- The vector of payoff functions is $u^S = (u_S^S, (u_j^S)_{j \in N \backslash S})$, where $u_S^S(t_S, t_{-S}) = \sum_{i \in S} u_i(t_S, t_{-S})$ is the payoff function of player S, and $u_j^S(t_S, t_{-S}) = u_j(t_S, t_{-S})$ is the payoff function of player $j \in N \backslash S$, for all $t_S \in T_S$ and $t_{-S} \in \times_{j \in N \backslash S} T_j$.

---

9. The lowercase letters n and s denote the cardinality of sets N and S, respectively.

Notice that if $(\tilde{t}_S, \tilde{t}_{-S}) = \tilde{t}$ is a Nash equilibrium of the *induced game* Γ^S, then $u_S^S(\tilde{t}_S, \tilde{t}_{-S}) = \sum_{i \in S} u_i(\tilde{t}_S, \tilde{t}_{-S}) \geq \sum_{i \in S} u_i(t_S, \tilde{t}_{-S})$ for all $t_S \in T_S$.[10] Thus, for each $S \subset N$, a Nash equilibrium of the induced game Γ^S assigns a payoff to coalition S, which it can obtain without cooperation from the remaining players. Notice that a Nash equilibrium of a strategic game Γ is a Nash equilibrium of each induced game $\Gamma^{(i)}$, $i \subset N$, and a Nash equilibrium of an induced game is defined more explicitly as follows.

Definition 4.5

Given a strategic game $\Gamma = (N, T, u)$ and a coalition $S \subset N, \tilde{t}$ is a Nash equilibrium of the induced game Γ^S if $\tilde{t} \in T$, $\sum_{i \in S} u_i(\tilde{t}) \geq \sum_{i \in S} u_i(t_S, \tilde{t}_{-S})$ for all $t_S \in T_S$, and $u_j(\tilde{t}) \geq u_j(t_j, \tilde{t}_{-j})$ for all $t_j \in T_j$ and all $j \in N \backslash S$.

An Alternative Definition of Strong Nash Equilibrium

Given a strategic game $\Gamma = (N, T, u)$, a strategy $\bar{t} \in T$ is a strong Nash equilibrium of the strategic game Γ if and only if it is a Nash equilibrium of every induced game $\Gamma^S, S \subset N$. Notice that in a Nash equilibrium of an induced game Γ^S, the coalition S has complete freedom in choosing its strategy. We can define an alternative concept by restricting the strategic choices of the coalition in the same manner as in a coalition-proof Nash equilibrium vis-à-vis a strong Nash equilibrium.[11] However, with restricted strategic choices, the payoff of a deviating coalition is generally lower, and as in the case of the coalition-proof Nash equilibrium, that would only lead to a Nash equilibrium of the strategic game (N, T, u). The result will be a concept that is weaker in the sense of assigning lower payoffs to the coalitions than the one introduced later. Therefore, I shall not pursue or apply the alternative concept in the rest of the book. Instead, in keeping with the spirit of our exercise to understand the outcome of free and unconstrained negotiations, I shall continue to assume that coalitions have complete freedom in choosing their joint strategies.

---

10. Note that a Nash equilibrium of an *induced game* may not necessarily be an equilibrium of the *original* game.

11. See Bernheim, Peleg, and Whinston (1987) and Moreno and Wooders (1996).

For games with transferable utility, a strengthening of the conditions that are sufficient for the existence of a Nash equilibrium (Proposition 4.1) is sufficient for the existence of a Nash equilibrium of an induced game, as the following proposition shows.

Proposition 4.2

Given a strategic game $\Gamma = (N, T, u)$, each induced game Γ^S, $S \subset N$, admits a Nash equilibrium if (1) each T_i is a compact and convex set of a Euclidean space, (2) each $u_i(t)$ is defined and continuous for all $t \in T$, and (3) each $u_i(t)$ is quasi-concave with respect to each $t \in T$.[12]

Notice that condition (3) in Proposition 4.2 is stronger than the corresponding condition in Proposition 4.1. Technically weaker conditions that ensure the existence of a Nash equilibrium of each induced game Γ^S are possible, but I do not pursue those here because the games we study in this book satisfy the stronger condition. Let s denote the cardinality of set S.

Proof of Proposition 4.2

Consider the strategic game consisting of coalition S as a single player with $\sum_{i \in S} u_i(t_S, t_{-S})$ as its payoff function and the remaining players in $N \backslash S$ with payoff functions $u_j(t_S, t_{-S})$ for each $j \in N \backslash S$. This is a game with $n - s + 1$ players with strategy sets T_S of player S and T_j of player $j \in N \backslash S$. Clearly, the so-defined game satisfies all the conditions for the existence of a Nash equilibrium as in Proposition 4.1, including condition (3). In particular, given the stronger condition (3), the payoff function $\sum_{i \in S} u_i(t_S, t_{-S})$ of player S is quasi-concave on its strategy set T_S, which is a compact and convex set of a Euclidean space. It is also defined and continuous over the joint strategy set $T_S \times_{j \in N \backslash S} T_j$. The proof now follows by noting that a Nash equilibrium of this modified game is a Nash equilibrium of the induced game Γ^S. ∎

---

12. That is, for every $\bar{t} \in T$, the set $\{t \in T : u_i(t) \geq u_i(\bar{t})\}$ is convex.

The existence of a Nash equilibrium of each induced game alone, however, is not sufficient for our purpose. Another equally important issue is that an induced game may admit multiple Nash equilibria. I resolve this issue by selecting for each coalition S a Nash equilibrium of the induced game Γ^S in which the payoff of coalition S is highest.[13] Though, as will be argued later, this makes the conditions for the existence of a nonempty γ-core more stringent, and other refinement criteria for selecting among the equilibria are possible,[14] I do not pursue those here because, as will be shown in the next chapter, each induced game of the environmental game admits a unique Nash equilibrium.[15] But if the induced game Γ^S does have multiple Nash equilibria, then any one of them with the highest payoff for S can be selected.[16] In this way, a unique payoff can be assigned to each coalition.

It is worth noting, however, that, unlike a Nash equilibrium and other familiar equilibrium concepts, a Nash equilibrium of an induced game of a strategic game is not a solution concept for the strategic game. Instead, it is an intermediate notion that leads to a solution concept, namely, the γ-core of a strategic game, to be introduced in section 4.2.3.

4.2. COALITIONAL FUNCTIONS

Let $w(S)$ denote the highest payoff that members of a coalition $S \subset N$ can jointly achieve using some strategies; that is,

$$w(S) = \max_{t_s \in T_S} \sum_{i \in S} u_i(t_S, t_{-S}).$$

---

13. That is clearly possible in games with compact (or finite) strategy sets and continuous payoff functions.

14. Refinements of the set of equilibria became a veritable cottage industry in the late 1980s and 1990s (see, e.g., Grossman and Perry 1986). I will return to this issue after I have formally introduced the concept of γ-core.

15. The same is true in the case of an oligopoly game (see, e.g., Chander 2010).

16. Such a payoff exists if the strategy sets are compact (or finite) and the payoff functions are continuous.

However, notice that $w(S)$ depends not only on the strategies of the players in coalition S but also on the strategies of the players not in S over which S has no control. This typical feature of strategic games requires that in order to determine the highest payoff achievable by a coalition, we must specify not only the strategies of its members but also of the players outside the coalition. One natural way suggested by the equilibrium concepts developed so far is to assume that the strategies of the players in $N \setminus S$ are the same as in a Nash equilibrium of the induced game Γ^S. The payoff $w(S)$ of coalition S may then be defined as its highest payoff in any Nash equilibrium of the induced game Γ^S. This is indeed how I define below the payoff achievable by a coalition. However, the issue of what payoff a coalition can achieve has been debated and treated differently in conventional cooperative game theory. I will therefore return to this in more detail later on. For the time being, I define the coalitional function associated with a strategic game independently of this issue.

Definition 4.6

A coalitional function[17] associated with a strategic game (N, T, u) is a scalar-valued function, w, which associates $w(S) \in R$ with each $S \subset N$. The coalitional function value for the empty coalition is zero; that is, $w(\phi) = 0$.

It is convenient to refer to the set of players and the coalitional function as a game, to be denoted by (N, w). The coalitional function w summarizes the power structure of the strategic game (N, T, u). This description of the game is called the *coalitional function form*. Once a representation in coalitional form has been specified, we can try to predict the outcome of negotiations among the players.

Definition 4.7

A coalitional game form of a strategic game (N, T, u) is a pair (N, w), where N is the set of players and w is a coalitional function.

Given a coalitional game (N, w), a payoff vector $x \in R^n$ is an *imputation* of the game if $\sum_{i \in N} x_i = w(N)$. Given an imputation x, we say that it

---

17. An alternative familiar term for this notion is "a characteristic function."

can be *improved upon* by a coalition S if $w(S) > \sum_{i \in S} x_i$, that is, if some coalition S can achieve on its own a higher payoff than assigned to it by the imputation x.

Definition 4.8

The core of a coalitional game (N, w) is the set of all imputations of the game that cannot be improved upon by any coalition.

Another way of stating this definition is that an imputation x is in the core of the game (N, w) if $\sum_{i \in S} x_i \geq w(S)$ for all coalitions $S \subset N$, that is, if no coalition can achieve a higher payoff on its own. This definition ensures that any core payoff vector x, being an imputation, is feasible for the *grand coalition* N and, thus, cannot be improved upon by N. Moreover, it introduces a stability requirement on the imputation x in the sense that no subcoalition S by acting on its own can achieve an aggregate payoff that is higher than the payoff that it receives under imputation x. Thus, a core imputation may be seen as a "strategically stable" outcome of negotiations on joint payoff maximization. The core of a game is *nonempty* or *exists* if there is at least one imputation that belongs to the core.

4.2.1. The α- and β-Coalitional Functions

I now return to the issue raised earlier, namely, what is the appropriate way to define the coalitional function associated with a strategic game. I consider first the two coalitional functions that have been used in conventional cooperative game theory for a long time. The first is based on the assumption that the players outside the coalition adopt those strategies that are least favorable to the coalition (von Neumann and Morgenstern 1953). Thus, the maximum payoff of a coalition $S \subset N$ is

$$w^\alpha(S) = \max_{t_S \in T_S} \min_{t_{N \setminus S} \in T_{N \setminus S}} \sum_{i \in S} u_i(t_S, t_{N \setminus S}).$$

In words, $w^\alpha(S)$ represents the highest payoff that coalition S can guarantee itself no matter what strategies are adopted by the players outside the coalition. For this reason, w^α is often referred to as the *maximin coalitional function*. In this formulation, coalition S moves first, choosing a strategy that maximizes its payoff, taking into account that $N\backslash S$ will move next after seeing its strategy and will choose a strategy that minimizes its payoff.

The second coalitional function that is also often used in traditional cooperative game theory is defined as follows:

$$w^\beta(S) = \min_{t_{N\backslash S} \in T_{N\backslash S}} \max_{t_S \in T_S} \sum_{i \in S} u_i(t_S, t_{N\backslash S}), S \subset N.$$

In words, $w^\beta(S)$ represents the maximum payoff that coalition S can be held down to no matter what strategies are adopted by its members. It is therefore often referred to as the *minimax coalitional function* as opposed to the *maximin coalitional function*. In this formulation, $N\backslash S$ moves first, choosing a strategy that minimizes the payoff of S and takes into account that S will move next after seeing its strategy and will choose a strategy that maximizes its payoff. Because coalition S can choose at least the same strategy that it chooses under the α-coalitional function, $w^\beta(S) \geq w^\alpha(S)$ for all S. I summarize this as the following proposition.

Proposition 4.3

The α- and β-characteristic functions of a strategic game satisfy $w^\alpha(S) \leq w^\beta(S)$ for all $S \subset N$.

The α- and β-coalitional functions lead to the α- and β-core concepts. These have been applied and contrasted to each other in various models with externalities by Scarf (1971), Starrett (1973), and Laffont (1977b) and in the public good context by Foley (1970), Roberts (1974), Kaneko (1977), Moulin (1987), and Chander (1993), among others. However, in the course of analyzing a global pollution model, Mäler (1989) concludes that both these core concepts are useless because when applied to his model, they result in too large a set of outcomes, actually the entire set of efficient allocations.[18] I illustrate this limitation of the α- and β-core concepts by applying them to a simple example borrowed from Chander (2007).

---

18. Also see Rosenthal (1971) for criticisms of the behavioral assumptions underlying the α- and β-cores.

4.2.2. A Limitation of the α- and β-Cores

Consider an economy consisting of three identical agents, that is, $N = \{1, 2, 3\}$. Let

$$y_i = e_i^{\frac{1}{2}}, i \in N, z = \sum_{i \in N} e_i \tag{4.1}$$

and

$$u_i(y_i, z) = y_i - z + \frac{1}{4}, i \in N, \tag{4.2}$$

where e_i and y_i denote emissions and output, respectively, of agent i, and z denotes the ambient pollution.

Let $T_i = \{e_i : e_i \geq 0\}, T = T_1 \times T_2 \times T_3$, and $u = (u_1, u_2, u_3)$. It is easily seen that the strategic game (N, T, u) has a dominant strategy equilibrium that induces the following state of the economy

$$\overline{e}_i = \frac{1}{4}; \ \overline{y}_i = \frac{1}{2}, i \in N; \ \overline{z} = \frac{3}{4} \ \text{ and } \ \overline{u}_i = 0, i \in N, \tag{4.3}$$

where $(\overline{e}_1, \overline{e}_2, \overline{e}_3) \in T$ are the unique dominant strategies.

Because the behavioral assumption underlying the α- and β-coalitional functions requires that the players outside a coalition must choose their strategies so as to maximin or minimax the payoff of the coalition, each player i not in the deviating coalition S must choose $e_i = \infty$ even though it has a dominant strategy $\overline{e}_i = 1/4$. Thus, recalling the definitions of the α- and β-coalitional functions, $e_i = \infty$ for each $i \in N \backslash S$ and $w^\alpha(S) = w^\beta(S) = -\infty$ for each $S \subset N, S \neq N$. Hence, every imputation of the game belongs to the α- and β-cores. But why should the players outside the coalition play maximin or minimax strategies when they each have a unique *dominant strategy*? This is clearly absurd.[19] Instead, it would be more reasonable if the players outside the coalition are assumed to choose their strategies so as to maximize their own payoffs rather than bother about how those would affect the payoff of the deviating coalition.

---

19. The α- and β-core concepts were conceived in the 1950s independently of models with public goods and/or externalities. But because of their successful applications in many spheres and lack of a more suitable concept, they were later also applied to models with public goods and/or externalities.

Because of these limitations of the α- and β-coalitional functions, the need for an alternative concept that is based on more plausible behavioral assumptions has often been expressed in the literature. In Harsanyi's (1977) words, "in a variable-sum game one would rather expect that each side would try to find a suitable compromise between trying to maximize the costs of a conflict to the other side and trying to minimize the costs of a conflict to itself—in other words, between trying to minimize the joint payoff of the opposing coalition and trying to maximize the joint payoff of their own coalition."

4.2.3. The γ-Coalitional Function

The above-discussed limitations of the α- and β-cores motivate an alternative coalitional function that seems to address the issue raised by Harsanyi and others. This alternative coalitional function is based on an assumption concerning the behavior of the nonmembers that is more plausible in that the players outside a deviating coalition are assumed to simply adopt their individually best reply strategies resulting in a Nash equilibrium of the corresponding induced game.

Definition 4.9

The γ-coalitional function of a strategic game (N, T, u) is the function

$$w^{\gamma}(S) = \sum_{i \in S} u_i(\tilde{t}), S \subset N,$$

where $\tilde{t} \in T$ is the Nash equilibrium of the induced strategic game Γ^S.

Under the α- and β-coalitional functions, coalition S and the players in $N \backslash S$ move sequentially: either coalition S moves first and the players in $N \backslash S$ move next after seeing the strategies of S or the players in $N \backslash S$ move first and coalition S moves next after seeing the strategies of $N \backslash S$. Under the γ-coalitional function, both S and the players in $N \backslash S$ move simultaneously as if coalition S and the players in $N \backslash S$ have decided to go their separate ways in pursuit of their own goals. If coalition S suffers a loss in payoff, it is incidental and not the intention of the players in $N \backslash S$. It amounts to noncooperation, but not to war on the deviating coalition by the outside players as under the α- and β-coalitional functions. We now show that the γ-core of a strategic game is indeed smaller.

Proposition 4.4

The γ-core of a general strategic game is a subset of the β-core, which is a subset of the α-core.

Proof of Proposition 4.4

In view of Proposition 4.3, we only need to show that $w^\beta(S) \leq w^\gamma(S), S \subset N$. Given $S \subset N$, let $(\tilde{t}_S, \tilde{t}_{-S})$ and $(\hat{t}_S, \hat{t}_{-S})$ be such that $w^\gamma(S) = \sum_{i \in S} u_i(\tilde{t}_S, \tilde{t}_{-S})$ and $w^\beta(S) = \sum_{i \in S} u_i(\hat{t}_S, \hat{t}_{-S})$. Let $t_S^*(t_{-S}) = \text{argmax}_{t_S \in T_S} \sum_{i \in S} u_i(t_S, t_{-S}), t_{-S} \in T_{-S}$. Then, by definition, $w^\beta(S) = \sum_{i \in S} u_i(t_S^*(\hat{t}_{-S}), \hat{t}_{-S}) \leq \sum_{i \in S} u_i(t_S^*(t_{-S}), t_{-S})$ for all $t_{-S} \in T_{-S}$. In particular, $\sum_{i \in S} u_i(t_S^*(\hat{t}_{-S}), \hat{t}_{-S}) \leq \sum_{i \in S} u_i(t_S^*(\tilde{t}_{-S}), \tilde{t}_{-S}) = w^\gamma(S)$. Hence, $w^\beta(S) \leq w^\gamma(S)$, and, therefore, $w^\alpha(S) \leq w^\beta(S) \leq w^\gamma(S), S \subset N$. ∎

Examples are easily constructed in which the inequalities established in the proposition are strict (e.g., the game shown later in figure 4.1). Because the proposition holds for a general strategic game, it can be applied to a variety of economic models.[20]

Because the proposition shows that $w^\beta(S) \leq w^\gamma(S)$ for all $S \subset N$, it follows from Proposition 4.3 that $w^\alpha(S) \leq w^\beta(S) \leq w^\gamma(S)$ for all $S \subset N$.[21]

Notice that in the class of strategic games Γ in which each induced game Γ^S, $S \subset N$, admits a unique Nash equilibrium, if a strategic game (N, T, u) has a strong Nash equilibrium, then the strong Nash equilibrium is unique. Furthermore, the core of the coalitional game (N, w^γ) is non-empty and consists of the *unique* imputation with payoffs equal to the unique strong Nash equilibrium payoffs. However, if the game (N, T, u) admits no strong Nash equilibrium, the core of the game (N, w^γ) may still be nonempty. Thus, the γ-core is a weaker concept than the strong Nash equilibrium, at least in the class of strategic games Γ in which each induced game Γ^S, $S \subset N$, has a unique Nash equilibrium.

---

20. For example, Proposition 2.1 concerning an oligopoly game in Lardon (2012) is a special case of this proposition.

21. This inequality also implies a consistency property of the γ-core in the sense that γ-core solutions are in general also α- and β-core solutions.

To illustrate the γ-core, I use the same three-agent example that was used earlier to illustrate the limitations of the α- and β-cores. Because the strategy $(\bar{e}_1, \bar{e}_2, \bar{e}_3) = (\frac{1}{4}, \frac{1}{4}, \frac{1}{4})$, as defined in (4.3), is the Nash equilibrium induced by either of the singleton coalitions {1}, {2}, and {3}, respectively, $\bar{u}_1 = w^\gamma(\{1\}) = \bar{u}_2 = w^\gamma(\{2\}) = \bar{u}_3 = w^\gamma(\{3\}) = 0$. I calculate the Nash equilibrium of the strategic games induced by coalitions of two players. Let $S = \{1, 2\}$. Define $(\tilde{e}_1, \tilde{e}_2, \tilde{e}_3)$ such that

$$\tilde{e}_1, \tilde{e}_2 = \operatorname{argmax}\left(e_1^{\frac{1}{2}} + e_2^{\frac{1}{2}} - 2e_1 - 2e_2 - 2\tilde{e}_3\right), \text{ and}$$

$$\tilde{e}_3 = \operatorname{argmax}\left(e_3^{\frac{1}{2}} - \tilde{e}_1 - \tilde{e}_2 - e_3\right).$$

Then,

$$\tilde{e}_1 = \tilde{e}_2 = \frac{1}{16}, \; \tilde{e}_3 = \frac{1}{4}, \; \tilde{u}_1 = \tilde{u}_2 = \frac{1}{8} \text{ and } \tilde{u}_3 = \frac{3}{8}. \tag{4.4}$$

The strategy $(\tilde{e}_1, \tilde{e}_2, \tilde{e}_3) = (\frac{1}{16}, \frac{1}{16}, \frac{1}{4})$ is the Nash equilibrium of the strategic game induced by coalition {1, 2}. Thus, $w^\gamma(\{1, 2\}) = \frac{1}{4}$. Similarly, we can assign payoffs to coalitions {2, 3} and {1, 3}. But because the players are identical, we have $w^\gamma(\{1, 2\}) = w^\gamma(\{2, 3\}) = w^\gamma(\{1, 3\}) = \frac{1}{4}$.

Let $e_1^*, e_2^*, e_3^* \equiv \operatorname{argmax}(e_1^{\frac{1}{2}} + e_2^{\frac{1}{2}} + e_3^{\frac{1}{2}} - 3e_1 - 3e_2 - 3e_3) + \frac{3}{4}$. Then, using the first-order conditions for maximization, it is seen that

$$e_i^* = \frac{1}{36}, \; y_i^* = \frac{1}{6}, \; i \in N; \; z^* = \frac{1}{12}; \text{ and } u_i^* = \frac{1}{3}; \; i \in N. \tag{4.5}$$

Thus, $w^\gamma(N) = 1$, and an imputation (x_1, x_2, x_3) of the game (N, w^γ) belongs to the core if and only if $x_1, x_2, x_3 \geq 0$, $\sum\limits_{j \in N\backslash\{i\}} x_j \geq \frac{1}{4}$ for each $i \in N$, and $\sum\limits_{j \in N} x_j = 1$. Clearly, the set of γ-core imputations is more deterministic than the set of α- and β-core imputations.

It is easily seen that the unique Nash equilibrium of the game in the above example is coalition-proof. This unique (coalition-proof) Nash equilibrium is not strong, as a strong Nash equilibrium must necessarily be efficient.

$$E$$

	Soccer match	Restaurant
Soccer match	+2, +2	+1, -1
Restaurant	-1, +1	+3/2, +3/2

A

FIGURE 4.1

A game with a nonempty γ-core

We consider another example to further illustrate the relationship between the concepts introduced so far.[22] The example has two players, A and E—Adam and Eve—who are to decide whether to watch a soccer match or go to a restaurant. Thus, the grand coalition $N = \{A, E\}$. The payoff matrix of the game is shown in figure 4.1.

Thus,

$$w^\alpha(\{A\}) = w^\alpha(\{E\}) = +1, w^\alpha(\{A, E\}) = +4,$$

$$w^\beta(\{A\}) = w^\beta(\{E\}) = +3/2, \ w^\beta(\{A, E\}) = +4, \ \text{and}$$

$$w^\gamma(\{A\}) = w^\gamma(\{E\}) = +2, w^\gamma(\{A, E\}) = +4.$$

$$\alpha\text{-core} = \{(x_1, x_2) : x_1 \geq +1, x_2 \geq +1, x_1 + x_2 = +4\},$$

$$\beta\text{-core} = \{(x_1, x_2) : x_1 \geq +3/2, x_2 \geq +3/2, x_1 + x_2 = +4\}, \text{and}$$

$$\gamma\text{-core} = \{(x_1, x_2) : x_1 \geq +2, x_2 \geq +2, x_1 + x_2 = +4\}.$$

---

22. This example was constructed and used by me in a course on public economics to explain the difference between the α-, β-, and γ-cores and to illustrate the notion of a focal-point equilibrium (Schelling 1960).

$$E$$

	Soccer match	Restaurant
Soccer match	+2, +5/2	+1, −1
Restaurant	−1, +1	+5/2, +2

(with H labeling the row player)

FIGURE 4.2

A game with an empty γ-core

It is seen that $w^{\alpha}(S) < w^{\beta}(S) < w^{\gamma}(S)$ for all $S \subset N$, and the game admits a strong Nash equilibrium. The γ-core consists of a unique imputation, and the γ-core payoffs are equal to the strong Nash equilibrium payoffs.

Notice that each induced game admits two Nash equilibria, and the γ-coalitional function is defined by selecting each time the Nash equilibrium that results in the highest payoff for the concerned coalition. However, as the next example in figure 4.2 shows, selecting among the equilibria is not always so obvious.

The α- and β-cores of the game shown in figure 4.2 are nonempty, and the game admits two (strong) Nash equilibria. On the one hand, the γ-core is empty, as by assumption we must select the Nash equilibrium that results in the highest payoff to the concerned coalition, but, on the other hand, it would be nonempty if the Nash equilibrium that results in the lowest payoff were selected every time instead. Another possibility is to first choose among the two Nash equilibria by intuitive criteria that are external to the game, as is often done in noncooperative game theory. For example, in a society in which women's preferences are given priority, the Nash equilibrium (soccer match, soccer match) is more likely and the coalitional payoffs may be defined accordingly. However, as noted earlier, the need for selecting among multiple Nash equilibria does not really arise in the games studied in this book as they all have a unique equilibrium.

4.2.4. Superadditivity and Efficiency

Definition 4.10

A coalitional game (N, w) is *superadditive* if $w(S \cup T) \geq w(S) + w(T)$ for every pair of coalitions S and T such that $S \cap T = \phi$.

In words, if two disjoint coalitions combine their forces, then the coalitions are able to obtain a payoff that is at least as high as what they could obtain without combining. Coalitional functions are often assumed to be superadditive as superadditivity of a coalitional function appears to be a natural condition that ensures that the grand coalition is an efficient coalition structure. However, in our framework, the grand coalition is an efficient coalition structure irrespective of whether or not the coalitional function is superadditive. That is because in a strategic game, the grand coalition can choose at least the same strategies as the players in any coalition structure. This general implication of my framework will be used throughout the book.

As will be seen in the next chapter, the grand coalition is the *only* efficient coalition structure in the environmental game,[23] but the γ-coalitional function is not superadditive.[24] However, superadditivity is neither necessary nor sufficient for the existence of the core of a coalitional game (see, e.g., Friedman 1990). A related property of coalitional games is convexity (Shapley 1971).

Definition 4.11

A coalitional game (N, w) is *convex* if $w(S) + w(T) \leq w(S \cup T) + w(S \cap T)$ for all $S, T \subset N$.

By definition of a coalitional game, $w(\phi) = 0$. Therefore, a coalitional game (N, w) is not convex unless it is superadditive. As will be shown in the next chapter, the γ-coalitional function corresponding to the environmental game is not superadditive and, therefore, not convex.

---

23. Thus, formation of coalition structures other than the grand coalition implies inefficiency.
24. This is perhaps also true in the case of the α- and β-coalitional functions, but they are not our focus.

4.3. PARTITION FUNCTIONS

Apart from its relationship with the strong Nash equilibria and the coalition-proof Nash equilibria, the γ-coalitional function is also related to the concept of a partition function introduced by Thrall and Lucas (1963) in which the worth of a coalition depends on the entire coalition structure and not just on the coalition in question. Ichiishi (1981) and Zhao (1996) showed how a strategic game can be converted into a partition function. This is as follows:

A *partition* of N is $P = \{S_1, S_2, \ldots, S_m\}$ such that $\bigcup_{j=1}^{m} S_j = N$ and for all $i \neq j$, $S_i \cap S_j = \phi$. A partition of N is also called a *coalition structure*. Given a strategic game $\Gamma = (N, T, u)$ and a coalition structure $P = \{S_1, \ldots, S_m\}$, let Γ^P denote the induced strategic game in which each coalition $S_i \in P, i = 1, \ldots, m$, acts as one single player.

Definition 4.12

Given a strategic game (N, T, u), $(\hat{t}_1, \ldots, \hat{t}_n) \in T$ is a Nash equilibrium of the strategic game Γ^P induced by the coalition structure $P = \{S_1, S_2, \ldots, S_m\}$, if
$$\hat{t}_{S_j} = \operatorname{argmax}_{i \in S_j} \sum u_i(t_{S_j}, \hat{t}_{-S_j}), j = 1, 2, \ldots, m.$$

To relate to the previous notions, let $[N]$ and $[N\backslash S]$ denote the finest/trivial partitions of N and $N\backslash S$, respectively. Then, a Nash equilibrium of the induced game Γ^S is a Nash equilibrium of the game Γ^P induced by the coalition structure $P = \{S, [N\backslash S]\}$ and vice versa.

Definition 4.13

A partition function representation of a strategic game (N, T, u) is the function
$$p(S_i, P) = \sum_{j \in S_i} u_j(\hat{t}), S_i \in P \in \wp,$$

where $\hat{t} \in T$ is a Nash equilibrium of the induced game Γ^P, and $\wp$ is the set of all partitions of N.

However, a partition function may not be well defined if a Nash equilibrium of the induced game corresponding to some coalition structure is not unique. If an induced game corresponding to a coalition structure has multiple Nash equilibria, then it is not clear which one should be selected for assigning payoffs to the coalitions in the coalition structure. If there is a Nash equilibrium of the induced game in which the payoff of *every* coalition in the coalition structure is highest compared to its payoff in any other Nash equilibrium of the induced game (as, e.g., in the game in figure 4.1), then the choice is clear. But that may not always be so, and there does not seem to be a general method for choosing between the alternative Nash equilibria of the induced game if that is indeed not so.[25] However, as is actually true in many applications, including those in this book, I assume that each induced game admits a unique Nash equilibrium and therefore the partition function is well defined. As will be seen, it is sometimes more convenient to work directly with the partition function representation rather than the strategic form of the game.

A partition function recognizes the fact that the payoff of a coalition depends on the entire coalition structure. It is a more general concept as it assigns a worth or payoff to each coalition in *each* coalition structure of which the coalition is a member.[26] The γ-coalitional function is a particular restriction of the partition function that assigns a worth or payoff to each coalition *only* in the coalition structure in which all except possibly the coalition are singletons.[27] However, as seen later, the γ-coalitional function may not inherit every property of the partition function.

Maskin (2003) introduces the following notion of superadditivity. A partition function p is *superadditive* if for any partition $P = \{S_1, \ldots, S_n\}$ and any $S_i, S_j \in P$,

$$p\left(\{S_i \cup S_j\}, P \setminus \{S_i, S_j\} \cup \{S_i \cup S_j\}\right) \geq p(S_i, P) + p(S_j, P).$$

---

25. Such a conceptual problem does not arise in the case of the γ-coalition function, as only one coalition in each coalition structure is to be assigned a payoff. Therefore, the payoff that is highest among all Nash equilibria of the induced game can be assigned to the coalition.

26. See Maskin (2003) and de Clippel and Serrano (2008) for some recent studies of games in partition function form. They are also sometimes referred to as games with externalities.

27. See Wooders and Page (2008) for additional discussion of this interpretation of the γ-coalitional function.

Maskin (2003) assumes the partition function to be superadditive but notes that this may be a strong assumption.[28] As the following example demonstrates, superadditivity of a partition function does not imply superadditivity of the corresponding γ-coalitional function, and the two are generally not comparable.[29]

Example 1

Consider the following symmetric three-player partition function: $N = \{1, 2, 3\}$, $p(\{i\}, \{\{1\}, \{2\}, \{3\}\}) = 4$ for $i = 1, 2, 3$; $p(\{j, k\}, \{\{j, k\}, \{i\}\}) = 9$ and $p(\{i\}, \{\{i\}, \{j, k\}\}) = 1$ for $\{i, j, k\} = N$; $p(N, \{N\}) = 12$.

The partition function in this example is superadditive in Maskin's sense, but the γ-coalitional function is not, because $w^\gamma(\{i\}) + w^\gamma(\{j, k\}) = p(\{i\}, \{\{1\}, \{2\}, \{3\}\}) + p(\{j, k\}, \{\{j, k\}, \{i\}\}) = 4 + 9 > w^\gamma(\{i, j, k\}) = 12$. The converse, that is, the γ-coalitional function is superadditive but the partition function is not, is also true. Though I do not include an explicit example here to substantiate this claim, it is straightforward to see why the two are not comparable.[30] More specifically, the superadditivity of the partition function concerns the impact of merger among two coalitions in the same coalition structure, whereas the superadditivity of the γ-coalitional function concerns merger among two coalitions in different coalition structures. The two notions are thus not comparable. In chapter 5, I show that neither the γ-coalitional function form nor the partition function form of the environmental game is *generally* superadditive.

For this reason, I will use below an alternative notion of superadditivity of a partition function that is more meaningful for my purpose and weaker than the stronger notions of superadditivity in Hafalir (2007) and de Clippel and Serrano (2008). Let (N, p) denote the partition function

---

28. Both Hafalir (2007) and de Clippel and Serrano (2008) introduce an even stronger notion of superadditivity.

29. Hafalir (2007) uses a similar example, but to demonstrate that superadditivity of a partition function does not imply efficiency of the grand coalition. I have modified his example such that the grand coalition is an efficient coalition structure, as it is indeed in my framework.

30. Because, in my framework, the grand coalition is efficient, the example should have at least four players.

game representation of the strategic game (N, T, u). Let $|S|$ denote the cardinality of set S.

Definition 4.14

The partition function (N, p) is partially superadditive if for any partition $P = \{S, [N \setminus S]\}$ and $\{S_1, \ldots, S_k\}$ such that $\bigcup_{i=1}^{k} S_i = S, |S_i| > 1,\ i = 1, \ldots, k$, $\sum_{i=1}^{k} p(S_i; P') \leq p(S; P)$ where $P' = P \setminus S \cup \{S_1, \ldots, S_k\}$.

Partial superadditivity of a partition function, as the term suggests, is weaker than the familiar notion of superadditivity, which requires that combining *any* arbitrary coalitions increases their total worth.[31] In contrast, partial superadditivity of a partition function requires that combining only *all* non-singleton coalitions increases their total worth. It is trivially satisfied by all partition functions with three players and also by all those with four players, as the grand coalition, by definition, is an efficient coalition structure.

4.3.1. A Class of Games with Nonempty γ-Cores

A number of studies have focused on symmetric games in which bigger coalitions in a coalition structure have lower per-member payoffs (see, e.g., Chander 2007; Funaki and Yamato 1999; Ray and Vohra 1997; Yi 1997) and the grand coalition is the efficient coalition structure. I show that the γ-cores of these games are nonempty. In particular, the imputation with equal payoffs is a γ-core payoff vector.[32] I do not explicitly state the strategic form of these games, as the result can be established generally by using their common property that larger coalitions in a coalition structure have lower per-member payoffs and the grand coalition is the efficient coalition structure. But it will be shown in the next chapter that the symmetric environmental game—which is a strategic game—also satisfies this property. We need some additional notation.

---

31. For a formal definition, see, for example, Hafalir (2007), who uses the term "full cohesiveness" for superadditivity.

32. This seems to be a case of a concept that has been implicitly used before it was formally defined.

A partition function game (N, p) is symmetric if for every partition $P=\{S_1,\ldots,S_m\},\left|S_i\right|=\left|S_j\right|\Rightarrow p(S_i,P)=p(S_j,P)$.[33]

Proposition 4.5

Let (N, p) be a symmetric partition function game such that for every partition $P=\{S_1,\ldots,S_m\}$, $p(S_i,P)/\left|S_i\right|\le p(S_j,P)/\left|S_j\right|$ if $\left|S_i\right|\ge\left|S_j\right|, i,j\in \{1,\ldots,m\}$. If the grand coalition is an efficient partition, then the game (N, p) admits a nonempty γ-core.

Proof of Proposition 4.5

Let $(x_1,\ldots,x_n)$ be the feasible payoff vector with equal shares, that is, $\sum_{i\in N} x_i = p(N,\{N\})$ and $x_i = x_j, i, j\in N$. We claim that $(x_1,\ldots,x_n)$ belongs to the γ-core.

We need to show that $\sum_{i\in S} x_i \ge p(S,\{S,[N\backslash S]\})$ for all $S\subset N$. Because the grand coalition is an efficient partition, $p(S,\{S,[N\backslash S]\})+\sum_{i\in N\backslash S} p(\{i\},[S,\{N\backslash S\}])\le p(N;\{N\})=\sum_{i\in N} x_i$. Because $p(S;\{S,[N\backslash S]\})/\left|S\right|\le p(\{i\},\{S,[N\backslash S]\})$, $i\in N\backslash S$, and $x_i = x_j, i, j\in N$, this inequality can hold only if $p(S,\{S,[N\backslash S]\})/\left|S\right|\le x_i$. Hence $p(S,\{S,[N\backslash S]\})\le\sum_{i\in S} x_i$, as $x_i = x_j, i, j\in N$. ∎

4.3.2. Normative Approaches and the γ-Core

Beside the concept of core, the other most important concept of cooperative game theory is the Shapley value (Shapley 1953). In recent years, the theory of the Shapley value has been extended to partition function games (see, e.g., de Clippel and Serrano 2008; Hafalir 2007; Maskin 2003). This raises the issue whether the Shapley value of the coalitional game (N, w^γ) is also a value of the more general partition function game (N, p). A result, de Clippel and Serrano (2008, proposition 3) show that a value of the more general partition function game (N, p) that satisfies the axioms of anonymity, efficiency, and weak marginality when they are suitably extended to a partition function is indeed the Shapley value of the corresponding

---

33. Remember that $\left|S\right|$ denotes the cardinality of set S.

characteristic function game (N, w^γ). Thus, the concept of γ-coalitional function is in a sense also supported by the axiomatic approach to cooperative game theory. It is well known that the Shapley value belongs to the core if the coalitional game is convex. But, as shown above, the coalitional game (N, w^γ) is not even superadditive, let alone convex. Therefore, in chapter 5, I propose a unique selection from among the core imputations that has a positive interpretation instead.

4.4. NONCOOPERATIVE FOUNDATIONS OF THE γ-CORE

A growing branch of the literature seeks to unify cooperative and noncooperative approaches to game theory through underpinning cooperative game-theoretic solutions with noncooperative equilibria, the "Nash program" for cooperative games. In the same vein, I now show that the γ-core imputations (a cooperative solution concept) of a general strategic game can be supported as equilibrium payoff vectors of a noncooperative game, and the grand coalition is the unique equilibrium outcome if and only if the γ-core of the game is nonempty.

As discussed in chapter 1, the noncooperative game for which the γ-core imputations are equilibrium outcomes is an intuitive description of how the players may agree to form a partition when they know in advance what their payoffs will be in each partition. It consists of infinite repetition of two stages. In the first stage of the two stages, which begins with the finest partition as the status quo, each player announces whether it wishes to stay alone or to form a nontrivial partition. In the second stage of the two stages, the players form a partition as per their announcements. The two stages are repeated if the outcome of the second stage of the two stages is the finest partition as in the status quo from which the game began in the first place. We need the definitions that follow.

4.4.1. Payoff Sharing Rules

A *payoff sharing rule* is a mapping $x :\rightarrow R^n$ that associates to each partition $P = \{S_1, \ldots, S_m\}$ a vector of individual payoffs $x(P) \in R^n$ such that

$\sum_{j \in S_i} x_j(P) = p(S_i; P), S_i \in P$.[34] A mapping $x :\to R^n$ is the *equal payoff sharing rule* if for each partition $P = \{S_1, \ldots, S_m\}$, $x_i(P) = x_j(P)$ for each $i, j \in S_k, k = 1, \ldots, m$. The equal payoff sharing rule, as in the class of symmetric games in Proposition 4.5, is a special case of the following general class of rules. A payoff sharing rule $x :\to R^n$ is *monotonic* if for each partition $P = \{S_1, \ldots, S_m\}$ and each coalition $S_i \in P$, $x_j(P) > (\leq) x_j(\{N\})$ for all $j \in S_i$ if and only if $p(S_i; P) > (\leq) \sum_{j \in S_i} x_j(\{N\})$.

A monotonic payoff sharing rule assigns higher (resp. lower) payoffs to each member of a coalition in a partition if the total payoff of the coalition is lower (resp. higher) in the grand coalition. In other words, a monotonic rule assigns payoffs to members of each coalition in a partition such that their individual payoffs are either all higher or all lower compared to their individual payoffs in the grand coalition. Thus, monotonic sharing rules ensure unanimity among the members of a coalition on the coalition's decision to leave or not to leave the grand coalition if they are farsighted and can foresee the resulting partition that will form subsequent to their departure from the grand coalition.

Definition 4.15

A payoff sharing rule $x :\to R^n$ is *proportional* if for each partition $P = \{S_1, \ldots, S_m\}$ and each coalition $S_i \in P, x_j(P) = x_j(\{N\}) \times [p(S_i; P) / \sum_{k \in S_i} x_k(\{N\})], j \in S_i$.

Proportional sharing rules are clearly monotonic, and the equal payoff sharing rule is clearly proportional and, therefore, monotonic. It will be convenient to denote a proportional sharing rule $x :\to R^n$ simply by a feasible payoff vector $(x_1, \ldots, x_n)$, meaning that for each partition $P = \{S_1, \ldots, S_m\}$ and each coalition $S_i \in P$, $x_j(P) = x_j \times [p(S_i; P) / \sum_{k \in S_i} x_k]$. Theorem 4.6 holds for all monotonic payoff sharing rules, but for the sake of a sharper proof I shall restrict myself to a proportional rule.

---

34. Thus, a payoff sharing rule enables the players to compare their individual payoffs in different partitions. In contrast, Hart and Kurz (1983) assume that the individual payoffs of the players are the same irrespective of which partition is formed and equal to the Shapley value allocations that the players would receive in the grand coalition.

4.4.2. An Infinitely Repeated Game

Because the partition function (N, p) is generated from a strategic game, the members of a coalition in a partition may unanimously decide to dissolve the coalition (i.e., not to give effect to the coalition). This possibility arises from the fact that the players in a coalition can always choose the same strategies that they would choose individually.[35] Thus, dissolving a coalition is equivalent to the players in the coalition choosing the same strategies that they would choose if they were all singletons, given the strategies of the remaining players. Accordingly, the noncooperative game introduced below allows for this possibility.

In what follows, I shall also assume, without loss of generality, that $p(\{i\}, [N]) > 0$ for all i, and the grand coalition is the unique efficient coalition structure. This means that if $(x_1, \ldots, x_n)$ is a γ-core imputation, then $x_i > 0$ for all i.

The noncooperative game, to be called an infinitely repeated game or simply repeated game, consists of infinitely repeated two-stages. The first stage of the two-stages begins from the finest/trivial partition $[N]$ as the status quo, and each player announces some nonnegative integer from 0 to n. In the second stage of the two-stages, all those players who announced the same positive integer in the first stage of the two-stages form a coalition and either give effect to the coalition or dissolve it. All those players who announced 0 remain singletons.[36] If the outcome of the second stage of the two-stages is not the finest partition, the game ends and the partition formed remains formed forever.[37] But if the outcome of the second stage of the two-stages is the finest partition as in the status quo from which the game began in the first place, the two-stages are

---

35. For instance, the United States signed the Kyoto Protocol but did not ratify it and thereby chose a strategy that was best for it individually.

36. Thus, no coalition of two or more players can be formed without the consent of all players, as any player can announce the same positive integer as the members of the coalition, and no player can be forced to form a coalition with any other player, as it can announce 0.

37. This is analogous to the rule in the infinite bargaining game of alternating offers (Rubinstein, 1982) in which the game ends if the players agree to a split of the pie, but continues, possibly ad infinitum, if no agreement is reached. It is also similar to the rule that coalition formation is irreversible (e.g., Compte and Jehiel 2010).

repeated, possibly ad infinitum, until some nontrivial partition is formed in a future round.[38] In either case, the players receive payoffs in each period in proportion to a prespecified feasible payoff vector $(x_1^*, \ldots, x_n^*)$. If no nontrivial partition is ever formed, the game continues forever—in other words, the players agree to disagree perpetually.

Notice that the payoff sharing game allows the players to form *any* partition and end the game; it does not rule out a priori any partition as a possible equilibrium outcome. The finest partition [N] is an outcome of the second stage of the two-stages if all players announce 0 in the first stage of the two-stages[39] or some players announce the same positive integers in the first stage but decide to dissolve the coalition(s) in the second stage of the two-stages. Because a nontrivial partition can be formed only with the agreement of all players, formation of a nontrivial partition is to be interpreted as an agreement among all players.[40] Also notice that the conditional repetition of the two stages does not imply stronger incentives to form a nontrivial partition. On the contrary, it may weaken them, because then the players stand to lose nothing by not forming a nontrivial partition in the current round as there will be opportunities to form it in a future round instead.

To describe the repeated game in more concrete terms, visualize the following story. All players meet in a negotiating room to decide on formation of coalitions knowing in advance what their payoffs will be in each resulting partition. They may form any nontrivial partition or they may all decide to stay alone (i.e., form the finest/trivial partition). If the players agree to form a nontrivial partition, the meeting ends, the players receive per-period payoffs according to the prespecified rule, and all leave the room. But if the players do not agree to form a nontrivial partition, the meeting and negotiations continue and nobody leaves the room until the players agree to form a nontrivial partition.

---

38. Because the game starts from the finest partition, not allowing repetition of the two-stages if the outcome of the second stage is again the finest partition as in the status quo, from which the game began in the first place, would be inconsistent.

39. Alternatively, all players may announce positive integers, but no two of the announced integers may be the same.

40. Because formation of a nontrivial partition depends on the strategies of each and every player, a nontrivial partition cannot be formed without the consent of all players.

Because the structure of the continuation game is exactly the same as that of the original game, I restrict myself to equilibria in stationary strategies of the repeated game. In fact, there is no loss in generality in restricting to equilibria only in stationary strategies. Accordingly, I characterize the equilibria of the repeated game by comparing only per-period payoffs of the players.

Theorem 4.6

If the partition function (N, p) is partially superadditive and the γ-core is nonempty, then any γ-core imputation $(x_1^*, \ldots, x_n^*)$ is an equilibrium payoff vector of the payoff sharing game.

Proof of Theorem 4.6

To obtain a sharper proof, I assume all inequalities to be strict. It will be made clear below that the proof also holds if the inequalities are weak. I show that in the payoff sharing game

(i)   to dissolve a coalition if it does not include all players is an equilibrium strategy of each player, and

(ii) the grand coalition N is an equilibrium outcome resulting in per-period equilibrium payoffs equal to $(x_1^*, \ldots, x_n^*)$.

The theorem is clearly true for $n = 2$. It will be useful to prove the theorem separately for $n = 3$ and $n > 3$.

Case $n = 3$.   I first show that (i) implies (ii) and then prove that the strategies in (i) are indeed equilibrium strategies as they imply (ii). Given strategies in (i) and players' responses to them, I derive a reduced form of the payoff sharing game as follows:

Given (i), let $(w_1, \ldots, w_n)$ be the per-period equilibrium payoff vector of the repeated game. (a) If in some period, all players do not announce the same positive integer or some player announces $i = 0$, then as the strategies in (i) require, any nontrivial coalition is dissolved and the outcome is the finest partition implying per-period payoffs of $(w_1, \ldots, w_n)$, as the continuation game is identical to the original game. (b) If in some period all players announce the same positive integer, then the outcome is the grand coalition, the game ends, and the per-period payoffs are (x_1^*, x_2^*, x_3^*).

Notice that if the grand coalition is indeed an equilibrium outcome of the game, then it may be realized in the first period itself. That is because the per-period payoffs of the players would be otherwise lower in the periods preceding the period in which the grand coalition is formed, as $x_i^* > p(i; \{1,2,3\})$, $i=1,2,3$ as (x_1^*, x_2^*, x_3^*) is a γ-core payoff vector.[41]

To resolve indifference on the part of players, I first solve the game with discounting and then take the limit as the discount rate goes to zero. Let $\delta < 1$ be the discount factor. There is no loss of generality in assuming that one of the players, say 3, chooses only between strategies $i=1$ and $i=0$. The same analysis holds if player 3 chooses instead between strategies $i=2$ or 3 and $i=0$. Given the strategies in (i), the payoff matrix of the reduced form of the repeated game is shown in figure 4.3.

To find a solution of the reduced form, consider first a mixed-strategy Nash equilibrium. Let q_1, q_2, q_3 be the probabilities assigned by the three players to the strategy $i=1$. Then, in equilibrium, each player, say 1, should be indifferent between strategies $i \neq 1$ and $i=1$. Therefore, $w_1 = q_2 q_3 \delta w_1 + (1-q_2 q_3)\delta w_1 = q_2 q_3 x_1^* + (1-q_2 q_3)\delta w_1$. If $x_1^* > \delta w_1$, then $i=1$ is the dominant strategy, and the resulting payoff is $w_1 = x_1^*$ (>0), confirming the inequality $x_1^* > \delta w_1$ for dominance. Thus, $i=1$ is indeed the dominant strategy of each player for each $\delta \to 1$, the grand coalition N is an equilibrium outcome, and (x_1^*, x_2^*, x_3^*) are the per-period equilibrium payoffs.[42]

I now prove that the strategies in (i) are indeed equilibrium strategies, as they imply (ii). Suppose in some period two players, say 2 and 3, announce $i=1$, but player 1 announces $i \neq 1$. Suppose further that in stage 2, players 2 and 3 give effect to the coalition 23 and do not dissolve it. Such a deviation from the strategies in (i) would lead to payoffs of $\left(\frac{x_2^*}{x_2^* + x_3^*}\right) p(23; \{23, 1\}) < x_2^*$ and $\left(\frac{x_3^*}{x_2^* + x_3^*}\right) p(23; \{23, 1\}) < x_3^*$ for players 2 and 3 (resp.), as $(x_1^*, \ldots, x_n^*)$ is a γ-core payoff vector, and, therefore, $x_2^* + x_3^* >$

---

41. To economize on commas and brackets whenever possible, I shall denote partitions $\{\{1\}, \{2\}, \{3\}\}$ by $\{1,2,3\}$, $\{\{123\}\}$ by $\{123\}$, and $\{i, \{jk\}\}$. by $\{i, jk\}$.

42. Because the equalities characterizing the mixed-strategy equilibrium also hold for $\delta = 1$, (x_1^*, x_2^*, x_3^*) is indeed the per-period equilibrium payoff vector in the limit.

Player 3

		$i = 1$		$i = 0$	
		Player 2		**Player 2**	
		$i = 1$	$i \neq 1$	$i - 1$	$i \neq 1$
Player 1	$i = 1$	x_1^*, x_2^*, x_3^*	$\delta w_1, \delta w_2, \delta w_3$	$\delta w_1, \delta w_2, \delta w_3$	$\delta w_1, \delta w_2, \delta w_3$
	$i \neq 1$	$\delta w_1, \delta w_2, \delta w_3$	$\delta w_1, \delta w_2, \delta w_3$	$\delta w_1, \delta w_2, \delta w_3$	$\delta w_1, \delta w_2, \delta w_3$

FIGURE 4.3

The payoff matrix of the reduced form of the repeated game

$p(23; \{23, 1\})$. However, if players 2 and 3 adhere to the strategies and thus dissolve the coalition, then the game will be repeated, and their payoffs from that, as shown above, will be x_2^* and x_3^*, which are higher. Thus, it is ex post optimal for both players 2 and 3 to dissolve the coalition, which player 1 *must* take into account when deciding its strategy.[43] This proves (i) as well.

Case $n > 3$. The extension to $n > 3$ follows from the fact that if the partition function game is partially superadditive, then any partition other than the finest partition has *at least* one non-singleton coalition whose payoff is lower compared to its payoff in the grand coalition. More formally, let $S_1, \ldots, S_m$ be the non-singleton coalitions in the partition $P = \{S_1, \ldots, S_m\}$. Let $S = \bigcup_{i=1}^k S_i$ and $P' = P \setminus \{S_1, \ldots, S_k\} \cup \{S\}$. Then, $\sum_{i=1}^k p(S_i; P) \leq p(S; P') < \sum_{i \in S} x_i^* (= \sum_{i=1}^k \sum_{j \in S_i} x_j^*)$, as (N, p) is partially superadditive and $(x_1^*, \ldots, x_n^*)$ is a γ-core payoff vector. This implies $p(S_i; P) < \sum_{j \in S_i} x_j^*$ for at least one

---

43. The argument here is not that players 2 and 3 can force player 1 to merge with them by threatening to dissolve the coalition (and thus deny player 1 the opportunity to free ride), but rather that given their strategies in (i) and the players' responses to it, such an action is ex post optimal for players 2 and 3, i.e., a subgame-perfect equilibrium strategy.

non-singleton coalition S_i of partition P. If the members of such a non-singleton coalition dissolve the coalition, then another coalition among the remaining non-singleton coalitions will similarly have a lower payoff than its payoff in the grand coalition, and so on. It is therefore ex post optimal for the members of *each* non-singleton coalition to dissolve their coalition so that the game is repeated and the grand coalition is formed. ∎

Clearly, the proof above also holds if the assumed inequalities are not strict and the discount factor $\delta = 1$, if, to resolve indifference on the part of players, we assume that the players strictly prefer to be members of the grand coalition rather than to be members of a coalition in a partition other than the grand coalition even if their payoffs are the same.

Because the grand coalition, by assumption, is the unique efficient partition, it follows that the equilibrium outcome of the payoff sharing game is efficient. In contrast, Ray and Vohra (1997) and Yi (1997) show that the grand coalition is not an equilibrium outcome, even though, as Proposition 4.5 shows, the γ-cores of their games are nonempty and the players' payoffs in each partition are proportional to a prespecified γ-core payoff vector.[44] The intuition for their contrasting result is the following: If the two-stages in a three-player game are to be played only once, then for a player i considering a unilateral deviation from the grand coalition, the coalition structure $\{i, jk\}$ rather than the finest partition $\{i, j, k\}$ is the strategically relevant partition, as the strategies of the other two players will not be oriented toward the finest partition if the two-stages are not to be repeated and the payoffs of the other two players are higher in the partition $\{i, jk\}$ than in the finest partition $\{i, j, k\}$, which is true especially in the case of superadditive partition function games. Therefore, if the payoff of player i in the partition $\{i, jk\}$ is higher than its γ-core payoff in the grand coalition, player i can benefit by deviating from the grand coalition, as that would lead to formation of the partition $\{i, jk\}$ and not the finest partition $\{i, j, k\}$. Hence, in three-player symmetric games, the three coalition structures with a pair and a singleton and not the grand coalition are the equilibrium outcomes if the payoff sharing game is limited to a single play of the two-stages.

---

44. Because these games are such that each partition includes at least one non-singleton coalition that is worse off, the theorem holds, though these games are not partially superadditive.

Theorem 4.6 does not show that the grand coalition is the unique equilibrium outcome of the payoff sharing game. For example, in the case of three-player games, the finest partition $\{1, 2, 3\}$ is not only the status quo but also an equilibrium outcome. However, this equilibrium is Pareto dominated, as (x_1^*, x_2^*, x_3^*), by hypothesis, is a γ-core payoff vector and thus $x_i^* > p(i; \{1, 2, 3\})$, $i = 1, 2, 3$. Therefore, this equilibrium may never be played, as the players know the structure of the game and know that their rivals know the structure of the game, and so on. Hence, if the players believe that perpetual disagreement is not a strategically relevant equilibrium outcome, their equilibrium strategies will be oriented toward formation of the grand coalition, and no partition other than the grand coalition can be sustained as an equilibrium outcome, because a unilateral deviation by a player from the grand coalition will lead to the finest partition and repetition of the two-stages as the other two players' strategies would remain focused on formation of the grand coalition in which their payoffs are higher. Therefore, the finest partition resulting in repetition of the two-stages, rather than any of the nontrivial partitions consisting of a pair and a singleton, is the strategically relevant partition for a player considering unilateral deviation from the grand coalition. Hence, the grand coalition is the only equilibrium outcome if the players believe that perpetual disagreement is not a strategically relevant equilibrium.[45] Clearly, this argument can be extended, as in the proof of Theorem 4.6, to the payoff sharing game with more than three players.

The converse of Theorem 4.6 is also true; that is, if the grand coalition is the unique equilibrium outcome of the payoff sharing game, then the equilibrium payoffs must be equal to a γ-core payoff vector. This too is more easily seen by focusing on the case of three-player games. If the grand coalition is the only equilibrium outcome, then as the players, despite having the opportunity, do not form a partition consisting of a pair and a singleton and end the game, the payoffs of the players in no pair must be higher than in the grand coalition. Furthermore, because the players do not form a partition consisting of a pair and a singleton, each player i can induce the finest partition by not agreeing to form the grand coalition

---

45. In Ray and Vohra (1997), the finest partition is a strategically relevant equilibrium, as in their formulation coalitions can never merge; they can only become finer.

and, therefore, obtain at least the payoff $p(i; \{1, 2, 3\})$. These conditions are exactly the same as those required of a γ-core payoff vector. Clearly, the argument can be extended to games with more than three players. To conclude, if the players believe and know that their rivals believe that perpetual disagreement is not a strategically relevant equilibrium and so on, then the grand coalition is the unique equilibrium outcome of the payoff sharing game if and only if the γ-core is nonempty.

5

ENVIRONMENTAL GAMES

I n this chapter, we will study the γ-core of a *specific* class of strategic games, to be called the environmental games. This brings into focus the strategic aspects of my basic flow model, introduced and discussed in chapters 2 and 3, and leads to additional interpretations and properties of the γ-core that are specific to environmental games.

The main question addressed is whether the γ-core of an environmental game is nonempty, which, as will be seen, is equivalent to the question whether voluntary negotiations among the countries involved can lead to cooperation and efficiency. The chapter provides both a general and a constructive proof for the existence (i.e., nonemptiness) of the γ-core of an environmental game and discusses some specific imputations of the γ-core that are of special interest. The nonemptiness of the γ-core of an environmental game is also interpreted as a generalization of the Coase theorem that applies to a situation in which the externalities are reciprocal and there is no authority empowered to assign and enforce pollution rights.

The chapter concludes with a discussion of an application of the γ-core to another international environmental problem; namely, the "seasonal haze" that afflicts at least three countries of Southeast Asia in years of drought and is in many respects similar to the climate change problem on a regional scale. Thus, it too can be formulated as an environmental game. In fact, the framework of an environmental game is eminently suitable for analyzing the regional haze problem, especially as haze is a flow and not a stock externality.

5.1 THE ENVIRONMENTAL GAME

Let $E_i = \{e_i : 0 \leq e_i \leq e^0\}$, $E = E_1 \times E_2 \times \cdots \times E_n$, and $u = (u_1, u_2, \ldots, u_n)$, where $u_i = g_i(e_i) - v_i(\sum_{j \in N} e_j)$ such that $g_i, v_i, i = 1, \ldots n$, and e^0 satisfy Assumptions 2.1–2.4. I shall refer to the strategic game $\Gamma \equiv (N, E, u)$ as the *environmental game*.

Notice that consumption good transfers are not a part of the strategy sets of the players, and the environmental game fits the standard definition of a strategic game. As seen in the preceding chapter, this formulation enables us to relate and interpret the γ-core concept in terms of other game theory concepts. In chapter 7, I introduce a dynamic version of the environmental game by assuming that emissions generate a durable stock that, rather than the emissions, causes damage; emissions *only* add to the stock.

5.1.1 The Nash Equilibrium of the Environmental Game

Let $(\bar{e}_1, \bar{e}_2, \ldots, \bar{e}_n)$ denote a Nash equilibrium of the environmental game $\Gamma = (N, E, u)$. Then, each $\bar{e}_i$, by definition, must maximize $g_i(e_i) - v_i(e_i + \sum_{j \in N \setminus \{i\}} \bar{e}_j)$. Thus, a Nash equilibrium $(\bar{e}_1, \bar{e}_2, \ldots, \bar{e}_n)$ must satisfy the first-order conditions

$$g_i'(\bar{e}_i) = v_i'(\bar{z}), \quad i = 1, 2, \ldots, n, \tag{5.1}$$

where $\bar{z} = \sum_{j \in N} \bar{e}_j$. Let $(\bar{u}_1, \bar{u}_2, \ldots, \bar{u}_n)$ denote the Nash equilibrium payoffs, that is, $\bar{u}_i = g_i(\bar{e}_i) - v_i(\bar{z})$, $i = 1, \ldots, n$.

Proposition 5.1

The environmental game $\Gamma = (N, E, u)$ admits a unique Nash equilibrium.

Proof of Proposition 5.1

Existence follows from Proposition 4.1, as (i) each player's strategy set E_i is a compact and convex subset of R, and (ii) each player's payoff function u_i is continuous and concave in its own strategies.

To prove uniqueness of the Nash equilibrium, suppose contrary to the assertion that the game has two Nash equilibria, say $(\bar{e}_1,\ldots,\bar{e}_n)$ and $(\tilde{e}_1,\ldots,\tilde{e}_n)$. If $\bar{z}=\sum_{i\in N}\bar{e}_i=\sum_{i\in N}\tilde{e}_i=\tilde{z}$, then strict concavity of each g_i and (5.1) imply $\bar{e}_i=\tilde{e}_i$ for all i, which is a contradiction. If $\bar{z}=\sum_{i\in N}\bar{e}_i>(<)\sum_{i\in N}\tilde{e}_i=\tilde{z}$, then strict concavity of each g_i, convexity of each v_i, and (5.1) imply $\bar{e}_i\leq(\geq)\tilde{e}_i$ for all i, which is again a contradiction. Hence, my supposition is wrong. ∎

Clearly, the noncooperative equilibrium discussed in chapter 3 is essentially the same as the unique Nash equilibrium of the environmental game, except that it is now described in game-theoretic terms. For this reason, I shall often refer to $(\bar{e}_1,\ldots,\bar{e}_n)$ as the unique Nash equilibrium or the noncooperative emission profile.

It is easily seen that if we fix the strategies of any subset of players, then the restricted game possesses essentially the same features as the original game Γ, and, therefore, the restricted game also admits a unique Nash equilibrium. Thus, there is no room for credible cooperation by any coalition of players. This means that the unique Nash equilibrium of Γ is coalition-proof.[1] This unique (coalition-proof) Nash equilibrium is not strong, as a strong Nash equilibrium must necessarily be efficient. But as seen from a comparison of (5.1) and the efficiency condition (3.3), it is not.

5.1.2. Nash Equilibria of the Induced Games

I next prove the existence and uniqueness of a Nash equilibrium for each induced game $\Gamma^S, S\subset N$. To that end, let $(\tilde{e}_1,\tilde{e}_2,\ldots,\tilde{e}_n)$ denote a Nash equilibrium of the induced game Γ^S. Then, $(\tilde{e}_i)_{i\in S}$ must maximize $\sum_{i\in S}[g_i(e_i)-v_i(\sum_{k\in S}e_k+\sum_{j\in N\backslash S}\tilde{e}_j)]$ for $\tilde{e}_j, j\in N\backslash S$, that maximize $g_j(e_j)-v_j(e_j+\sum_{i\in N\backslash\{j\}}\tilde{e}_i)$. The first-order conditions imply that $(\tilde{e}_1,\tilde{e}_2,\ldots,\tilde{e}_n)$ must satisfy

$$g_i'(\tilde{e}_i)=\sum_{k\in S}v_k'(\tilde{z}), i\in S \tag{5.2}$$

---

1. Thus, my comment in footnote 2 in the preceding chapter also can be interpreted to mean that the current negotiations on climate change are an attempt to overcome the inefficiency of a coalition-proof Nash equilibrium.

and

$$g'_j(\tilde{e}_j) = v'_j(\tilde{z}), \, j \in N \setminus S, \tag{5.3}$$

where $\tilde{z} = \sum_{i \in N} \tilde{e}_i$.

Proposition 5.2

Each induced game Γ^S, $S \subset N$, of the environmental game $\Gamma = (N, E, u)$ admits a unique Nash equilibrium $(\tilde{e}_1, \tilde{e}_2, \ldots, \tilde{e}_n)$ such that $\tilde{e}_i \geq \overline{e}_i, i \in N \setminus S$, but $\sum_{i \in N} \tilde{e}_i \leq \sum_{i \in N} \overline{e}_i$, where $(\overline{e}_1, \overline{e}_2, \ldots, \overline{e}_n)$ is the Nash equilibrium of the environmental game Γ.

In words, in each induced game Γ^S, $S \subset N$, the Nash equilibrium emissions of no player outside the coalition S are lower than the Nash equilibrium emissions of the player in the game Γ, but the total Nash equilibrium emissions of all players in the induced game are not higher. This implies that the total emissions of the players in coalition S are not higher and the total emissions of those outside S are not lower. Intuitively, when a coalition forms, its members take into account the effect of their emissions on the other members of the coalition and hence reduce their emissions, whereas the players outside the coalition acting alone have no such incentive. On the contrary, given the reductions in the total emissions of the members of coalition S, their incentive to free ride and emit more becomes stronger and, therefore, they may increase their emissions.

Proof of Proposition 5.2

The existence of a Nash equilibrium of each induced game $\Gamma^S, S \subset N$, follows from Proposition 4.2, as the strategy set of each player is a compact and convex set of R, and the payoff function of each player is continuous and concave in the strategies of *all* players. Uniqueness follows from the fact that the first-order conditions (5.2) and (5.3), by the same arguments as in the proof of Proposition 5.1, admit a unique solution. I prove the remaining part of the proposition.

Let $\tilde{z} = \sum_{i \in N} \tilde{e}_i$ and $\overline{z} = \sum_{i \in N} \overline{e}_i$. I first show that $\tilde{z} \leq \overline{z}$. Suppose contrary to the assertion that $\tilde{z} > \overline{z}$. In that case, by Assumption 2.2, we have $v_i'(\tilde{z}) \geq v_i'(\overline{z})$ for each $i \in N$. Comparing (5.2) and (5.1), it follows that

$$\sum_{j \in S} v_j'(\tilde{z}) = g_i'(\tilde{e}_i) \geq g_i'(\overline{e}_i) = v_i'(\overline{z}) \text{ for all } i \in S, \text{ and}$$

$$v_i'(\tilde{z}) = g_i'(\tilde{e}_i) \geq g_i'(\overline{e}_i) = v_i'(\overline{z}) \text{ for all } i \subset N\backslash S.$$

From the strict concavity of each function g_i, it follows that $\tilde{e}_i \leq \overline{e}_i$ for each $i \in N$. But this contradicts my supposition that $\tilde{z} > \overline{z}$, as $\tilde{z} = \sum_{i \in N} \tilde{e}_i$ and $\overline{z} = \sum_{i \in N} \overline{e}_i$. Hence, $\tilde{z} \leq \overline{z}$.

Finally, because $\tilde{z} \leq \overline{z}$, $v_i'(\tilde{z}) \leq v_i'(\overline{z})$ for each $i \in N\backslash S$. From a comparison of (5.1) and (5.3) and the strict concavity of each g_i, it follows that $\tilde{e}_i \geq \overline{e}_i$ for each $i \in N\backslash S$.∎

It may be noted that Proposition 5.2 shows only that $\sum_{i \in S} \tilde{e}_i \leq \sum_{i \in S} \overline{e}_i$. Though this proposition plays a crucial role in the proof for the existence (i.e., nonemptiness) of the γ-core of the environmental game, it does not show that $\tilde{e}_i \leq \overline{e}_i$ for each $i \in S$. I introduce below a number of alternative sufficient conditions for this stronger result to hold.

5.2. THE γ-COALITIONAL FUNCTION OF THE ENVIRONMENTAL GAME

We are now well prepared to introduce the γ-coalitional function of the environmental game.

Definition 5.1

The coalitional function representation of the environmental game $\Gamma = (N, E, u)$ is the function

$$w^\gamma(S) = \sum_{i \in S}\left[g_i(\tilde{e}_i) - v_i\left(\sum_{j \in N} \tilde{e}_j \right)\right] = \sum_{i \in S} \tilde{u}_i, S \subset N, \qquad (5.4)$$

where $(\tilde{e}_1, \tilde{e}_2, \ldots, \tilde{e}_n)$ is the unique Nash equilibrium of the induced game Γ^S and $\tilde{u}_i = g_i(\tilde{e}_i) - v_i(\sum_{j \in N} \tilde{e}_j), i = 1, \ldots, n$.

I shall refer to $w^\gamma(S)$ as the *worth* of coalition S. The definition implies that $w^\gamma(N) = \max_{e_1,\ldots,e_n}[\Sigma_{i \in N}\, g_i(e_i) - \Sigma_{i \in N}\, v_i(\Sigma_{j \in N}\, e_j)] = \Sigma_{i \in N}\, g_i(e_i^*) - \Sigma_{i \in N}\, v_i(\Sigma_{i \in N}\, e_i^*)$ where $(e_1^*,\ldots,e_n^*)$ is the unique efficient emission profile as characterized by (3.3), (3.4), and Proposition 3.2. This means that every distribution of the worth of the grand coalition is equivalent to choosing an efficient consumption profile. The definition of $w^\gamma(N)$ also means that the grand coalition is an efficient coalition structure, because the players as members of the grand coalition can choose at least the same strategies that they can as members of coalitions in any coalition structure.

5.2.1. An Interpretation of the γ-Coalitional Function

As in the definition of the γ-coalitional function of a general strategic game, the worth of a coalition in the environmental game is equal to its Nash equilibrium payoff in the game induced by the coalition. Unlike the α- and β-coalitional functions, a deviating coalition does not presume that its complement would follow max-min or min-max strategies. It, however, presumes that the complement $N \setminus S$ would break apart into singletons. In the extreme case, it means that each individual player expects that a deviation by it alone will be enough to precipitate disintegration of the remaining coalition of $N-1$ players into singletons. As will be seen, this "all-or-none" expectation is central to the definition and existence of the γ-core of the environmental game.[2]

In the context of environmental games, I justify this assumption by showing below that if the players outside the deviating coalition S form one or more non-singleton coalitions, then the worth of S is generally higher.[3] Thus, the assumption is equivalent to granting the deviating coalition S a

---

2. There is ample empirical evidence in support of the "all-or-none" expectation. For instance, the Comprehensive Test Ban Treaty (CTBT) on nuclear testing can come into force *only if all* current and potential nuclear powers sign it. No non-singleton coalition that does not include all current and potential nuclear powers has been formed despite many years of negotiations, and none other than the grand coalition is expected to be formed.

3. As mentioned earlier, I offer in chapter 6 a further justification for this assumption in terms of an intuitive noncooperative game of coalition formation in which the players are free to decide whether to stay alone or to merge with other players.

certain degree of pessimism in the sense that it presumes the formation of that coalition structure that is the worst from its point of view. This is pessimism of a different sort: it is not about the strategies that will be adopted by the outside players subsequent to a deviation by a coalition (as in the case of α- and β-coalitional functions), but about the coalitions that will be formed by them. In other words, the uncertainty regarding the coalition structure that will form subsequent to a deviation by a coalition is resolved by assuming that the deviating coalition presumes formation of that coalition structure that is worst from its point of view. This is seen as follows:

Given the environmental game $\Gamma = (N, E, u)$, let $(\hat{e}_1, \hat{e}_2, \ldots, \hat{e}_n)$ denote the Nash equilibrium of the induced game Γ^P, where $P = \{S_1, S_2, \ldots, S_m\}$ is a partition of N. Then, the first-order conditions imply

$$g_i'(\hat{e}_i) = \Sigma_{k \in S_j} v_k'(\Sigma_{h \in N} \hat{e}_h), i \in S_j, j = 1, 2, \ldots, m. \qquad (5.5)$$

Let $\hat{u}_i = g_i(\hat{e}_i) - v_i(\Sigma_{j \in N} \hat{e}_j), i \in N$, denote the corresponding payoffs of the players. The existence and uniqueness of a Nash equilibrium of the induced game Γ^P follows from similar arguments as in the proofs of Propositions 4.2 and 5.1, as the strategy set of each player (i.e., each coalition in the partition) is a compact and convex set of R, the payoff function is continuous and concave in the strategies of all players, and the solution of (5.5) is unique.

Proposition 5.3

Let $(\hat{e}_1, \hat{e}_2, \ldots, \hat{e}_n)$ and $(\tilde{e}_1, \tilde{e}_2, \ldots, \tilde{e}_n)$ denote the Nash equilibria of the induced games Γ^P and Γ^{P^j}, where $P = \{S_1, S_2, \ldots, S_m\}$ and $P^j = \{S_j, [N \setminus S_j]\}$, respectively.[4] Then, $\Sigma_{i \in S_j} \tilde{u}_i \leq \Sigma_{i \in S_j} \hat{u}_i$, where $\tilde{u}_i = g_i(\tilde{e}_i) - v_i(\Sigma_{j \in N} \tilde{e}_j)$ and $\hat{u}_i = g_i(\hat{e}_i) - v_i(\Sigma_{j \in N} \hat{e}_j)$.

Proof of Proposition 5.3

It is sufficient to show that $\Sigma_{i \in N} \hat{e}_i \leq \Sigma_{i \in N} \tilde{e}_i$, but $\tilde{e}_i \leq \hat{e}_i$ for all $i \in S_j$. Suppose contrary to the assertion that $\Sigma_{i \in N} \hat{e}_i > \Sigma_{i \in N} \tilde{e}_i$. Then, because each g_i is

---

4. Recall that $[N \setminus S_j]$ denotes the finest partition of the set $N \setminus S_j$.

strictly concave, and each v_i is convex, the first-order conditions (5.2), (5.3) for $S = S_j$ and (5.5) imply $\hat{e}_i \leq \tilde{e}_i$ for all $i \in N$. But this contradicts my supposition that $\sum_{i \in N} \hat{e}_i > \sum_{i \in N} \tilde{e}_i$. Hence, $\sum_{i \in N} \hat{e}_i \leq \sum_{i \in N} \tilde{e}_i$. Because each g_i is strictly concave, each v_i is convex, and $\sum_{i \in N} \hat{e}_i \leq \sum_{i \in N} \tilde{e}_i$, comparing the first-order conditions (5.2), (5.3) for $S = S_j$ and (5.5), it follows that $\hat{e}_i \geq \tilde{e}_i$ for all $i \in S_j$. Because $\sum_{i \in N} \hat{e}_i \leq \sum_{i \in N} \tilde{e}_i$ and $\hat{e}_i \geq \tilde{e}_i$ for all $i \in S_j$, it follows that $\hat{u}_i \geq \tilde{u}_i$ for all $i \in S_j$. ∎

The γ-Core and Coalition Formation

Though the γ-core assumption that the players outside a deviating coalition act individually and do not form any non-singleton coalitions is, as noted earlier, a meaningful behavioral assumption, it may appear arbitrary and an unreasonable forecast of what coalitions may actually form. Conceivably, a deviating coalition may be able to guess better than simply presume formation of the coalition structure that is worst from its perspective. It may be able to predict more accurately the coalition structure that would actually form subsequent to its deviation and not just presume that the complement would break apart into singletons.[5] The next chapter addresses this issue in more detail and shows that breaking apart into singletons is actually a subgame-perfect equilibrium strategy of the players outside a deviating coalition in a repeated game of coalition formation in which the players are free to stay alone or merge with other players. In other words, the players in the complement of a deviating coalition indeed have incentives to break apart into singletons, and, therefore, the assumption underlying the γ-core is justified.

---

5. The extant literature on coalition formation itself does not have much to say on this as there is no general characterization result that shows which coalition structures may form upon deviation by a coalition, especially if the countries are not identical. Thus, assuming that the players in the complement form a coalition of their own or some other coalition structure is as questionable as the assumption that they form singletons.

5.2.2. Superadditivity of the γ-Coalitional Function

Before proving that the environmental game (N, E, u) admits a nonempty γ-core, I show by an example that the γ-coalitional function is not generally superadditive.[6]

Example 1

The game has three players with production and damage functions:

$$g_1(e_1) = 6e_1 - \frac{1}{2}e_1^2; \; v_1(z) = \frac{1}{2}z^2,$$

$$g_2(e_2) = 3e_2 - \frac{1}{2}e_2^2; \; v_2(z) = z, \; \text{and}$$

$$g_3(e_3) = 3e_3 - \frac{1}{2}e_3^2; \; v_3(z) = z.$$

The three players are identical except for the damage function of player 1. Using the first-order conditions, the Nash equilibrium strategies and payoffs are $\bar{e}_1 = 1, \bar{e}_2 = 2, \bar{e}_3 = 2, \bar{z} = 5; \bar{u}_1 = -7, \bar{u}_2 = \bar{u}_3 = -1$, respectively. Let $S_1 = \{2\}$, $S_2 = \{3\}$, and $S = S_1 \cup S_2 = \{2,3\}$. Then, $w^\gamma(S_1) = w^\gamma(S_2) = -1$.

Using the first-order conditions, the Nash equilibrium strategies and the payoffs of the induced game corresponding to coalition S $(= S_1 \cup S_2)$ are $\tilde{e}_1 = 2, \tilde{e}_2 = 1, \tilde{e}_3 = 1, \tilde{z} = 4; \tilde{u}_1 = 2, \tilde{u}_2 = -\frac{3}{2}, \tilde{u}_3 = -\frac{3}{2}$, respectively. Thus, $w^\gamma(S_1 \cup S_2)(= -3) < w^\gamma(S_1) + w^\gamma(S_2)(= -2)$, which proves that the coalitional function w^γ is not superadditive.

Because $p(S_1 \cup S_2, \{S_1 \cup S_2, \{1\}\}) = w^\gamma(S_1 \cup S_2) < p(S_1, \{S_1, S_2, \{1\}\}) + p(S_2, \{S_1, S_2, \{1\}\}) = w^\gamma(S_1) + w^\gamma(S_2)$, the example also shows that the partition function representation p of the environmental games may not be superadditive either.

---

6. Eyckmans (1997) also shows that the γ-coalitional function is not superadditive, but by simulations of a parameterized example.

The reason why players 2 and 3 are worse off if they form a coalition than if they stay apart is that player 1 increases its emissions, which partly offsets the emission reductions made by the coalition of 2 and 3. Thus, the example highlights the complex nature of the externality problem studied in this book.

Though it is customary in conventional cooperative game theory to take the coalitional function to be superadditive, superadditivity is neither necessary nor sufficient for the existence of the core of a coalitional game (see, e.g., Friedman 1990). It is also well known that convex coalitional games are superadditive and have nonempty cores (Shapley 1971). But the γ-coalitional function, as shown, is not even superadditive, let alone convex. Thus, we need to verify whether the γ-core of the environmental game is nonempty or not.

5.3 NONEMPTINESS OF THE γ-CORE

Bondareva (1963) and Shapley (1967) show that the core of a coalitional game with transferable utility is nonempty if and only if the game is balanced. The Bondareva–Shapley theorem has since become the standard procedure to check whether the core of a coalitional game is nonempty. It uses the following notion of balanced weights.[7]

5.3.1 Balanced Collection of Weights

Let C denote the set of all coalitions, and let $C_i = \{S \in C : i \in S\}$ denote the subset of all those coalitions of which i is a member. A vector $(\delta_S)_{S \in C}$ such that $\delta_S \in [0,1]$ for all $S \in C$ is a *balanced collection of weights* if for all $i \in N$, $\sum_{S \in C_i} \delta_S = 1$.

One interpretation of a balanced collection of weights is that for a coalition S to be active for the fraction of time δ_S, each of its members must be active in S for the same fraction of time δ_S. Suppose each player

---

7. See Helm (2000, 2001) for a lucid discussion of this notion.

has 1 unit of time available, then the condition $\sum_{S \in C_i} \delta_S = 1$ for all $i \in N$ means that the total time of each player is used up as a member of the various coalitions. In other words, the vector $(\delta_S)_{S \in C}$ is a balanced collection of weights if it represents a feasible allocation of time of the players.

5.3.2 The Bondareva–Shapley Theorem

A coalitional game (N, w) with transferable utility has a nonempty core if and only if $\sum_{S \in C} \delta_S w(S) \leq w(N)$ for every balanced collection of weights $(\delta_S)_{S \in C}$.

To interpret, if coalition S is active for the fraction of time δ_S, it yields the payoff $\delta_S w(S)$. The Bondareva–Shapley theorem means that the core of a game is nonempty if there is no feasible allocation of the time of the players that yields a higher aggregate payoff than what the grand coalition can achieve.

Chander and Tulkens (1997) show that the coalitional game (N, w^γ) has a nonempty core if the damage functions are linear or satisfy other similar assumptions. Helm (2001) proves the same under more general assumptions.[8] The proof of the next theorem closely follows Helm.

Theorem 5.4

The coalitional game (N, w^γ) has a nonempty core.

Before proceeding with the proof of this theorem, I take note of a useful accounting identity: Given an emission profile $(\bar{e}_1, \bar{e}_2, \ldots, \bar{e}_n)$, for any balanced collection of weights $(\delta_S)_{S \in C}$, we have $\sum_{j \in N} \bar{e}_j = \sum_{S \in C} \delta_S \sum_{j \in S} \bar{e}_j$. Therefore,

$$\sum_{S \in C \backslash C_i} \delta_S \sum_{j \in S} \bar{e}_j = \sum_{j \in N} \bar{e}_j - \sum_{S \in C_i} \delta_S \sum_{j \in S} \bar{e}_j = \sum_{S \in C_i} \delta_S (\sum_{j \in N} \bar{e}_j - \sum_{j \in S} \bar{e}_j)$$

$$= \sum_{S \in C_i} \delta_S \sum_{j \in N \backslash S} \bar{e}_j.$$

8. However, the proofs in both Chander and Tulkens (1997) and Helm (2001) use Proposition 5.2 above.

Proof of Theorem 5.4

Let C be the set of all coalitions and $(\delta_S)_{S \in C}$ a balanced collection of weights. In view of the Bondareva–Shapley theorem, I need to show that $w^\gamma(N) \geq \sum_{S \in C} \delta_S w^\gamma(S)$.

Let $(e_1(S), e_2(S), \ldots, e_n(S))$ denote the unique Nash equilibrium of the game induced by coalition S. For each $i \in N$, let $e_i(C) = \sum_{S \in C_i} \delta_S e_i(S)$. Then, $e_i(C)$ is the weighted average of the equilibrium emission of player i as a member of the various coalitions. By concavity,

$$g_i\left(e_i(C)\right) = g_i\left(\sum_{S \in C_i} \delta_S e_i(S)\right) \geq \sum_{S \in C_i} \delta_S g_i\left(e_i(S)\right), \qquad (5.6)$$

as $\delta_S \in [0,1]$ and $\sum_{S \in C_i} \delta_S = 1$. By Proposition 5.2, $\sum_{j \in S} \overline{e}_j \geq \sum_{j \in S} e_j(S)$ and $\overline{e}_j \leq e_j(S)$ for all $j \in N \backslash S$. Substituting in the accounting identity noted above,

$$\sum_{S \in C \backslash C_i} \delta_S \sum_{j \in S} e_j(S) \leq \sum_{S \in C_i} \delta_S \sum_{j \in N \backslash S} e_j(S). \qquad (5.7)$$

Because each damage function v_i is increasing and convex,

$$v_i\left(\sum_{j \in N} e_j(C)\right) = v_i\left(\sum_{j \in N} \sum_{S \in C_j} \delta_S e_j(S)\right) = v_i\left(\sum_{S \in C} \delta_S \sum_{j \in S} e_j(S)\right)$$

$$= v_i\left(\sum_{S \in C_i} \delta_S \sum_{j \in S} e_j(S) + \sum_{S \in C \backslash C_i} \delta_S \sum_{j \in S} e_j(S)\right)$$

$$\leq v_i\left(\sum_{S \in C_i} \delta_S \sum_{j \in S} e_j(S) + \sum_{S \in C_i} \delta_S \sum_{j \in N \backslash S} e_j(S)\right) \text{ (using (5.7))}$$

$$= v_i\left(\sum_{S \in C_i} \delta_S \left(\sum_{j \in S} e_j(S) + \sum_{j \in N \backslash S} e_j(S)\right)\right)$$

$$= v_i\left(\sum_{S \in C_i} \delta_S \sum_{j \in N} e_j(S)\right)$$

$$\leq \sum_{S \in C_i} \delta_S v_i\left(\sum_{j \in N} e_j(S)\right) \text{ (by convexity).} \qquad (5.8)$$

Finally, because $(e_1(C), e_2(C), \ldots, e_n(C))$ is not necessarily equal to the unique efficient emission profile $(e_1^*, e_2^*, \ldots, e_n^*)$,

$$w^\gamma(N) \geq \sum_{i \in N} g_i\big(e_i(C)\big) - \sum_{i \in N} v_i\Big(\sum_{j \in N} e_j(C)\Big)$$

$$\geq \sum_{i \in N} \sum_{S \in C_i} \delta_S g_i\big(e_i(S)\big) - \sum_{i \in N} \sum_{S \in C_i} \delta_S v_i\Big(\sum_{j \in N} e_j(S)\Big) \text{(using (5.6) and (5.8))}$$

$$= \sum_{S \in C} \delta_S \sum_{i \in S} g_i\big(e_i(S)\big) - \sum_{S \in C} \delta_S \sum_{i \in S} v_i\Big(\sum_{j \in N} e_j(S)\Big)$$

$$= \sum_{S \in C} \delta_S \Big(\sum_{i \in S} g_i\big(e_i(S)\big) - v_i\Big(\sum_{j \in N} e_j(S)\Big)\Big)$$

$$= \sum_{S \in C} \delta_S w^\gamma(S),$$

by definition of the γ-coalitional function.[9] ∎

5.4. SOME SPECIFIC γ-CORE IMPUTATIONS

By definition, the coalitional game (N, w^γ) has a nonempty core if there exists an *imputation* $(a_1, a_2, \ldots, a_n)$ such that $\sum_{i \in S} a_i \geq w^\gamma(S)$ for all $S \subset N$. Thus, a more direct and constructive proof for the existence of a nonempty γ-core is to demonstrate that a specific imputation of the game belongs to the core of the coalitional game (N, w^γ). I demonstrate that some specific imputations can indeed be shown to belong to the γ-core if the damage functions satisfy additional conditions. Furthermore, such imputations have economic interpretations of their own and, as will be seen, are of

---

9. Because, as shown, the γ-core is in general a subset of the α- and β-cores, the theorem also proves, by default, that the α- and β-cores of the environmental game are also nonempty.

special interest as they can be used to find γ-core imputations in simulation models of climate change.[10]

5.4.1. Linear Damage Functions

Assumption 5.1

For all $i \in N$, $v_i(z) = \pi_i z$, $\pi_i > 0$; that is, the disutility or the damage functions are linear, and the willingness to pay for the environment of each country i is constant and equal to π_i.

As can be seen from the first-order conditions (5.1) for a Nash equilibrium, an important consequence of the linearity assumption is that the Nash equilibrium is a dominant strategy equilibrium. Furthermore, if the damage functions are linear, then, as seen from (5.3), in the game induced by a coalition, the Nash equilibrium strategies of the players outside the coalition are the same as in the Nash equilibrium of the environmental game.

Let $(x_1^*, \ldots, x_n^*, z^*)$ denote the efficient consumption profile in which

$$x_i^* = g_i(\bar{e}_i) - \frac{\pi_i}{\sum\limits_{j \in N} \pi_j}\left(\sum_{j \in N} g_j(\bar{e}_j) - \sum_{j \in N} g_j(e_j^*)\right)$$

$$= g_i(e_i^*) + \left(g_i(\bar{e}_i) - g_i(e_i^*)\right)$$

$$- \frac{\pi_i}{\sum\limits_{j \in N} \pi_j}\left(\sum_{j \in N} g_j(\bar{e}_j) - \sum_{j \in N} g_j(e_j^*)\right), i = 1, \ldots, n, \qquad (5.9)$$

where $(e_1^*, e_2^*, \ldots, e_n^*)$ and $(\bar{e}_1, \bar{e}_2, \ldots, \bar{e}_n)$ are the unique efficient and Nash equilibrium emission profiles, respectively, and, thus, $z^* = \sum_{i \in N} e_i^*$. In this efficient consumption profile, each country gets an amount of the private good that is equal to its Nash equilibrium output minus a weighted share of the difference between the total Nash equilibrium and the total efficient outputs of all countries. In other words, the cost

---

10. See, for example, Eyckmans and Tulkens (2003).

of reducing the ambient pollution from the Nash equilibrium level $\bar{z}$ to z^* is shared by each country in proportion to its willingness to pay, π_i, for the environment.[11]

Let $a_i^* = x_i^* - \pi_i z^*$, where x_i^* is as defined in (5.9). Then, $\sum_{i \in N} a_i^* = w^\gamma(N)$; that is, $(a_1^*, a_2^*, \ldots, a_n^*)$ is an imputation of the coalitional game (N, w^γ). I show that it belongs to the core of (N, w^γ), and hence the core of (N, w^γ) is nonempty.

Theorem 5.5

If the damage functions are linear, that is, satisfy Assumption 5.1, then the imputation $(a_1^*, a_2^*, \ldots, a_n^*)$ constructed from the efficient consumption profile defined by (5.9) belongs to the core of the game (N, w^γ).

Proof of Theorem 5.5

Suppose contrary to the assertion that the imputation $(a_1^*, a_2^*, \ldots, a_n^*)$ does not belong to the core of the game (N, w^γ). Then, by definition, there exists a coalition $S \subset N$ and a joint strategy $(\tilde{e}_1, \ldots, \tilde{e}_n)$ such that $(\tilde{e}_1, \ldots, \tilde{e}_n)$ is the Nash equilibrium of the game induced by S and $\sum_{i \in S} g_i(\tilde{e}_i) - \sum_{i \in S} \pi_i \bar{z} > \sum_{i \in S} x_i^* - \sum_{i \in S} \pi_i z^*$. Consider now an alternative imputation $(\hat{x}_1, \ldots, \hat{x}_n, z^*)$ where

$$\hat{x}_i = g_i(\tilde{e}_i) - \frac{\pi_i}{\sum_{j \in N} \pi_j} \left(\sum_{j \in N} g_j(\tilde{e}_j) - \sum_{j \in N} g_j(e_j^*) \right), i \in N$$

and $\hat{a}_i = \hat{x}_i - \pi_i z^*, i \in N$. Clearly, $(\hat{a}_1, \ldots, \hat{a}_n)$ is also an imputation of the game. I show that $(\hat{a}_1, \ldots, \hat{a}_n)$ strictly dominates $(a_1^*, a_2^*, \ldots, a_n^*)$; that is, $\sum_{i \in N} \hat{a}_i > \sum_{i \in N} a_i^*$.

Because the payoff functions are linear, the first-order conditions (5.1) and (5.3) imply $\tilde{e}_i = \bar{e}_i$ for all $i \in N \backslash S$. Furthermore, by comparing (3.3)

---

11. The idea to share the cost of public good provision in proportion to each agent's willingness to pay goes back to Chander (1993).

and (5.2), linearity of the payoff functions and strict concavity of g_i imply $\bar{e}_i \geq \tilde{e}_i \geq e_i^*$ for all $i \in S$. Thus, $\sum_{i \in N} g_i(\tilde{e}_i) \leq \sum_{i \in N} g_i(\bar{e}_i)$. Because $\tilde{e}_i = \bar{e}_i$ for all $i \in N \setminus S$ and $\bar{e}_i \geq \tilde{e}_i$ for all $i \in S$,

$$\hat{x}_i = g_i(\tilde{e}_i) - \frac{\pi_i}{\sum\limits_{j \in N} \pi_j}\left(\sum_{j \in N} g_j(\tilde{e}_j) - \sum_{j \in N} g_j(e_j^*)\right)$$

$$\geq g_i(\bar{e}_i) - \frac{\pi_i}{\sum\limits_{j \in N} \pi_j}\left(\sum_{j \in N} g_j(\bar{e}_j) - \sum_{j \in N} g_j(e_j^*)\right) = x_i^* \text{ for all } i \in N \setminus S.$$

This means that $\hat{a}_i \geq a_i^*$ for all $i \in N \setminus S$. I next show that $\sum_{i \in S} \hat{a}_i > \sum_{i \in S} a_i^*$. By definition,

$$\sum_{i \in S} \hat{a}_i = \sum_{i \in S} g_i(\tilde{e}_i) - \frac{\sum_{i \in S} \pi_i}{\sum_{i \in N} \pi_i}\left(\sum_{i \in N} g_i(\tilde{e}_i) - \sum_{i \in N} g_i(e_i^*)\right) - \sum_{i \in S} \pi_i z^*$$

$$= \sum_{i \in S} g_i(\tilde{e}_i) - \frac{\sum_{i \in S} \pi_i}{\sum_{i \in N} \pi_i}\left(\sum_{i \in N} \pi_i z^* - \sum_{i \in N} \pi_i \tilde{z}\right.$$

$$\left. + \left(\sum_{i \in N} g_i(\tilde{e}_i) - \sum_{i \in N} g_i(e_i^*)\right)\right) - \sum_{i \in S} \pi_i \tilde{z}$$

$$> \sum_{i \in S} g_i(\tilde{e}_i) - \sum_{i \in S} \pi_i \tilde{z}, \text{ as } g_i(\tilde{e}_i) - g_i(e_i^*) < \sum_{j \in N} \pi_j(\tilde{e}_i - e_i^*), i \in N,$$

by strict concavity of g_i and efficiency condition (3.3), which implies $\sum_{j \in N} \pi_j = g_i'(e_i^*), i \in N$.

Because, by supposition, $\sum_{i \in S} g_i(\tilde{e}_i) - \sum_{i \in S} \pi_i \tilde{z} > \sum_{i \in S} x_i^* - \sum_{i \in S} \pi_i z^* = \sum_{i \in S} a_i^*$, this implies $\sum_{i \in S} \hat{a}_i > \sum_{i \in S} a_i^*$. Thus, as required, I have shown that $(\hat{a}_1, \ldots, \hat{a}_n)$ dominates the imputation $(a_1^*, a_2^*, \ldots, a_n^*)$; that is, $\sum_{i \in N} \hat{a}_i > \sum_{i \in N} a_i^*$. But this contradicts that $(a_1^*, a_2^*, \ldots, a_n^*)$ is an imputation of the game. Hence, my supposition is wrong, and the imputation $(a_1^*, a_2^*, \ldots, a_n^*)$ belongs to the core of the game (N, w^γ). ∎

Affine Damage Functions

It is straightforward to generalize Theorem 5.5 to the case when the damage functions are affine but not linear; that is, $v_i(z) = \pi_i z + b_i, \pi_i > 0, i \in N$.

The proof of Theorem 5.5 still holds except that the calculations and the imputation now carry an extra constant term. In particular, the imputation $a_i^* = x_i^* - \pi_i z^* - b_i$, $i = 1, \ldots, n$, where each x_i^* is as defined in (5.9), belongs to the γ-core.

Corollary to Theorem 5.5

If the damage functions are affine, that is, $v_i(z) = \pi_i z + b_i$, $\pi_i > 0$, $i = 1, \ldots, n$, then the imputation $(a_i^*, \ldots, a_n^*)$ where $a_i^* = x_i^* - \pi_i z^* - b_i$ belongs to the core of the game (N, w^γ).

As noted above, the γ-core imputations with an explicit form can be particularly useful for computing γ-core imputations in simulation models, beside their theoretical interest. Thus, I consider a few other γ-core imputations that also have explicit forms. As seen above, linearity of the damage functions implies simpler strategic interaction between the players. If the damage functions are not linear or affine, then it may no longer be the case that $\tilde{e}_i \leq \bar{e}_i$ for each $i \in S$, where $\tilde{e}_i$ are the Nash equilibrium emissions in the game induced by coalition S and $(\bar{e}_1, \bar{e}_2, \ldots, \bar{e}_n)$ is the unique Nash equilibrium profile of the environmental game. Because these inequalities play a role in the proof of Theorem 5.5, I look for other conditions under which they may continue to hold.

5.4.2. Nonlinear Damage Functions

Assumption 5.2

The damage functions v_i, $i = 1, \ldots, n$ are such that for each $S \subset N$, $S \neq N$, $|S| \geq 2$, we have $\sum_{i \in S} v_i'(z^*) \geq v_j'(\bar{z})$, $j \in S$, where $\bar{z}$ and z^* correspond to the Nash equilibrium and the efficient emission profiles, respectively.

In words, the assumption requires that the willingness to pay of no country should fall "too much" as the total pollution is reduced from the Nash equilibrium level $\bar{z}$ to the efficient level z^*. This is a weaker assumption in the sense that it is satisfied if the damage functions are linear. As can be easily checked, it is also satisfied by a general class of quadratic damage functions.

Lemma 5.6

For each $S \subset N$, let $(\tilde{e}_1, \tilde{e}_2, \ldots, \tilde{e}_n)$ be the Nash equilibrium of the game induced by coalition S. Then, $\tilde{e}_i \leq \overline{e}_i$ for each $i \in S$ if the damage functions $v_i, i = 1, \ldots, n$ satisfy Assumption 5.2.

Proof of Lemma 5.6

I claim that $\sum_{i \in N} \tilde{e}_i \geq \sum_{i \in N} e_i^*$. Suppose contrary to the claim that $\sum_{i \in N} \tilde{e}_i < \sum_{i \in N} e_i^*$. By convexity, $v_i'(z^*) \geq v_i'(\tilde{z})$ for all $i \in N$. From (3.3), (5.2), and (5.3), it follows that $g_i'(e_i^*) \geq g_i'(\tilde{e}_i)$ for all $i \in N$. Because each g_i is strictly concave, $e_i^* \leq \tilde{e}_i$ for each $i \in N$. But this contradicts my supposition. Therefore, $\sum_{i \in N} \tilde{e}_i \geq \sum_{i \in N} e_i^*$ and, by convexity, $\sum_{i \in S} v_i'(\tilde{z}) \geq \sum_{i \in S} v_i'(z^*)$. Using Assumption 5.2, $\sum_{i \in S} v_i'(\tilde{z}) \geq v_j'(\tilde{z})$ for all $j \in S$. Thus, comparing (5.1) and (5.2), $g_j'(\tilde{e}_j) \geq g_j'(\overline{e}_j)$ for all $j \in S$. Because each g_j is strictly concave, $\tilde{e}_j \leq \overline{e}_j$ for all $j \in S$. ∎

Let

$$x_i^* = g_i(\overline{e}_i) - \frac{v_i'(z^*)}{\sum\limits_{j \in N} v_j'(z^*)} \left(\sum_{j \in N} g_j(\overline{e}_j) - \sum_{j \in N} g_j(e_j^*) \right), \qquad (5.10)$$

and $a_i^* = x_i^* - v_i(z^*), i \in N$. Then, $(a_1^*, a_2^*, \ldots, a_n^*)$ is an imputation of the game (N, w^γ) and has the same economic interpretation as the imputation constructed from (5.9).

Theorem 5.7

If the damage functions $v_i, i = 1, \ldots, n$, satisfy Assumption 5.2, the imputation $(a_1^*, a_2^*, \ldots, a_n^*)$ defined from the efficient consumption profile (5.10) belongs to the core of the game (N, w^γ).

Proof of Theorem 5.7

Suppose contrary to the assertion that the imputation $(a_1^*, a_2^*, \ldots, a_n^*)$ does not belong to the core of the game (N, w^γ). Then there exists a coalition $S \subset N$ and a joint strategy $(\tilde{e}_1, \ldots, \tilde{e}_n)$ such that $(\tilde{e}_1, \ldots, \tilde{e}_n)$ is the Nash equilibrium of the game induced by S and $\sum_{i \in S} g_i(\tilde{e}_i) - \sum_{i \in S} v_i(\tilde{z}) > \sum_{i \in S} x_i^* - \sum_{i \in S} v_i(z^*)$.

Consider now an alternative imputation of the game, which is defined as follows. Let

$$\hat{x}_i = g_i(\tilde{e}_i) - \frac{v_i'(z^*)}{\sum\limits_{j \in N} v_j'(z^*)}\left(\sum_{j \in N} g_j(\tilde{e}_j) - \sum_{j \in N} g_j(e_j^*)\right), i \in N$$

and $\hat{a}_i = \hat{x}_i - v_i(z^*), i \in N$. Clearly, $(\hat{a}_1, \ldots, \hat{a}_n)$ is also an imputation of the game. I show below that $(\hat{a}_1, \ldots, \hat{a}_n)$ strictly dominates $(a_1^*, a_2^*, \ldots, a_n^*)$; that is, $\sum_{i \in N} \hat{a}_i > \sum_{i \in N} a_i^*$. By definition,

$$\sum_{i \in N \backslash S} \hat{x}_i = \sum_{i \in N \backslash S} g_i(\tilde{e}_i) - \sum_{i \in N \backslash S} \frac{v_i'(z^*)}{\sum\limits_{j \in N} v_j'(z^*)}\left(\sum_{j \in N} g_j(\tilde{e}_j) - \sum_{j \in N} g_j(e_j^*)\right)$$

$$= \sum_{i \in N \backslash S} g_i(\bar{e}_i) - \sum_{i \in N \backslash S} \frac{v_i'(z^*)}{\sum\limits_{j \in N} v_j'(z^*)}\left(\sum_{j \in N} g_j(\tilde{e}_j) - \sum_{j \in N} g_j(e_j^*)\right)$$

$$+ \left(\sum_{i \in N \backslash S} g_i(\tilde{e}_i) - \sum_{i \in N \backslash S} g_i(\bar{e}_i)\right)$$

$$+ \sum_{i \in N \backslash S} \frac{v_i'(z^*)}{\sum\limits_{j \in N} v_j'(z^*)}\left(\sum_{j \in N} g_j(\bar{e}_j) - \sum_{j \in N} g_j(\tilde{e}_j)\right)$$

$$= \sum_{i \in N \backslash S} x_i^* + \left(\sum_{i \in N \backslash S} g_i(\tilde{e}_i) - \sum_{i \in N \backslash S} g_i(\bar{e}_i)\right)$$

$$+ \sum_{i \in N \backslash S} \frac{v_i'(z^*)}{\sum\limits_{j \in N} v_j'(z^*)}\left(\sum_{j \in N} g_j(\bar{e}_j) - \sum_{j \in N} g_j(\tilde{e}_j)\right)$$

$$= \sum_{i \in N \backslash S} x_i^* + \left[\left(\sum_{i \in N \backslash S} g_i(\tilde{e}_i) - \sum_{i \in N \backslash S} g_i(\bar{e}_i)\right)\right.$$

$$\left. - \sum_{i \in N \backslash S} \frac{v_i'(z^*)}{\sum\limits_{j \in N} v_j'(z^*)}\left(\sum_{j \in N \backslash S} g_j(\tilde{e}_j) - \sum_{j \in N \backslash S} g_j(\bar{e}_j)\right)\right]$$

$$- \sum_{i \in N \backslash S} \frac{v_i'(z^*)}{\sum\limits_{j \in N} v_j'(z^*)}\left(\sum_{j \in S} g_j(\tilde{e}_j) - \sum_{j \in S} g_j(\bar{e}_j)\right).$$

Because Proposition 5.2 implies $\tilde{e}_i \geq \bar{e}_i$ for all $i \in N \setminus S$ and, by Lemma 5.6, $\tilde{e}_i \leq \bar{e}_i$ for all $i \in S, \sum_{i \in N \setminus S} \hat{x}_i > \sum_{i \in N \setminus S} x_i^*$, we have $\sum_{i \in N \setminus S} \hat{a}_i > \sum_{i \in N \setminus S} a_i^*$. I next show that $\sum_{i \in S} \hat{a}_i \geq \sum_{i \in S} a_i^*$. By definition,

$$
\sum_{i \in S} \hat{x}_i = \sum_{i \in S} g_i(\tilde{e}_i) - \sum_{i \in S} \frac{v_i'(z^*)}{\sum_{j \in N} v_j'(z^*)} \left(\sum_{j \in N} g_j(\tilde{e}_j) - \sum_{j \in N} g_j(e_j^*) \right)
$$

$$
\geq \sum_{i \in S} g_i(\tilde{e}_i) - \sum_{i \in S} \frac{v_i'(z^*)}{\sum_{j \in N} v_j'(z^*)} \left(\sum_{i \in N} v_i'(z^*) \left(\sum_{j \in N} \tilde{e}_j - \sum_{j \in N} e_j^* \right) \right),
$$

as the functions g_j are concave and efficiency condition (3.3) implies $g_j'(e_j^*) = \sum_{i \in N} v_i'(z^*)$ for each $j \in N$. Thus, $\sum_{i \in S} \hat{x}_i - \sum_{i \in S} v_i'(z^*) z^* \geq \sum_{i \in S} g_i(\tilde{e}_i) - \sum_{i \in S} v_i'(z^*) \tilde{z}$, which in turn implies $\sum_{i \in S} \hat{x}_i - \sum_{i \in S} v_i(z^*) \geq \sum_{i \in S} g_i(\tilde{e}_i) - \sum_{i \in S} v_i(\tilde{z}) - (\sum_{i \in S} v_i(z^*) + \sum_{i \in S} v_i(\tilde{z}) - \sum_{i \in S} v_i'(z^*)(z^* - \tilde{z})) \geq \sum_{i \in S} g_i(\tilde{e}_i) - \sum_{i \in S} v_i(\tilde{z})$, as each v_i is convex.

This shows, as required, that $(\hat{a}_1, \ldots, \hat{a}_n)$ strictly dominates imputation $(a_1^*, a_2^*, \ldots, a_n^*)$. But this contradicts that $(a_1^*, a_2^*, \ldots, a_n^*)$ is an imputation of the coalitional game (N, w^γ). Hence, my supposition is wrong, and, thus, the imputation $(a_1^*, a_2^*, \ldots, a_n^*)$ defined from (5.10) belongs to the core of the game (N, w^γ). ∎

Identical Damage Functions

Another case in which the transfers or a γ-core imputation of the environmental game can be stated explicitly is that of damage functions that are nonlinear in general but identical across the players; that is, $v_i = v_j$ for all $i, j \in N$. If the damage functions are identical, then Lemma 5.6 holds because (5.1), (5.2), and (5.3), which characterize Nash equilibrium and Nash equilibrium of the game induced by a coalition S, imply that $g_i'(\bar{e}_i) = g_j'(\bar{e}_j)$ and $g_i'(\tilde{e}_i) = g_j'(\tilde{e}_j)$ for all $i, j \in S$. Because $\sum_{i \in S} \tilde{e}_i \leq \sum_{i \in S} \bar{e}_i$ by Proposition 5.2, $\tilde{e}_i \leq \bar{e}_i$ for at least one i in S; that is, $g_i'(\tilde{e}_i) \geq g_i'(\bar{e}_i)$ for *at least one* $i \in S$. But this implies $g_i'(\tilde{e}_i) \geq g_i'(\bar{e}_i)$ *for all* $i \in S$ from equalities (5.1) and (5.2) and the assumption that the damage functions are identical. That is, $\tilde{e}_i \leq \bar{e}_i$ for all $i \in S$. In view of that, it is straightforward to see that the proof of Theorem 5.7 applies to this case as well. I summarize this result as Theorem 5.8.

Theorem 5.8

If the damage functions of the agents are identical, that is, $v_i = v_j$ for all $i, j \in N$, then the imputation $(a_1^*, a_2^*, \ldots, a_n^*)$ defined from the efficient consumption profile (5.10) belongs to the core of the game (N, w^γ).

As noted in the previous chapter, a number of studies have focused on partition function games with identical players. It was shown that the imputation with equal payoffs belongs to the γ-cores of these games. I show that the same is also true for the environmental game with identical players.

Identical Players

Recall that $(e_1^*, \ldots, e_n^*)$ denotes the unique efficient emission profile and in the absence of transfers implies utilities $u_i^* = g_i(e_i^*) - v_i(z^*), i \in N$. If the agents are identical, that is, $v_i = v_j$ and $g_i = g_j$ for all $i, j \in N$, the efficiency conditions (3.3) imply that $e_i^* = e_j^*$ and $u_i^* = u_j^*$ for all $i, j \in N$.

Theorem 5.9

If the agents are identical, that is, $v_i = v_j$ and $g_i = g_j$ for all $i, j \in N$, then the imputation $(u_1^*, u_2^*, \ldots, u_n^*)$ assigning equal payoffs to all players belongs to the core of the game (N, w^γ) and $w^\gamma(S) < \sum_{i \in S} u_i^*$ for all $S \subset N, S \neq N$.

Proof of Theorem 5.9

Let $(\hat{e}_1, \hat{e}_2, \ldots, \hat{e}_n)$ denote the Nash equilibrium of the game induced by coalition S, and let $(\hat{u}_1, \hat{u}_2, \ldots, \hat{u}_n)$ be the Nash equilibrium payoffs, that is, $\hat{u}_i = g(\hat{e}_i) - v(\sum_{j \in N} \hat{e}_j)$. Because $v_i = v_j$ and $g_i = g_j$ for all $i, j \in N$ and each g_i is strictly concave, the first-order conditions (5.2) and (5.3) imply $\hat{e}_i < \hat{e}_j$ for $i \in S$ and $j \in N \backslash S$. Thus, $\hat{u}_i < \hat{u}_j$ for each $i \in S$ and $j \in N \backslash S$. If $S \neq N$, $(\hat{e}_1, \hat{e}_2, \ldots, \hat{e}_n)$ is not equal to the unique efficient emission profile $(e_1^*, \ldots, e_n^*)$ and thus $\sum_{i \in N} \hat{u}_i < \sum_{i \in N} u_i^*$. Hence, $\sum_{i \in S} u_i^* > \sum_{i \in S} \hat{u}_i = w^\gamma(S)$, as $\hat{u}_i < \hat{u}_j$ for each $i \in S$ and $j \in N \backslash S$. Because $(u_1^*, u_2^*, \ldots, u_n^*)$ is an imputation and $w^\gamma(S) < \sum_{i \in S} u_i^*$ for all $S \subset N, S \neq N$, it follows that $(u_1^*, u_2^*, \ldots, u_n^*)$ belongs to the core of the game (N, w^γ). ∎

Corollary to Theorem 5.9

For each $i \in S, \hat{u}_i < u_i^*$.

The corollary follows from the fact that $\sum_{i \in S} \hat{u}_i = w^\gamma(S) < \sum_{i \in S} u_i^*$ and the players are identical, that is, $u_i^* = u_j^*$ and $\hat{u}_i = \hat{u}_j$ for all $i, j \in S$.

5.5. THE γ-CORE AND THE COASE THEOREM

We now interpret the existence of a nonempty γ-core of the environmental game as a generalization of the Coase theorem (Coase 1960). The Coase theorem has two parts: (1) an assignment of well-defined pollution rights and bargaining among the parties involved would lead to efficiency, and (2) the initial assignment of pollution rights does not matter from an efficiency perspective.[12]

As noted earlier, Coase took for granted the existence of a government or authority that can assign and enforce pollution rights. He did not foresee a problem such as climate change when there is no such supranational authority. In such situations, efficiency cannot be achieved unless the countries themselves voluntarily agree to an allocation of rights.[13] I claim that because the γ-core of the environmental game is nonempty, the countries, in the absence of a supranational authority, will themselves assign rights that are mutually acceptable and achieve efficiency. Emptiness of the γ-core of the environmental game would have implied the opposite.

In order to see this, I interpret (5.9) as follows: (i) each country i arrogates to itself the right to pollute up to its Nash equilibrium emissions $\bar{e}_i$, which it can enforce without the cooperation of other countries, and accordingly it must be compensated if it is required to reduce its pollution from $\bar{e}_i$ to the efficient level e_i^* by an amount that is equal to its

---

12. See Cooter (1987) among others for this interpretation of Coase (1960).
13. The issue is not confined to the problem of climate change alone. First, there are many other similar international pollution problems (e.g., depletion of the ozone layer). Second, in many parts of the world (as in the Middle Ages), authorities who could allocate and enforce pollution rights are often missing. One may wonder whether efficiency can be achieved in such situations by voluntary negotiations among the parties involved.

cost $g_i(\overline{e}_i) - g_i(e_i^*)$ of doing that, and (ii) because each country i is also a pollutee and benefits from pollution reduced by all countries, it must in turn share the total cost $\sum_{j \in N}(g_j(\overline{e}_j) - g_j(e_j^*))$ of reducing pollution from $(\overline{e}_1, \ldots, \overline{e}_n)$ to $(e_1^*, \ldots, e_n^*)$ in proportion to its willingness to pay. Because the resulting imputation belongs to the γ-core, no country or group of countries can be better off by refusing to accept the assignment of pollution rights $(\overline{e}_1, \overline{e}_2, \ldots, \overline{e}_n)$ and sharing the total cost of reducing pollution from $(\overline{e}_1, \overline{e}_2, \ldots, \overline{e}_n)$ to $(e_1^*, e_2^*, \ldots, e_n^*)$.

Clearly, the principle underlying (5.9) is that the pollutee, and not the polluter, pays. This is not a matter of choice, but comes from the fact that in the absence of a supranational authority that can assign and enforce pollution rights, the independent sovereign countries will not voluntarily agree to any assignment of rights unless they all, individually or as coalitions, stand to gain from the assignment of rights. The nonemptiness of the γ-core and the fact that the imputation resulting from (5.9) belongs to the γ-core implies that it is indeed possible to assign rights such that all countries would voluntarily accept them. Put another way, it also means that not *all* arbitrary assignments of pollution rights would be voluntarily acceptable to the countries unless they lead to a γ-core imputation.[14]

In other words, it means that in the absence of an authority that can assign and enforce pollution rights, part (1) of the Coase theorem still holds, but part (2) does not. In one sense, it is a generalization of the Coase theorem, as it shows that even if there is no authority to assign and enforce pollution rights, the parties involved will themselves assign rights to each other and achieve efficiency.[15] But at the same time it also shows that part (2) of the Coase theorem does not hold as the parties may not voluntarily accept any arbitrary assignment of pollution rights, and therefore not *every* arbitrary assignment of rights may lead to efficiency.[16] Thus, in the absence of an authority to allocate rights, the initial assignment of rights *does* matter from an efficiency perspective. An additional

---

14. See Chander (2011) for an illustrative example.

15. I argue later in chapter 9 that the Kyoto Protocol was really an attempt to negotiate such rights.

16. Part (2) of the Coase theorem has also been questioned by others (see, e.g., Aivazian and Callen 1981 and the reply thereto by Coase 1981).

interpretation of (5.9) reinforces this argument. Let $(e_1^0, e_2^0, \ldots, e_n^0)$ be such that for each $i \in N$,

$$(e_i^0 - e_i^*)\Sigma_{j \in N} \pi_j = \left(g_i(\overline{e}_i) - g_i(e_i^*)\right) - \frac{\pi_i}{\Sigma_{j \in N} \pi_j}\Sigma_{j \in N}\left(g_j(\overline{e}_j) - g_j(e_j^*)\right). \quad (5.11)$$

The right-hand side of this equality is equal to the compensation that country i receives for reducing its pollution from the Nash equilibrium level $\overline{e}_i$ to the efficient level e_i^* minus what it pays toward the total cost of all countries for reducing pollution from levels $(\overline{e}_1, \overline{e}_2, \ldots, \overline{e}_n)$ to $(e_1^*, e_2^*, \ldots, e_n^*)$. The net result is a receipt or a payment. The left-hand side of the equality implies that this receipt or payment is equal to what country i would get if it has a pollution right equal to e_i^0 that can be traded at a price $\Sigma_{j \in N} \pi_j$. Indeed, by definition of $(e_1^0, e_2^0, \ldots, e_n^0)$, $\Sigma_{i \in S} e_i^0 = \Sigma_{i \in S} e_i^*$ and by the efficiency conditions (3.3), $\Sigma_{j \in N} \pi_j = g_i'(e_i^*) = g_j'(e_j^*)$, $i, j = 1, 2, \ldots, n$. Therefore, $e_i^* = \text{argmax}_{e_i}[g_i(e_i) + \Sigma_{j \in N} \pi_j(e_i^0 - e_i)]$. In words, $\Sigma_{j \in N} \pi_j$ is the competitive price of pollution permits at which the countries may buy or sell their pollution permits to maximize their payoffs, and demand for pollution permits is equal to supply. Comparing (5.9) and (5.11), it follows that the resulting after-trade payoffs of the countries are same as in the γ-core imputation $(a_1^*, a_2^*, \ldots, a_n^*)$.

This γ-core property of the after-trade payoffs implies that if a country or group of countries refuses the assignment of tradable pollution rights $(e_1^0, e_2^0, \ldots, e_n^0)$, then it will be worse off. To conclude, there exists a specific assignment of pollution rights such that each country would voluntarily accept it and, as part (1) of the Coase theorem implies, achieve an efficient outcome. Part (2) of the Coase theorem, however, does not hold as not every assignment of rights may be acceptable to all countries and lead to efficiency. That is so because the resulting after-trade payoffs for every assignment of rights may not have the γ-core property, and thus not every arbitrary assignment of rights may be voluntarily acceptable to all countries which is a prerequisite for the markets for rights to operate and lead to efficiency.

Though this latter interpretation, which relies on competitive markets in pollution rights, relates more directly to the Coase theorem, it blurs, in comparison, the distinction between each country's dual role as both a polluter and a pollutee. However, neither of these interpretations is restricted to the environmental game with linear damage functions and/or the specific imputation implied by (5.9). The latter interpretation applies to every γ-core imputation and even if the damage functions are not linear. In other words, every γ-core imputation is equivalent to a mutually acceptable allocation of pollution rights which are tradable on a competitive international market. This is seen as follows.

Proposition 5.10

Let $a_i^* = x_i^* - v_i(z^*)$, $i=1,\ldots,n$ be a γ-core imputation. Then there exist emission rights $(e_1^0,\ldots,e_n^0)$ that can be traded on an international market such that the after-trade payoffs of the countries are equal to their γ-core payoffs $a_i^*, i=1,\ldots,n$.

Proof of Proposition 5.10

Let $e_i^0, i=1,\ldots,n$ be such that $(e_i^0 - e_i^*)\sum_{i=1}^{n} v_j'(z^*) + g_i(e_i^*) = x_i^*$. I claim that if each country i is assigned e_i^0 pollution permits that can be traded at an international market, then the competitive price is equal to $\sum_{i=1}^{n} v_j'(z^*)$. Because, by definition, $\sum_{i=1}^{n} x_i^* = \sum_{i=1}^{n} g_i(e_i^*)$, we have $\sum_{i=1}^{n} e_i^0 = \sum_{i=1}^{n} e_i^*$. Furthermore, if the international price is equal to $\sum_{i=1}^{n} v_j'(z^*)$, then each e_i^* is the solution of the payoff maximization problem $\max_{e_i} (e_i^0 - e_i)\sum_{i=1}^{n} v_j'(z^*) + g_i(e_i)$ of each country i, as, by (3.3), $\sum_{i=1}^{n} v_j'(z^*) = g_i'(e_i^*)$. Thus, $\sum_{i=1}^{n} v_j'(z^*)$ is the competitive international price as the demand for permits equals supply, and the resulting payoffs of the countries are equal to their γ-core payoffs $a_i^*, i=1,\ldots,n$. ∎

Proposition 5.10 shows that each γ-core imputation is equivalent to an allocation of emission rights that can be traded on a competitive international market. This is a useful property of the γ-core, as sometimes making direct transfers can be sensitive to domestic politics in the countries making the transfers. For example, the U.S. government has no authority to commit finances without congressional approval but has authority to sign agreements regarding pollution rights and participate in international markets.[17]

Though the γ-core is smaller than the conventional α- and β-cores, the multiplicity of the γ-core imputations still leaves enough room for indeterminacy. Because this indeterminacy is prevalent among the cores of coalitional function games in general, the cooperative game theory has addressed it by proposing normative criteria for selecting a unique imputation from among the many core imputations. Adopting a normative criteria, however, may not be consistent with the approach in this book, which assumes the countries to be self-interested and strategic and rarely, if at all, motivated by considerations of equity and fairness. Thus, I propose a new selection criterion that is motivated by a positive consideration. This comes from the fact noted above that making direct income transfers may be sensitive to domestic politics, and, therefore, to facilitate an agreement, the transfers, as far as possible, should be kept at the minimum. This means the imputation that minimizes transfers between countries may be selected from among the set of all γ-core imputations, that is, the one that minimizes $\sum_{i=1}^{n}(g_i(e_i^*) - x_i)^2$ over the set of all γ-core imputations $(a_1, \ldots, a_n)$ where, by definition, $a_i = x_i - v_i(z^*)$, $i = 1, \ldots, n$, and $\sum_{i \in N} x_i = \sum_{i \in N} g_i(e_i^*)$. This minimization problem is well defined and has a unique solution, as the set of γ-core imputations is compact and convex and the objective function is continuous and strictly convex.[18]

Diagrammatic Exposition of the γ-Core

Analogous to the diagrammatic exposition of the Coase theorem in chapter 3, figure 5.1 illustrates the γ-core in the case of three countries.[19] In this illustration, country 1 is a polluter *only*, that is, $v_1(z) = 0$ for all $z \geq 0$, and

---

17. Indeed, the U.S. government could sign the Paris Agreement without Congressional approval, though the Agreement has provision for trade in emissions.
18. To be precise, the γ-core is a closed and convex set, as, by definition, it satisfies a system of *weak* linear inequalities. It is compact, as the x_i's are, by definition, bounded above.
19. I restrict discussion to three-player games because they are minimally sufficient for studying coalitional behavior and it seems difficult, if not impossible, to diagrammatically illustrate the γ-core of games with more than three players.

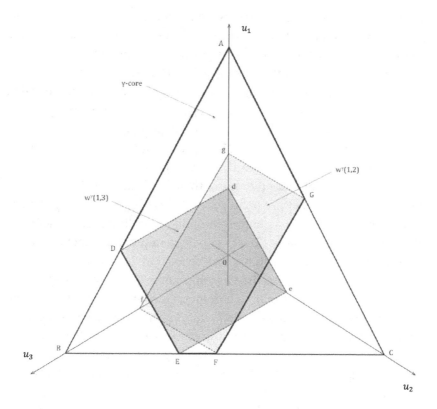

FIGURE 5.1

The γ-core of an economy with one polluter and two pollutees

countries 2 and 3 are pollutees *only*, that is, $v_2(z), v_3(z) > 0$ for all $z > 0$ and $g_2(e_2) = g_3(e_3) = 0$ for all $e_2, e_3 \geq 0$. The origin O of the polyhedral $OABC$ represents the unique Nash equilibrium payoffs/utilities (assumed to be nonnegative) of the three countries.

Because an imputation can be in the γ-core only if it assigns payoffs that are at least as high as the Nash equilibrium payoffs of the players, the two-dimensional simplex ABC (i.e., the face of the polyhedral $OABC$) represents the set of all imputations that could possibly be in the γ-core (i.e., that part of the utility frontier that includes the γ-core). The utility frontier is a two-dimensional simplex, as, by assumption, the utility/payoff functions of the three countries are quasi-linear. The line *fg* represents the utility frontier of coalition {1,3}. It is shown to be a straight line because the utility functions are quasi-linear. The line *FG* is the projection of the line *fg* onto the simplex *ABC*. Therefore, the points on the simplex *ABC* that are to the left of the line *FG* represent imputations that cannot be improved upon by coalition {1,3}. Similarly, line *de* represents the utility frontier of coalition {1,2}, and the points on the simplex *ABC* that are to the right of the line *DE* represent imputations that cannot be improved upon by coalition {1,2}. In contrast, coalition {2,3} cannot achieve utilities that add up to more than the sum of their Nash equilibrium payoffs, as they are pollutees only. Accordingly, no imputations in the simplex *ABC* can be improved upon by coalition {2,3}. Therefore, the points in the polygon *ADEFG* represent imputations that cannot be improved upon by any coalition, and hence they represent the γ-core.

If the payoff functions satisfy Assumption 5.2, the γ-core imputation $(a_1^*, a_2^*, \ldots, a_n^*)$ implied by (5.10) is a point on the line segment *EF*. Though the polluting country 1 reduces its pollution, its payoff remains the same as at the Nash equilibrium confirming my interpretation earlier that (5.10) implicitly assigns the pollution rights to the polluter.[20] But the entire ecological surplus is assigned to the two pollutees, 2 and 3, though that is not so in every γ-core imputation.

Figure 5.2 similarly displays the γ-core when countries 1 and 2 are polluters only and country 3 is a pollutee only. If the payoff functions satisfy Assumption 5.2, the γ-core imputation $(a_1^*, a_2^*, \ldots, a_n^*)$ corresponding to (5.10) is point *A*.[21] The entire surplus is assigned to the pollutee only and none to the two polluters, though that is not so in every γ-core imputation.

---

20. In contrast, the compensation mechanism (Varian 1994) implicitly assigns pollution rights to the pollutee and thus leads to an imputation that does not belong to the γ-core. In fact, the compensation mechanism confirms my assertion in chapter 3 that if the agents are sufficiently heterogeneous, then the outcome of a mechanism that, by definition, rules out transfers between agents cannot be both individually rational and efficient.

21. Diagrams similar to figures 5.1 and 5.2 also appear in Chander and Tulkens (2008).

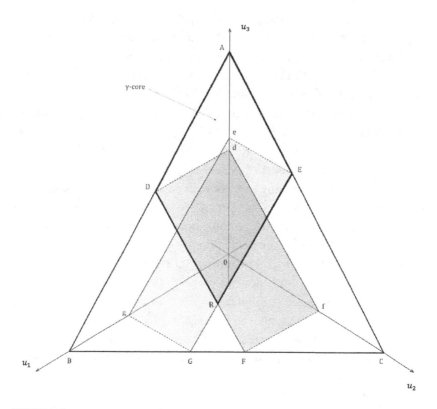

FIGURE 5.2

The γ-core of an economy with two polluters and one pollutee

5.6. THE SEASONAL HAZE IN SOUTHEAST ASIA

It was mentioned in chapter 1 that applications of the game-theoretic approach proposed in this book are not restricted to the climate change problem alone, and it can also be applied to other similar problems of international environmental externalities. One such problem is the "seasonal haze" that afflicts three countries in Southeast Asia in years of drought (for a lucid description and analysis of the problem, see Quah 2002). The major cause of this haze is forest, bush, and field fires to clear and prepare land for agriculture in Indonesia. Besides Indonesia itself, Singapore and Malaysia are the other two most affected countries.[22] The haze has led to poor air quality in these countries for a number of years now and has become an almost annual occurrence. While the region experienced its first episode of haze in 1982, it was viewed as a serious problem only in 1997 when Singapore, parts of Malaysia, and southern Thailand were affected by it for almost three months with widespread concerns about its impacts on health, environment, and economy. In 2002, the Southeast Asian nations agreed to implement measures to prevent the forest fires that lead to haze. But the agreement was not effective, and the haze continues to be a problem with major crises arising subsequently in 2005, 2007, and 2013 when successively higher levels of air pollution were observed in Singapore and Malaysia. Despite a recent regional agreement to curb the haze, 2015 was yet another year of severe haze because of the drought caused by the El Niño weather pattern and its damage cost has been estimated to be US$16 billion or more for Indonesia alone.

Clearly, the Southeast Asian haze problem is an environmental game, especially as it is a flow and not a stock externality. Moreover, it is reasonable to assume that the damages from haze are convex functions of the amount of haze, and the production function for the related activities in Indonesia, which is the only polluter (as well as a pollutee), is a concave function of the amount of haze.

The estimates for damages from the Southeast Asian haze for the affected countries vary, but a conservative estimate made by the Asian Development Bank was US$9 billion for the 1997 episode. No similar estimates for the cost of controlling and preventing fires that lead to haze are available, but a rough estimate can be obtained by taking into account the average number of hectares that are cleared by fire each year and the average cost of clearing a hectare by nonburning methods. An educated guess for these costs is approximately US$1.2 billion, much smaller than the estimated US$9 billion in damage. Thus, controlling the Southeast Asian haze and distributing the cost of doing so is implicitly equivalent to sharing an ecological surplus of US$7.8 billion among the three primarily affected countries: Indonesia, Malaysia, and Singapore.

---

22. See Siddiqui and Quah (2004) for estimates of the damage.

A seminal study by Siddiqui and Quah (2004) has estimated the relative effects of haze on Indonesia, Malaysia, and Singapore to be 93.8 percent, 5.1 percent, and 1.1 percent, respectively. Thus, if I apply the rule enunciated in (5.7) or (5.11), Indonesia, Malaysia, and Singapore must contribute 93.8 percent, 5.1 percent, and 1.1 percent, respectively, of the abatement costs, which, as noted above, have been estimated to be US$1.2 billion. In absolute terms, this may require Indonesia, Malaysia, and Singapore to contribute approximately US$1.125 billion, US$61.2 million, and US$13.2 million a year, respectively. The bulk of the abatement cost has to be borne by Indonesia as it is the most affected by haze. Contributions of US$61.2 million and US$13.2 million by Malaysia and Singapore are small compared to the damages they suffer. But a contribution of US$1.125 billion a year may seem to be a huge sum for a developing country such as Indonesia, though this is relatively much smaller than the damage of approximately US$8.447 billion a year that it will be able to avoid by controlling and preventing the fires.

Though an agreement such that all affected countries will have incentive to honor it is theoretically possible, there seem to be practical difficulties in implementing such an agreement. This is because Indonesia has a highly decentralized governance structure with five different levels of government. This often leads to conflict and passive opposition to the central government's policies and programs. While the Environment and Forestry Ministry of Indonesia's central government has the power to make national policies regarding land and forest use, the authority to enforce them is with head of regency or head of city. The latter are entrusted with the responsibility for generating income for local development and have little incentive to conserve forests in accordance with the national policies, as they see land and forest resources as a major source of profits and local revenues (Jotzo 2013). The financial and political benefits from allowing business to proceed as usual exceed those from following national policies decided by authorities thousands of miles away who have limited capacity to verify what is actually happening on the ground (Henda 2016).[23]

---

23. Also see Purnomo and Shantiko (2015) for a detailed study of how the decentralized governance structure may impede implementation of Indonesia's central government policies regarding control of haze.

In addition, obtaining local authority by winning local elections is highly lucrative as it provides access to central government funds and power to hand out licenses to businesses in the mining and agriculture sectors. Thus, the local authorities often buy votes for their reelection by giving land (or right to use land) to local voters—often through the village heads who then distribute it among local residents. In confirmation of this practice, Purnomo (2015) notes that over the past decade, forest fires tended to spike prior to and just after local elections. Weak enforcement of laws generally provides further support to this practice.

Besides causing poor air quality in Southeast Asia, the fires in Indonesia also contribute significantly to climate change. In fact, they are the largest single contributor to Indonesia's greenhouse gas emissions, and in 1997, the haze over the three-month period contained more carbon than the carbon emissions from all of Europe during the same period. However, this fact has not received much international attention. Because Indonesia has committed to reduce its greenhouse gas emissions by 26% to 41% with international assistance by 2020, it would make perfect sense for Indonesia to promote clearing land by nonburning methods as part of its "intended nationally determined contributions" required under the Paris Agreement.

Thus, despite two regional agreements and strong incentives to tackle the Southeast Asian haze, Indonesia has not been successful in tackling it because of its complex internal political economy (see Chander 2017b).

6

COALITION FORMATION GAMES

The γ-core, introduced and studied in the two preceding chapters, involves concepts from both cooperative and noncooperative approaches to game theory. On the one hand, it is a stronger concept than the conventional α- and β-cores, as $w^\alpha(S) \le w^\beta(S) \le w^\gamma(S)$ for all $S \subset N$, and, on the other hand, it is a weaker concept than that of a strong Nash equilibrium in a general class of strategic games that includes the environmental games, among others. Furthermore, the γ-core imputations, as shown in Theorem 4.6, can be supported as equilibrium outcomes of a repeated game of coalition formation. Though, from its form and properties, the γ-core is an appealing concept that is applicable to the environmental games, a question that arises is why this and not some other core concept (from among the many that are possible) is more compelling. For example, Maskin (2003) proposes a core concept in which the worth of a coalition S is determined by assuming that it faces the complementary coalition $N \backslash S$, which is sometimes referred to as the δ-core.[1]

One justification for taking the players in the complement to be singletons is that it leads to a concept that is related and can be nicely compared with the other concepts of cooperation in a strategic game, namely, the strong and the coalition-proof Nash equilibria, in which also the players in the complement are taken to be singletons. Another justification is that it is a meaningful expectation on the part of a coalition leaving the

---

1. Unlike my approach, however, Maskin (2003) introduces his core concept in the primitive framework of a partition function game, and thereby he abstracts away from the interesting strategic interactions that are behind the payoffs of the coalitions.

grand coalition, which in the face of uncertainty regarding the coalition structure that may form in the complement may assume the one that is worst from its perspective would form. This is because, as shown in Proposition 5.3 in the previous chapter, a deviating coalition's payoff is lowest when the players in the complement form singletons,[2] and, thus, it is not an arbitrary but a conservative behavior on the part of a deviating coalition that is *assumed*. While that does seem to justify the assumption as a meaningful behavioral assumption in a general class of games—called environmental games—the question still remains whether the players in the complement would *actually* form singletons. This chapter addresses this important question, among others.

Notice that the question subtly changes the strategic choices available to the players from those of choosing emission levels to those of forming different coalitions. Accordingly, Chander (2007) considers a coalition formation game that starts from the finest coalition structure (i.e., all coalitions are singletons) as the status quo, and the players choose whether to stay alone or to form coalitions with other players. For reasons that will be made clear later, the game may be repeated infinitely many times. It is then shown, as assumed in the definition of the γ-core, that if a coalition other than the grand coalition is formed by some players, then forming singletons is actually a subgame-perfect equilibrium strategy of the remaining players. In other words, the assumption underlining the γ-core that when a coalition deviates, the rest of the players break apart into singletons, actually has a strong theoretical justification and is not arbitrary. Furthermore, as will be shown, the grand coalition is an equilibrium coalition structure, and the equilibrium payoff vector is a γ-core imputation. Though a similar result (Theorem 4.6) was proved in chapter 4 for a general strategic game, the analysis that follows below exploits the special structure of the environmental game and provides additional insights, especially as the environmental game, as shown, is not partially superadditive, and thus Theorem 4.6, though closely related, does not really apply.

---

2. The environmental game is not the only game that has this property. A number of other games including models of oligopoly and pure public goods (Ray and Vohra 1997) also satisfy it. Yi (1997) classifies games in which formation of a coalition raises the payoff of other coalitions as games with "positive externalities."

A related branch of literature (see, e.g., Ray and Vohra 1997; Yi 1997) also considers coalition formation games with transferable utility in which, as in my formulation, the grand coalition is the unique efficient coalition structure.[3] The coalition formation game introduced later is similar as in this previous literature except that the game may be repeated infinitely many times. As will be seen, this has important consequences and leads to contrasting results. For comparison purposes, I first assume, as is common in this branch of the literature, that the players are identical and then discuss how the analysis may be extended to the case in which they are not.[4]

To further highlight the assumption that is central to the γ-core and motivate the repeated game of coalition formation, this chapter begins with a motivating example with three players. It is shown that if a coalition that does not include all players forms, then it is ex post optimal for the remaining players to act as if they were singletons and precipitate a repetition of the game. It then extends the analysis to games with n identical players and discusses how it can be further extended to the case in which the players are not necessarily identical. The chapter ends with a review of the related literature and concluding remarks.

6.1. RATIONALE FOR THE γ-CORE

To motivate the analysis, I return to the three-player example that was used to illustrate the limitations of the α- and β-coalitional functions in chapter 4. But this time it is to justify the assumption underlying the γ-core.

6.1.1. A Motivating Example

I first check whether the imputation $(u_1^*, u_2^*, u_3^*) = (1/3, 1/3, 1/3)$ [see (4.5)] belongs to the γ-core. Because the players are identical, we need to

---

3. Thus, formation of coalition structures other than the grand coalition implies inefficiency.

4. Though the analysis is carried out in the framework of the environmental game, informed readers should be able to recognize that it also applies to models of symmetric oligopoly and public goods (Ray and Vohra 1997; Yi 1997).

consider only two types of deviations; namely, a deviation by a coalition of any two players, say {1, 2}, and a deviation by a coalition of any single player, say {3}.

Let $(\tilde{e}_1, \tilde{e}_2, \tilde{e}_3)$ be such that $\tilde{e}_1, \tilde{e}_2 = \mathrm{argmax}(e_1^{\frac{1}{2}} + e_2^{\frac{1}{2}} - 2e_1 - 2e_2 - 2\tilde{e}_3)$ and $\tilde{e}_3 = \mathrm{argmax}(e_3^{\frac{1}{2}} - \tilde{e}_1 - \tilde{e}_2 - e_3)$. Then,

$$\tilde{e}_1 = \tilde{e}_2 = \frac{1}{16}, \ \tilde{e}_3 = \frac{1}{4}, \ \tilde{u}_1 = \tilde{u}_2 = \frac{1}{8}, \text{ and } \tilde{u}_3 = \frac{3}{8}. \tag{6.1}$$

The strategies $(\tilde{e}_1, \tilde{e}_2, \tilde{e}_3)$ represent the Nash equilibrium of the game induced by coalition {1, 2}. By comparing the payoffs of the deviating coalition {1, 2} under the strategies $(\tilde{e}_1, \tilde{e}_2, \tilde{e}_3)$ and (e_1^*, e_2^*, e_3^*), it is seen that coalition {1, 2} cannot gain by deviating from the grand coalition, as $\tilde{u}_1 + \tilde{u}_2 < u_1^* + u_2^*$.

Now consider deviation by {3}, which brings into play the assumption underlying the γ-coalitional function. If $S = \{3\}$ deviates, it believes that $N \backslash S = \{1, 2\}$ will break apart into singletons and the resulting equilibrium will be the Nash equilibrium of the game induced by the singleton coalition {3} and, therefore, result in the same payoffs as in the Nash equilibrium of the game; that is, a payoff of 0 [see (4.3)], less than its payoff of 1/3 in the grand coalition. Thus, coalition {3} cannot gain by leaving the grand coalition. Similarly, no other coalition can gain by deviating. This proves that the imputation $(u_1^*, u_2^*, u_3^*) = (1/3, 1/3, 1/3)$ belongs to the γ-core of the game.

Turning now to the rationale, why should coalition {1, 2} break apart into singletons when {3} deviates from the grand coalition? The stability of the grand coalition and the γ-core imputation $(1/3, 1/3, 1/3)$ depends critically on the answer to this question, because {3} would gain from its deviation if coalition {1, 2} were not to break apart as its payoff will then be 3/8 compared to its payoff of 1/3 in the grand coalition.

Let us first consider the possible arguments against the breaking up of {1, 2}: if {3} deviates and {1, 2} does not break apart, then the resulting equilibrium and the corresponding payoffs of its members will be 1/8 each [by (6.1) or (4.4)], which, as seen from (4.3), are higher than their payoffs of 0 each if it were to break apart and induce the finest coalition structure {{1}, {2}, {3}}. Thus, both members of coalition {1, 2} would be worse off if {1, 2} breaks apart. However, this argument relies on the implicit assumption that either coalition {1, 2} thinks that after it breaks apart the finest coalition structure {{1}, {2}, {3}} would be the final coalition structure or coalition {1, 2} is myopic and it does not consider what may happen after it breaks apart and takes the finest coalition structure {{1}, {2}, {3}} to be the strategically relevant coalition structure for its decision to deviate or not. In either case, coalition {1, 2} believes that it will be worse off if it breaks apart, as that would result in the finest coalition structure {{1}, {2}, {3}} and in a lower payoff. Let me elaborate on the two arguments.

As to the former, Ray and Vohra (1997) assume that coalitions are farsighted but can only split and not merge. More specifically, if coalition {1, 2} breaks apart, then the coalition structure {{1}, {2}, {3}} would indeed be the final coalition structure, as singleton coalitions cannot split and, by assumption, they cannot merge either. Therefore, coalition {1, 2} will not break apart, as that will lead to formation of the finest coalition structure. But if coalition {1, 2} does not break apart, coalition {3} would deviate, as it would then be better off. Hence, it can be claimed that the coalition structure {{1, 2}, {3}} is stable, but the grand coalition {1, 2, 3} is not.

As to the latter, the grand coalition is also not stable in terms of the coalition formation games in d'Aspremont and Gabszewicz (1986), Carraro and Siniscalco (1993), and Barrett (1994a), among others, who implicitly assume that the coalitions are myopic in the sense that coalitions do not take into account the possibility that there may be further deviations or mergers after their deviation. More specifically, coalition {1, 2} will not break apart because it is myopic and thinks that it will be worse off if it does, as that would lead to the finest coalition structure {{1}, {2}, {3}} in which it is worse off. Thus, under their assumption too coalition structure {{1, 2}, {3}} is stable but the grand coalition is not.

Now suppose that coalitions are not only farsighted but can also merge.[5] In particular, it means that coalition {1, 2} takes into account the possibility that there may be further deviations after it breaks apart and induces the finest coalition structure {{1}, {2}, {3}}. For instance, after it breaks apart, all three singleton coalitions in the resulting finest coalition structure may merge and form the grand coalition {1, 2, 3}, as that would give them

---

5. Diamantoudi and Sartzetakis (2006) show that allowing coalitional mergers alone is not sufficient to rule out the inefficient equilibrium coalition structures obtained in Ray and Vohra (1997).

higher payoffs of 1/3 each compared to their payoffs of 0 each in the finest coalition structure {{1}, {2}, {3}} [by (4.3) and (4.5)]. Thus, contrary to the two arguments discussed above, it might be ex post optimal for the far-sighted coalition {1, 2} to break apart and induce the temporary coalition structure {{1}, {2}, {3}}, if it believes that that would lead to formation of the grand coalition later on and, therefore, to payoffs for its members that are strictly higher than if it did not break apart, as the payoffs of players 1 and 2 in the coalition structure {{1, 2}, {3}} are 1/8 each compared to their payoffs of 1/3 each in the grand coalition [by (6.1) or (4.4) and (4.5)]. In other words, if the players believe that the three singletons in the finest partition {{1}, {2}, {3}} will merge and form the grand coalition, then coalition {1,2} in the coalition structure {{1, 2}, {3}} will indeed break apart. Thus, coalition {3} cannot gain by leaving the grand coalition, as coalition {1, 2} will break apart because that will only temporarily lead to the finest coalition structure {{1}, {2}, {3}} but eventually to formation of the grand coalition. Thus, the grand coalition is stable but the coalition structure {{1, 2}, {3}} is not.

What emerges from this discussion is that if the players expect the grand coalition to form then it will indeed form and will be stable, provided the γ-core is nonempty.

Notice the role played in this argument by the fact that $(u_1^*, u_2^*, u_3^*) = (1/3, 1/3, 1/3)$ is a γ-core imputation, and, therefore, members of coalition {1, 2} are worse off in the coalition structure {{1, 2}, {3}} than in the grand coalition. If this were not so, coalition {1, 2} will not have the incentive to break apart even if the players believe that the grand coalition will form from the finest coalition structure {{1}, {2}, {3}}.

To put the arguments in more formal terms, coalition {3} will not leave the grand coalition, as there exists a "chain" of deviations that terminates at a coalition structure in which the payoff of {3} is not strictly higher compared to its payoff in the grand coalition {1, 2, 3}. This "chain," which may be compactly written as $\{\{1, 2, 3\}\} \overset{\{3\}}{-} \{\{1, 2\}, \{3\}\} \overset{\{1, 2\}}{-\!\!\!-} \{\{1\}, \{2\}, \{3\}\} \overset{\{1\}, \{2\}, \{3\}}{-\!\!\!-\!\!\!-} \{\{1, 2, 3\}\}$, where the deviating coalition(s) at each step of the chain is (are) shown over the arrows, is such that the "final" payoffs of all deviating coalitions, except {3}, in the chain [i.e., {1, 2} and {1}, {2} and {3}], are strictly higher than in the coalition structures to which they belong at the time of their deviation, and the "final" payoff of coalition {3} that makes the initial deviation is the same and not strictly higher. To be precise, the "final" payoffs of the members of coalition $\{1, 2\}$ in this chain are $1/3$ each, which are strictly higher than their payoffs of $1/8$ each in the coalition structure $\{\{1, 2\}, 3\}$ at the time of deviation by their coalition, and the "final" payoffs of singleton coalitions $\{1\}$, $\{2\}$, and $\{3\}$ are $1/3$ each, which are strictly higher than their payoffs of 0 each in the coalition structure $\{\{1\}, \{2\}, \{3\}\}$ at the time of their merger/deviation. Only the payoff of coalition $\{3\}$, which is responsible for the initial deviation, is the same and not strictly higher than its payoff in the coalition structure at the time of its deviation. It is easily seen that if any other coalition deviates from the grand coalition, then there exists similarly a chain such that the final payoff of the deviating coalition is the same and not strictly higher. Thus, the grand coalition is stable if coalitions are farsighted and conservative in the sense that a coalition prefers not to leave the grand coalition unless its "final" payoff is strictly higher in *all* possible "chains" that may form upon its deviation.[6]

6.1.2. Coalition Formation as a Noncooperative Game

The above discussion leads us to the question: Which coalition structures may form from the finest coalition structure $\{\{1\}, \{2\}, \{3\}\}$ if the coalitions are farsighted and can merge? This is clearly a more fundamental question than the question of which coalition structures are stable. As argued in the preceding section, the stability of the grand coalition depends critically on the answer to this question.[7] Thus, we need to resolve it first. I do so in the context of the motivating example above and then extend it to all symmetric environmental games. I also discuss extensions to environmental games with heterogeneous players.

---

6. The notion of "farsighted and conservative" coalitions used presently is conceptually the same as in the definitions of farsighted stable sets in Harsanyi (1974) and Ray and Vohra (2015). Also, see Mauleon and Vannetelbosch (2004).

7. As our coalition formation game starts from the finest coalition structure, it would also be the equilibrium coalition structure if, as in Ray and Vohra (1997), we assume that coalitions cannot merge.

The question has been dealt with in the extant literature in the framework of a two-stage noncooperative game of coalition formation that begins from the finest coalition structure $P^0 = \{\{1\}, \{2\}, \{3\}\}$ as the status quo. In stage 1 of the game, each player acting independently proposes to either cooperate with other players (i.e., announce C) or to not cooperate (i.e., announce NC). All those players who announce C form a coalition, and the rest of players, if any, form singletons.[8] In stage 2, the players choose their actions/emissions to maximize payoffs of their coalitions, given the actions/emissions of other players.

What are the equilibrium coalition structures in this game? The game has at least three pure-strategy asymmetric equilibria in which exactly two players cooperate and the remaining player gets to "free ride," resulting in payoffs that are equal to the Nash equilibrium payoffs in the game induced by the coalition of the two players.[9] But these equilibria are questionable on at least two counts. First, they mean that every possible coalition structure except the grand coalition is an equilibrium outcome because, as will be seen later, the finest coalition structure is also an equilibrium outcome of this game. How do we then choose between these equilibria? There is really no way to coordinate which player "gets" to free ride and which other two "have" to cooperate. A large literature on free riding in acceptance of corporate takeover bids argues that in symmetric games, only symmetric mixed-strategy equilibria are relevant (see, e.g., Bagnoli and Lipman 1988; Holmstrom and Nalebuff 1992). If we believe in this argument, then the three asymmetric pure-strategy equilibria must be ignored, and we may consider only symmetric mixed-strategy equilibria in which who gets to free ride is left to chance, and everyone has an equal chance to be the free rider.[10] Remember that not

---

8. In the three-agent case, restricting the choice of strategies to C and NC does not rule out a priori formation of any coalition structure. A player can join any coalition (i.e., announce C) or decide to stay alone (i.e., announce NC). However, in the n-agent case later, I allow more general strategies in order to ensure the same.

9. The number of such asymmetric equilibria depends on the number of players in the game and can be very large.

10. See Crawford and Haller (1990) for a *theoretical* argument in favor of symmetric mixed-strategy equilibria in symmetric games.

only are the players identical, but also the game starts from the finest coalition structure, which is symmetric as well.[11] Second, because the grand coalition is the unique efficient coalition structure, each of these equilibria, as Maskin (2003) notes, can be sustained ex post only if each player has the ability to commit to not merge with other players even when a merger would result in higher payoffs for everyone. I will return to this issue again below.

Notice that in the simultaneous-move two-stage game of coalition formation described above, deviations by players cannot be countered with subsequent moves by their opponents—they must be accepted by them as fate accompli. Notice also that the game has a symmetric equilibrium in pure strategies in which each player announces NC in stage 1 and chooses the same action in stage 2 as in the Nash equilibrium of the environmental game. But if the game results in this equilibrium, then everyone stands to gain if the game is played all over again. Thus, considering conditional repetitions of the game seems appropriate.[12] As will be shown, doing so dramatically changes the set of equilibria.

This is because conditional repetitions of the two-stage game make it possible for the players to counter the moves of their opponents with subsequent moves. For example, players can now choose actions that force a repetition of the game rather than maximize the current payoff of their coalition. To be more specific, the members of a two-player coalition can now force a repetition of the game by choosing the same actions/emissions in stage 2 as they would in the Nash equilibrium. Such actions by members of a two-player coalition in stage 2 are effectively equivalent to "dissolving" their coalition and act as if they were singletons resulting in effectively the same outcome as in the Nash equilibrium and, thus, resulting in a repetition of the game.

---

11. In this I follow Holmstrom and Nalebuff (1992), who use a similar argument in favor of symmetric mixed-strategy equilibria in a context that also involves free riding and identical players.

12. Also, because the two-stage game starts from the finest coalition structure, it would be inconsistent to not allow repetition of the game if the equilibrium outcome is the finest coalition structure.

6.1.3. An Efficient Equilibrium of the Repeated Game

I show that in the infinitely repeated two-stage game:

(a) to dissolve a coalition if it does not include all players is a subgame-perfect equilibrium strategy of each player; and
(b) the grand coalition is an equilibrium outcome.

I first establish (b) assuming (a), and then establish (a) given that it implies (b). Because the structure of the game is the same at the beginning of each repetition, I only need to consider equilibria in stationary strategies and per-period payoffs of the players. Let w_i denote the equilibrium per-period payoff of the repeated game of player i. To obtain a sharper characterization, I solve for the equilibrium of the repeated game first with discounting and then take the limit as the discount rate goes to zero.[13] Let β denote the discount factor; that is, $\beta = 1/(1+r)$, where r is the discount rate. The payoff matrix of the reduced game is as follows: if all players choose C, the grand coalition is formed, and each player gets $1/3$ by (4.5). If any player chooses NC, then, in view of the strategies in (a), all players choose the same emissions as in the Nash equilibrium, and the game is repeated next period. The equilibrium payoff of player i from this will be w_i starting one period later, which has a discounted present value of βw_i in the current period (figure 6.1).

We now find a solution to this reduced form of the repeated game. Let p_1, p_2, p_3 be the probabilities assigned by the three players to the C strategy. Then, in equilibrium, each player, say 3, should be indifferent between choosing strategies C and NC. Thus,

$$w_3 = p_1 p_2 \frac{1}{3} + (1 - p_1 p_2)\beta w_3$$
$$w_3 = p_1 p_2 \beta w_3 + (1 - p_1 p_2)\beta w_3$$

13. Actually, assuming a positive discount rate is merely a tie-breaking rule. My proofs below still hold if I adopt instead the alternative tie-breaking rule that the players strictly prefer to be members of the grand coalition than to be members of a coalition in a coalition structure other the grand coalition even if their payoffs are the same.

Player 3

Players	C	NC
Both Player 1 and 2 choose C	$\dfrac{1}{3}, \dfrac{1}{3}, \dfrac{1}{3}$	$\beta w_1 \beta w_2, \beta w_3$
Player 1 and/or 2 choose NC	$\beta w_1 \beta w_2, \beta w_3$	$\beta w_1 \beta w_2, \beta w_3$

FIGURE 6.1

The reduced form of the repeated game

Clearly, C is a dominant strategy for player 3 if $\beta w_3 < 1/3$. Thus, C is a dominant strategy of each player i if $\beta w_i < 1/3$, and the resulting equilibrium payoffs are $w_i = 1/3$, $i = 1, 2, 3$, which in turn confirm the condition for dominance, that is, $\beta w_i < 1/3$ for each i, as $\beta < 1$. Thus, announcing C is a dominant strategy of each player. This means the grand coalition is a dominant stationary strategy equilibrium outcome, and the per-period payoffs of the players are a γ-core imputation. It may be noted that this *symmetric* equilibrium is obtained without imposing the restriction that the equilibrium should be symmetric. I will return to this point again at the end of the chapter.

I now prove (a), because, as shown, (a) implies (b). Suppose some player, say 3, tries to test the credibility of the strategies described in (a) by announcing NC in stage 1 when both players 1 and 2 announce C in stage 1; then if players 1 and 2, contrary to their equilibrium strategies, give effect to their coalition in stage 2 [i.e., choose their emissions according to (6.1) or (4.4)], they will each get a payoff of 1/8. But if they stick to their equilibrium strategies, the game will be repeated, and each of them will get $\beta w_i = \beta(1/3) > 1/8$ for β close enough to 1.

The argument is not that players 1 and 2 can force player 3 to join their coalition by threatening that they will choose the same emission levels as in the Nash equilibrium (and thereby deny it the opportunity to free ride) unless it joins their coalition, but that such an action is ex post optimal for players 1 and 2 if the players' strategies are as in (a). Hence, this is a credible strategy of players 1 and 2 that player 3 cannot ignore when choosing its strategy in stage 1.

We have thus found an efficient symmetric equilibrium of the repeated game and shown that the grand coalition is an equilibrium coalition structure. However, this equilibrium is not unique.

6.1.4. Alternative Symmetric Equilibria

The repeated game has at least one other symmetric mixed-strategy equilibrium, which is defined by an alternative set of strategies. In this case, the strategy of each player who announces C in stage 1 is to form a coalition with all those players who have also announced C and choose emissions in stage 2 to maximize the payoff of their coalition, which would result in payoffs as in (6.1) or (4.4); if two or more players announce NC in stage 1, then the game is repeated in the next period.

Given the set of these strategies, let w_i again denote the equilibrium payoff of player i. Let p_1, p_2, p_3 be the probabilities assigned by the three players to the C strategy. Then, in equilibrium, each player, say 3, should be indifferent between the strategies C and NC. Thus,

$$w_3 = \frac{1}{3} p_1 p_2 + \frac{1}{8} \left[p_1 (1 - p_2) + (1 - p_1) p_2 \right] + (1 - p_1)(1 - p_2) \beta w_3$$

$$w_3 = \frac{3}{8} p_1 p_2 + \beta w_3 \left[p_1 (1 - p_2) + (1 - p_1) p_2 + (1 - p_1)(1 - p_2) \right].$$

Because the equilibrium must be symmetric, $p_1 = p_2 = p_3 \equiv p$. Solving for the equilibrium p and w_3 from these two simultaneous equations, it is easily seen that they have a unique solution $p > 0$, and

$$\beta w_3 = \frac{1}{8} - \frac{p}{48(1 - p)},$$

implying that the equilibrium payoff of no player is higher than 1/8, which is smaller than the payoff of 1/3 under the efficient equilibrium discussed in the preceding section.

Given this solution of the reduced game, I verify the ex post optimality of the players' strategies. Suppose all players happen to have chosen C, then as per their strategies the players form the coalition and their payoffs are 1/3 each by (4.5). If any two players, say 1 and 2, choose C and player 3 chooses NC, then as shown players 1 and 2 together get strictly less than $2\beta(1/8)$ if they dissolve their coalition, and thereby precipitate another round of play, but 2(1/8) if they maximize their joint payoff by (6.1), and, thus, the game is not repeated. Because $\beta < 1$, we have found an equilibrium of the repeated game that is symmetric but not efficient.

Notice that $p_1 = p_2 = p_3 = 0$ is also an equilibrium, because if all three players announce NC, then no individual player can gain by changing its strategy to C—it needs at least one more player to announce C. The repeated game does not seem to have any other symmetric equilibrium, and both the alternative symmetric equilibria are Pareto dominated by the efficient equilibrium. I now discuss how to select among the three symmetric equilibria.

6.1.5. A Focal-Point Equilibrium

When a game admits multiple equilibria, we must consider whether the players might be able to coordinate on a particular equilibrium by using information that is abstracted away from the game. Schelling (1960) argues that if the players' expectations can converge on one of the equilibria, then they may all expect it and hence make it happen, like a self-fulfilling prophecy. According to Schelling, an equilibrium is a *focal point* if it has some property that conspicuously distinguishes it from all other equilibria of the game, and such an equilibrium is more likely to be observed.

As noted earlier, the efficient equilibrium indeed has one such property, namely that it Pareto dominates the other two inefficient equilibria, as the payoff of each player in the efficient equilibrium is strictly higher than in either of the inefficient equilibria. Because each player has full knowledge about the structure of the game, each can independently find out that the efficient equilibrium offers higher payoffs to everyone. Each player should then think that everyone else must have done the same calculations and conclude that the efficient equilibrium is to be favored by everyone. Therefore, each player should think that all will think that . . . each player will play the strategy corresponding to the efficient equilibrium.

It might be possible to prevent the inefficient equilibria in some other ways, too. Because each player has full knowledge about the structure of the game, the players may voluntarily agree to introduce a unanimity clause before the game begins, which is similar to that in the case of the nuclear test ban treaty mentioned in footnote 2 in chapter 5.[14]

To conclude this section, if we believe, as in the large literature on free riding in acceptance of corporate takeover bids, that in games with identical players only the symmetric mixed-strategy equilibria are relevant, then it follows from Schelling's focal-point argument that formation of the grand coalition is the *only* relevant equilibrium outcome of the repeated game.

6.2. EXTENSIONS TO *n*-PLAYER GAMES

I first extend the analysis in the preceding section to the environmental game with n identical players. Because the players are identical, let $v \equiv v_i = v_j$ and $g \equiv g_i = g_j$ for all $i, j \in N$. Let $(e_1^*, \ldots, e_n^*)$ denote the unique efficient emissions profile. Then the efficiency conditions (3.3) can be rewritten as

$$g'(e_i^*) = nv'(z^*) \qquad (6.2)$$

and consequently $e_i^* = e_j^*$, $i, j \in N$. Let

$$u_i^* = g(e_i^*) - v(z^*), i \in N. \qquad (6.3)$$

Then, $u_i^* = u_j^*$, for $i, j \in N$.

In the three-player coalition formation game above, the players were allowed to choose between only two strategies: C and NC. For games with more than three players, such a restriction would amount to allowing

---

formation of at most one non-singleton coalition. Because we should not rule out a priori formation of any coalition structure, I assume that each player can now choose more generally from among the set of $n+1$ strategies consisting of 0 and integers 1 to n. The players who announce the same positive integer form a coalition, but the players who choose 0 remain singletons. Thus, the players can choose any coalition structure $\{S_1, S_2, \ldots, S_m\}$, $m \leq n$, and obtain the Nash equilibrium payoffs of the game induced by the coalition structure.

As in the three-player game, it is straightforward to see that announcing the same integer is a dominant strategy of each player in the reduced game if all players follow the strategy to dissolve a coalition if it does not include all players. The resulting equilibrium payoff to each player i is u_i^*. Thus, I only need to show that such strategies are ex post optimal as in the three-player environmental game above. This means I must show that members of a coalition other than the grand coalition cannot gain by not dissolving their coalition. As before, let $|S|$ denote the cardinality of set S.

Lemma 6.1

Let $P = \{S_1, S_2, \ldots, S_m\}$ be some coalition structure with $P \neq \{N\}$. Then the individual payoffs of the members of the largest coalition in the Nash equilibrium of the game induced by P are lower than their payoffs as members of the grand coalition;[15] that is, $\tilde{u}_i < u_i^*$ for all $i \in S_k$ such that $|S_k| \geq |S_j|$ for all j.

Proof of Lemma 6.1

Let $(\tilde{e}_1, \tilde{e}_2, \ldots, \tilde{e}_n)$ denote the Nash equilibrium of the game induced by the coalition structure $P = \{S_1, S_2, \ldots, S_m\}$. Because the players are identical, the first-order conditions [by (5.5)] imply

$$g'(\tilde{e}_i) = |S_j|\, v'\!\left(\sum_{k \in N} \tilde{e}_k \right), i \in S_j,\ j = 1, 2, \ldots, m. \tag{6.4}$$

---

15. As Yi (1997) shows, many other games, including models of oligopoly and public goods, also have this property.

Because g is strictly concave, $\tilde{e}_i < \tilde{e}_j$ if $i \in S_k$ and $j \in S_l$ with $|S_k| > |S_l|$. Let $\tilde{z} = \sum_{j \in N} \tilde{e}_j$ denote the total emissions corresponding to the Nash equilibrium of the game induced by the coalition structure $\{S_1, S_2, \ldots, S_m\}$. Then, $\tilde{u}_i \equiv g(\tilde{e}_i) - v(\tilde{z}) < \tilde{u}_j \equiv g(\tilde{e}_j) - v(\tilde{z})$ if $i \in S_k$ and $j \in S_l$ with $|S_k| > |S_l|$, because as shown $\tilde{e}_i < \tilde{e}_j$. Thus, the payoffs of the members of larger coalitions are lower. By definition, the efficiency condition (6.2) implies $\sum_{i \in N} \tilde{u}_i < \sum_{i \in N} u_i^*$ if $P \neq \{N\}$. This implies $\tilde{u}_i < u_i^*$ for all $i \in S_k$ if $|S_k| \geq |S_j|$ for all $j = 1, 2, \ldots m$, as the players are identical and as shown $\tilde{u}_i \equiv g(\tilde{e}_i) - v(\tilde{z}) < \tilde{u}_j \equiv g(\tilde{e}_j) - v(\tilde{z})$ if $i \in S_k$ and $j \in S_l$ with $|S_k| > |S_l|$. ∎

Suppose now that the players attempt to test the credibility of the players' strategies to dissolve their coalition if it does not include all players. Let $\{S_1, S_2, \ldots, S_m\} \neq \{N\}$ be some coalition structure. Assume without loss of generality that $|S_1| \leq |S_2| \leq \cdots \leq |S_m|$. Consider the finite sequence of coalition structures $P^m \equiv \{S_1, S_2, \ldots, S_m\}$, $P^{m-1} \equiv \{S_1, S_2, \ldots, S_{m-1}, [S_m]\}$, $P^{m-2} \equiv \{S_1, S_2, \ldots, S_{m-2}, [S_{m-1}], [S_m]\}, \ldots, P^1 \equiv \{S_1, [S_2], \ldots, [S_m]\}$. This sequence of coalition structures is obtained if the largest coalition in each subsequent coalition structure breaks apart into singletons, starting from the largest coalition S_m in P^m. Let $\tilde{u}_i^k, i \in S_k, k = 1, 2, \ldots, m$, be the corresponding sequence of payoffs; that is, $\tilde{u}_i^k, i \in S_k$, are the Nash equilibrium payoffs of the members of the largest coalition in the game induced by the coalition structure $\{S_1, S_2, \ldots, S_k, [S_{k+1}], \ldots, [S_m]\}$. Then, by Lemma 6.1, $\tilde{u}_i^k < u_i^*$ for all $i \in S_k, k = 1, 2, \ldots, m$. Therefore,

$$\tilde{u}_i^k < \beta u_i^* \text{ for all } i \in S_k, k = 1, 2, \ldots, m, \tag{6.5}$$

for β sufficiently close to 1.

I begin with a partition $P^1 = \{S_1, [N \setminus S_1]\}$, $S_1 \neq N$, and then extend it by induction to any partition $P = \{S_1, S_2, \ldots, S_m\}$. Suppose in partition $P^1 = \{S_1, [N \setminus S_1]\}$ some players do not dissolve their coalition S_1 when the rest of the players have announced 0. Then their payoffs will be $\tilde{u}_i^1, i \in S_1$. But if they dissolve their coalition the game will be repeated, and each of them will get a payoff of u_i^* starting one period later, which has a discounted present value of βu_i^* in the current period. Because, by (6.5), $\beta u_i^* > \tilde{u}_i^1, i \in S_1$, for β sufficiently close to 1, it is ex post optimal for the members of S_1 to dissolve their coalition (i.e., break apart into singletons).

Consider next the partition $P^2 = \{S_1, S_2, [N \backslash S_1 \cup S_2]\}$. If members of S_2 dissolve their coalition, then the coalition structure is effectively equivalent to $\{S_1, [N \backslash S_1]\}$, and, thus, as shown, the members of S_1 will also dissolve their coalition. Similarly, if members of S_1 dissolve their coalition, then the coalition structure is effectively equivalent to $\{S_2, [N \backslash S_2]\}$, and, as shown, the members of S_2 will also dissolve their coalition. Suppose neither members of S_2 nor members of S_1 dissolve their coalitions. Then, as $|S_2| \geq |S_1|$, the payoffs of members of S_2 are $\tilde{u}_i^2 < u_i^*$, $i \in S_2$, by (6.5). But if the members of S_2 dissolve their coalition, then, as shown, so will the members of S_1 and the game will be repeated, from which the payoff of each member i of coalition S_2 is $\beta u_i^* > \tilde{u}_i^2$. Hence, the players in S_2 will dissolve their coalition and so will the players in S_1.

It follows, by induction, that members of the coalitions in the coalition structure $\{S_1, S_2, \ldots, S_m\}$ will dissolve their coalitions one-by-one, starting from the largest coalition S_m, until the finest coalition structure P^0 is reached.

This proves that dissolving a coalition that does not include all players is an ex post optimal strategy of each player, and formation of the grand coalition is an equilibrium outcome of the repeated game. It is straightforward to show, as in the three-player environmental game, that this equilibrium strictly Pareto dominates the inefficient symmetric mixed-strategy equilibria. Their inefficiency comes from the fact that in either of the symmetric mixed-strategy equilibria, the probability of forming the grand coalition is strictly less than one, and, therefore, the payoff to each player is lower than in the symmetric efficient equilibrium. By the focal-point argument, it follows that the grand coalition is the only relevant equilibrium outcome of the repeated game.

6.2.1. Incentive to Disintegrate

Coalitional incentive to disintegrate seems to have been mostly ignored in the extant literature on coalition formation, perhaps because either the players are assumed to be myopic or mergers are ruled out by assumption. Because disintegration of a coalition in the context of the environmental game imposes negative externalities on other coalitions, it can weaken coalitions' incentives to deviate from the grand coalition. Thus, farsighted coalitions may disintegrate in order to discourage deviations. The following proposition confirms this intuition.

Given $P = \{S_1, S_2\}$, let $P' = \{S_1, [S_2]\}$ denote the coalition structure that is obtained if S_2 breaks apart into singletons. Because the Nash equilibrium payoff of each player in the game induced by a coalition structure depends on the entire coalition structure, let $u_i(P)$ denote player i's Nash equilibrium payoff in the game induced by the partition P.

Proposition 6.2

Let $P = \{S_1, S_2\}$ be some coalition structure such that $|S_2| > 1$, and let $P' = \{S_1, [S_2]\}$. Then, $u_i(P') < u_i(P)$ for all $i \in S_1$.

Proof of Proposition 6.2

Let $(\tilde{e}_1, \tilde{e}_2, \ldots, \tilde{e}_n)$ and $(e'_1, e'_2, \ldots, e'_n)$ be the Nash equilibrium strategies of the games induced by the partitions P and P', respectively. Let $\tilde{z} = \sum_{i \in N} \tilde{e}_i$ and $z' = \sum_{i \in N} e'_i$. I claim that $z' > \tilde{z}$. Suppose not; that is, let $z' \leq \tilde{z}$. Then, from (6.4), given strict concavity of g and convexity of v, $e'_i \geq \tilde{e}_i$ for each $i \in S_1$, and $e'_i > \tilde{e}_i$ for each $i \in S_2$, as $|S_2| > 1$. But this contradicts our supposition that $z' \leq \tilde{z}$. Hence, $z' > \tilde{z}$ and $e'_i \leq \tilde{e}_i$ for each $i \in S_1$, which implies $u_i(P') < u_i(P)$ for $i \in S_1$. ∎

Examples are easily constructed such that $u_i(P) > u_i^* > u_i(P')$ for all $i \in S_1$, which means that disintegration of coalition S_2 may reverse the incentives of members of coalition S_1 to leave the grand coalition and form coalition S_1. This is indeed true in the three-player environmental game above for $S_2 = \{1, 2\}$ and $S_1 = \{3\}$.

If disintegration of coalitions results in the finest coalition structure, then, as shown, it will lead to formation of the grand coalition. This provides members of farsighted coalitions—who have lower payoffs compared to their payoffs in the grand coalition—incentives to disintegrate, as that will lead to the finest coalition structure and, therefore, eventually to formation of the grand coalition.

6.2.2. Stability of the Grand Coalition

I first clarify what is meant by a stable coalition structure. A coalition structure $P = \{S_1, S_2, \ldots, S_m\}$ is *stable* if no player (or group of players) can strictly improve its final payoff by leaving a coalition in the coalition structure.[16]

Theorem 6.3

The grand coalition N is a stable coalition structure.

Proof of Theorem 6.3

Suppose a coalition S leaves the grand coalition N. There are two possibilities: either $|S| \leq |N \backslash S|$ or $|S| > |N \backslash S|$. Let $P = \{S, N \backslash S\}$ and $P' = (S, [N \backslash S])$.

Consider first $|S| \leq |N \backslash S|$. Then, in view of Lemma 6.1, $u_i(P) < u_i^*$ for all $i \in N \backslash S$ and, by Proposition 6.2, $u_i(P') < u_i^*$ for all $i \in S$. This means that, given the coalition structure $P = \{S, N \backslash S\}$, the members of $N \backslash S$ stand to gain if they break apart, as that would then induce the members of S also to break apart and lead to repetition of the game and formation of the grand coalition. This proves that coalition S cannot strictly improve its payoff by breaking away from N.

If $|S| > |N \backslash S|$, then by interchanging S and $N \backslash S$ and applying the same argument as in the preceding paragraph, S will break apart first, followed by the breaking up of $N \backslash S$, and finally repetition of the game and formation of N. This again means that S cannot strictly improve its payoff by breaking away from N.

This leaves out the possibility that rather than breaking up into singletons, $N \backslash S$ or S may break up into two or more non-singleton coalitions. However, there is no loss of generality in ignoring this possibility, because

---

16. There are two related but different stability concepts in game theory. One concerns the stability of imputations against actions by deviating players as in the conventional cooperative game theory, and the other concerns the stability of coalition structures as in the more recent coalition formation theory.

as shown, that would only lead to intermediate coalition structures in which the coalitions will continue to break apart into singletons until the finest coalition structure is reached, the game is repeated, and the grand coalition is formed. ▪

The analysis above suggests that the coalition structures that may form from the *finest* coalition structure determines which coalition structures are stable. This prompted me to consider coalition formation as a repeated game.

6.2.3. Heterogeneous Players

It can be argued that the analysis in the chapter so far depends on the assumption of identical players. In particular, certain equilibrium coalition structures could be ruled out on the ground that only symmetric mixed-strategy equilibria need be considered in games with identical players. Such a simplification is no longer possible if the players are heterogeneous. This is indeed true, but only if there exist at least some inefficient asymmetric equilibria. However, there does not seem to be any result in the extant literature that proves existence and characterizes equilibrium coalition structures in a general class of games with heterogeneous players. Aside from n-player games, there do not seem to be any existence and characterization results under reasonable assumptions even for general three-player games with heterogeneous players. In fact, the focus of the coalition formation theory seems to have been on demonstrating that the grand coalition may not form even in games with identical players. But what are the equilibrium coalition structures if not the grand coalition? This question seems to have been considered by Yi (2003), who concludes at the end of his comprehensive survey on coalition formation games:

> I have restricted my survey to the case of ex-ante symmetric players. This symmetric assumption is common to many papers on the endogenous formation of coalitions. When heterogeneous players are introduced, the analysis is typically done in a setting with small number of players or only one non-degenerate coalition is allowed. . . . But so far there is not much

result on the existence and characterization of stable coalition structures among heterogeneous players. (See Ray and Vohra 1999 for an important contribution to the existence question.) The difficulty arises from the fact that when players are not symmetric, it is no longer possible to identify a coalition by its size. How are the coalitional payoffs divided among the members? One way to begin the analysis in this direction might be to assume just two types of players and see if equilibrium coalitions consist of the same type or of different types and how the coalitional payoffs are divided among members. Recently, Belleflame (2000) made a start in examining stable coalition structures among heterogeneous players, albeit under the restrictive assumption that at most two coalitions can form. (Yi 2003, p. 120)

These comments by Yi should clarify the significance of Theorem 4.6 in chapter 4, which not only proposes how to divide coalitional payoffs among the members, but also shows that the grand coalition is an equilibrium coalition structure in *general* partially superadditive strategic games.

6.3. OTHER APPROACHES TO COALITION FORMATION

In a seminal paper, Ray and Vohra (1997) introduce a general approach to coalition formation and characterize equilibrium coalition structures in a number of games with identical players.[17] Their approach, though of an independent interest, has not been applied for characterization of equilibrium coalition structures in games with heterogeneous players. The assumption that coalitions cannot merge and the players are identical makes their approach unsuitable for applications to the environmental game with heterogeneous players. As illustrated in the three-player game above, ruling out mergers of coalitions biases the coalition formation game against formation of the grand coalition, and, as noted in chapter 1,

---

17. As shown in chapter 4, these games admit nonempty γ-cores.

heterogeneity of the players is the rule rather than the exception in the context of climate change.

6.3.1. Sequential Games of Coalition Formation

Bloch (1996) studies "endogenous" formation of coalitions in a sequential game with an "exogenously" given order of play. He also assumes that the players are identical and coalitional payoffs are distributed according to an "exogenously" given fixed rule. His approach is, therefore, not applicable to the environmental game not only because the players are assumed to be identical, but also because the order of play is decided exogenously. Let me elaborate on the latter. For instance, if China, the European Union, and the United States were to negotiate an agreement on climate change, then who will decide the order of play? Each one of them may reason that a higher (free-rider's) payoff (as Bloch shows) can be obtained by sitting back and letting the other two form a coalition. However, if each one of them reasons in the same way, the game would develop into a war of attrition (as negotiations on climate change indeed seemed to be at one time) in which each one waits for the others to form a coalition with the hope of free riding on them. Such a war of attrition is a repeated (and not a sequential) game in which each and *every* player decides in each period whether to stay alone or to form a coalition with other players. One way to get around the problem raised by an exogenously given order of play that has been considered in the literature is that the order of play may be randomized. But would the countries really agree to time their moves in the negotiations on climate change by rolling dice? Moreover, if the countries are not identical, why would they agree that each of them should have an equal chance to be the first mover and enjoy, if it is lucky and only if it is lucky, the free-rider's payoff?

Maskin (2003) notes that existence of inefficient coalition structures as equilibria in the sequential games of coalition formation involves yet another assumption; namely, that players can commit to refrain from joining a coalition. Let me elaborate on this too.

Maskin (2003) uses an example of a three-player partition function game. It represents a simple free-rider problem created by a public good that can be produced by each coalition of two players.[18] More formally, the set of agents is $N = \{a, b, c\}$, and the partition function is

$$v(N) = 24;$$

$$v\big(\{a,b\}, \{\{a,b\}, \{c\}\}\big) = 12, v\big(\{a,c\}, \{\{a,c\}, \{b\}\}\big)$$
$$= 13, v\big(\{b,c\}, \{\{b,c\}, \{a\}\}\big) = 14,$$

$$v\big(\{i\}, \{\{i\}, \{j,k\}\}\big) = 9 \text{ for all } i, j, k \in N,$$

$$v\big(\{i\}, \{\{i\}, \{j\}, \{k\}\}\big) = 0 \text{ for all } i, j, k \in N.[19]$$

Maskin assumes that each player has an equal chance to be the first mover and shows that whatever is the order of play, a coalition structure of the form $(\{i\}, \{j, k\})$ is a subgame-perfect equilibrium outcome of a sequential bargaining game. Let us see why.

Consider, for example, the natural order a, b, and c. There are two possibilities. If a does not form a coalition with b, then it has to compete with b to induce c to form a coalition with it. Player a may reason that if it lets c join b, then its payoff will be 9. Therefore, it will be willing to pay c no more than 4 (= 13 − 9). But b is happy to pay (a little more than) 4 and form a coalition with c because that will result in a payoff of 10 (= 14 − 4) to it, which is more than the payoff of 9 that it would get if it were to let c join a instead. On the other hand, if a forms a coalition with b, then it will have to pay 10 to b because, as noted above, b can obtain a payoff of 10 by not forming a coalition with a. Thus, a can get a payoff of only 2 by forming a coalition with b. Furthermore, a can induce c to also join the coalition by giving a payoff of 9 to c resulting in a total payoff of only 5 (= 24 − 10 − 9) for a. Hence, the coalition structure $\{\{a\}, \{b, c\}\}$ is a subgame-perfect equilibrium outcome of the sequential bargaining game with payoffs of (9, 10, 4). Because the sum of these payoffs is less than 24, it follows that the equilibrium outcome is inefficient.

---

18. A similar example was first used in Ray and Vohra (1999).

19. Notice that the players are not identical, but just barely.

As Maskin notes, this result depends on the implicit assumption that a's decision not to form a coalition with b is irreversible. In the absence of such a commitment by $\{a\}$, coalition $\{b,c\}$ in a subsequent move can sign up player a and achieve payoffs that are higher for everyone.[20] For example, they may subsequently agree to payoffs of (9.5, 10.25, 4.25): a γ-core imputation! But if a's decision not to form a coalition with b is reversible, then c may refuse in the first place to form a coalition with b in the hope that a and b will then be forced to form a coalition, and it will be able to free ride resulting in a payoff of 9 for itself. Thus, by its irreversible commitment not to merge with b, a is able to force b and c to form a coalition and give itself the opportunity to free ride.[21] But if c is not convinced of a's commitment, then it may not form a coalition with b in the hope that it will force a and b to form a coalition instead, and it will get the free-rider's payoff of 9. Thus, if players cannot commit to refrain from forming coalitions with other players, the game develops into a war of attrition in which each player waits for the other two to form a coalition in the hope of free riding on them. Such a war of attrition is clearly a repeated game without discounting in which each player decides every moment whether to cooperate or not, and, by assumption, there is no cost of waiting.[22] This leads yet again to the repeated game of coalition formation studied above.

6.3.2. Internal-External Stability of Coalition Structures

One of the oldest problems of game theory, namely, stability of coalition structures or more generally coalition formation, has been studied in the

---

20. Hafalir (2007) similarly argues that in an environment in which side payments are allowed, formation of inefficient coalition structures is implausible because a Pareto superior allocation would be available to the players. He proposes a sequential game of coalition formation with the property that all subgame-perfect equilibria result in an efficient coalition structure, which in our case is the grand coalition.

21. Because each player is to be a first mover in some random order of play, this actually amounts to assuming that *every* player can commit not to form a coalition with other players. Thus, the bus may stay in the garage, and no one will have the opportunity to free ride.

22. This is true in many real-life externality problems. For example, the negotiations on climate change or free world trade appear to be repeated games, though not without cost of waiting.

context of a symmetric oligopoly.[23] The concept of a stable cartel introduced in d'Aspremont and Gabszewicz (1986) seems to have been the first contribution on this. These authors study stability of a cartel (i.e., one nondegenerate coalition) against single-player deviations.

The concept of stability against deviations by single players has been subsequently imported into the literature on what has come to be known as models of international environmental agreements (IEAs). The notable contributors include Carraro and Siniscalco (1993), among others. They could extend the concept of a stable cartel to a flow model of climate change because it has a similar mathematical structure as a symmetric oligopoly. According to these authors, a coalition of size s is stable if it satisfies a property of *internal stability*, ensuring that no member of the coalition can benefit by leaving the coalition and becoming a singleton, and a property of *external stability*, ensuring that no single player outside the coalition can benefit by joining the coalition.

In my terminology, this translates into the following game: Each of the n symmetric players can announce either 1 or 0. Thus, the strategy set of each player i is

$$T_i = \{1, 0\}, i \in N.$$

The outcome of this game is a strategy profile $t = (t_1, \ldots, t_n)$ that determines the coalition structure. The structure consists of a coalition $S \subset N$ of size $s = \sum_{i \in N} t_i$ and of $n - s$ players who remain singletons. Note that for all $i \in S$, $t_i = 1$, and for all $i \in N \backslash S$, $t_i = 0$. Thus, $\{S, [N \backslash S]\}$ is the coalition structure induced by strategy t. Each strategy of the players induces a unique coalition structure that consists of a coalition S of size s and singleton coalitions of all remaining players. The internal-external stability concept then simply means: A coalition structure induced by t^* is *stable* if and only if t^* is a Nash equilibrium of the game.[24] Carraro and Siniscalco

---

23. Ray and Vohra (1999, 2001) study the same problem in an economy with a public good.

24. Thoron (1998) generalizes this concept of stability by considering instead strong and coalition-proof equilibria of the game. This leads to what she calls strong and coalition-proof stability. But as she continues to assume that the coalition structure consists of at most one non-singleton coalition, the strategy sets in her game are the same.

(1993) show that a coalition structure of size $1 < s < n$ is stable and s is "small," meaning that only small coalitions can form. This result has been claimed to prove that international negotiation on climate change will result in an inefficient outcome!

The internal-external stability approach suffers from several limitations. First, the strategy sets are very restrictive. In contrast, the strategy sets of the players in the coalition formation game above are much richer (see section 6.2). More specifically, the strategy set of each player i is

$$T_i = \{0, 1, \ldots, n\}, i \in N.$$

That is, each player can choose from among a set of $n+1$ strategies consisting of 0 and numbers 1 to n. The players who announce the same number form a coalition, but the players who choose 0 remain singletons. Thus, the players can induce any coalition structure $(S_1, S_2, \ldots, S_m)$, $m \leq n$, in which any number of coalitions may be non-singletons.

Second, in the internal-external stability approach, the players are assumed implicitly to be myopic. Each deviating player assumes that the remaining coalition structure will remain unaffected subsequent to its deviation. It does not take into account that there may be further deviations subsequent to its deviation, resulting in an entirely different coalition structure than the one it presumed at the time of its deviation. Indeed, Diamantoudi and Sartzetakis (2006) show that if players are farsighted and take into account that there may be further deviations subsequent to their deviation, then the "small" coalition result of internal-external stability analysis no longer holds and near efficiency obtains.

Third, the rather strong assumption of identical players makes the internal-external stability analysis unsuitable for application to the climate change problem where heterogeneity rather than symmetry is the rule. In fact, the internal-external stability analysis has not been extended even to a model with three *heterogeneous* players. This seems to be due to the following difficulty (also raised more generally by Yi 2003): when the players are identical there is an appealing way to split a coalition's worth among its members; namely, to split it equally. Given the equal split rule, the players know in advance whether they would be better off by joining or leaving a coalition. But if the players are heterogeneous, what should be the split of the coalitional worth among its members? Unless a split is specified for each coalition, the players would not know what would be their payoffs if they join a coalition or what payoff they would forgo if they leave a coalition. Because the stability of a coalition depends on how the coalitional worth is split, any meaningful result concerning the stability of a coalition that may form can be obtained only by determining simultaneously the distribution of the worth within the coalition.[25] This significantly complicates the internal-external stability analysis and makes it hard to characterize the equilibrium coalition structure if the players are not identical. Assuming identical players is the easy way out.

6.4. CONCLUDING REMARKS

This chapter provides a game-theoretic argument for the γ-core assumption that when a coalition deviates from the grand coalition, the remaining players form singletons. It supplements Theorem 4.6 and completes my attempts to justify the γ-core concept. In the rest of the book, I consider applications of this concept to alternative game formulations of climate change including dynamic games and a trade model.

---

25. For example, if the rule is to assign the entire coalitional payoff to one of the members of the coalition, say the one with the highest index, then only singleton coalitions are stable, which shows that arbitrary choices of the payoff sharing rule can lead to absurd outcomes.

7

DYNAMIC ENVIRONMENTAL GAMES

I t was noted in chapter 2 that the natural phenomenon of *accumulation* of greenhouse gases (GHGs) as a stock pollutant in the atmosphere adds a time dimension to climate change in an essential way. However, in the chapters that followed, I limited the discussion to a flow (i.e., a static) model, and accordingly no mention of time was made in any of the formulations. In addition to the intrinsic interest of the flow model in the preceding chapters, expositional convenience motivated this choice so as to enable me to present the issues and their solutions in the simplest terms.

This chapter explores whether and how the analysis thus far can be reformulated in terms that recognize the role of time, and characterizes agreements to tackle climate change in this enlarged (i.e., dynamic) context. To a large extent, the structure of the exposition parallels the one followed in the preceding chapters. Thus, connections can easily be made with the concepts and analysis in those chapters. However, reformulating climate change as a dynamic model is not just for the sake of generality; newer important issues arise that have a bearing on climate policy. The chapter pursues these new issues with full force in a manner that is conceptually not inconsistent with the preceding chapters.

The dynamic model of climate change introduced and studied in this chapter incorporates the fact that while production and emissions occur at specific periods of time, environmental damage extends over very long periods of time because of the durability of the GHGs released and accumulated as a stock in the atmosphere. Thus, future damage enters the picture in an essential way. In this intertemporal setting, economic and game theories combined show that, as in the flow model, the externalities created by the stock pollutant imply that the spontaneous noncooperative equilibria in the dynamic model do not meet the criterion of efficiency, and policies to correct for this can be formulated in terms of a unique efficient emissions time profile and a stream of transfers that induce cooperation.

To be specific, it is shown that there exists a *unique* time profile of emissions (and, therefore, of the GHG stock) that leads to an intertemporally efficient outcome. This means efficiency cannot be achieved unless the countries emit according to this unique efficient emissions time profile. However, the costs and benefits of doing so differ over time and across countries, unless they are identical. Thus, as in the static model of climate change in the preceding chapters, transfers between countries to balance the costs and benefits of controlling climate change that can induce them to emit according to the unique efficient time profile are necessary. However, an important new issue arises in that the stream of transfers must be such that no coalition of countries will have incentives to choose emissions that are different from those in the unique efficient emissions time profile, not only in the current state (i.e., the level of GHG stock) but also in every state that may possibly occur in the future. I interpret the unique efficient emissions time profile and a stream of transfers with this property as a *subgame-perfect* cooperative agreement. To be precise, an agreement is subgame perfect if no coalition of countries has incentive to withdraw from it in *any* subgame.

From a methodological perspective, the chapter brings together two of the most important solution concepts in game theory: the subgame-perfect Nash equilibrium of a noncooperative game and the core of a cooperative game—a link between the two is apparently missing in the extant literature.

Mathematically as well as from a policy perspective, it will be seen that specification of the time horizon as finite or infinite matters.[1] Quite different results obtain depending on whether the time horizon is finite or infinite. Thus, the infinite time horizon version of the game is analyzed side by side so that differences in the results and their policy implications can be contrasted and highlighted.

---

1. In chapter 4, the symbol *t* was used to denote a joint strategy of the players. In this chapter, it is used to denote time. This should not cause any confusion.

7.1. THE DYNAMIC MODEL

Dynamic models are more intricate than static ones, in terms of both economics and game theory. To prevent complexities of the economics from blurring the game-theoretic arguments, I shall make the presentation without introducing accumulation of capital. Including capital accumulation could permit me to explore connections between climate change and economic growth—an important topic that the dynamic model below does not allow me to address. But as my main objective is to highlight and channelize the strategic forces behind climate change, this is left for another occasion.

Time is discrete, indexed by $t = 1, \ldots, T$, where the time horizon T is finite but may approach infinity. The variables $x_{it} \geq 0$ and $y_{it} \geq 0$ denote the consumption and production (resp.) of the composite private good of country i in period t. As in the basic flow model in chapter 2, the amounts x_{it} and y_{it} may differ because transfers between countries are permitted. The variables $e_{it} \geq 0$ and $z_{it} \geq 0$ denote (resp.) the amount of GHGs emitted by country i and the GHG stock in period t. While x_{it}, y_{it}, and e_{it} are flow variables, z_t is a stock variable that evolves overtime according to equation (7.2) below.

The output of the composite private good and emissions of each country i are related according to the equation $y_{it} = g_i(e_{it})$, where g_i is the production function. Each country i suffers damage from climate change and derives utility from the private good consumption in *each* period t according to the utility function $u_i(x_{it}, z_t) = x_{it} - v_i(z_t)$, where x_{it} is the private good consumption and $v_i(z_t)$ is the damage or disutility function. Thus, as in the basic flow model in chapter 2, utility is transferable unit for unit.

I assume that the production and the utility functions satisfy the same assumptions as in the basic flow model in chapter 2. That is, the production function $g_i(e_{it})$ of each country i is strictly increasing and strictly concave, and the damage function $v_i(z_t)$ is strictly increasing and convex or linear [i.e., $g_i'(e_{it}) > 0, g_i''(e_{it}) < 0, v_i'(z_t) > 0$, and $v_i''(z_t) \geq 0$]. Thus, the derivative $g_i'(e_{it})$ is the marginal abatement cost or the marginal product of emissions e_{it} of country i in period t, and the derivative $v_i'(z_t)$ is the marginal damage of country i due to the GHG stock z_t in period t. I assume that the marginal damage $v_i'(z), z \geq 0$ of each country i is bounded above. Analogous to Assumption 2.4, I also assume that for all $z \geq 0$ and each country i, there exists an $e^0 > 0$ such that $g_i'(e^0) \leq v_i'(e^0 + z)$ and $\lim_{e_i \to 0} g_i'(e_i) = \infty > \sum_{j \in N} v_j'(z)$. These assumptions imply that it is never optimal to emit more than e^0 or emit nothing and marginal damage is bounded above. As in the basic flow model, these assumption are just to ensure that the emissions e_{it} of country i in period t, whether chosen cooperatively or noncooperatively, are such that $0 < e_{it} \leq e^0$, $t = 1, \ldots, T$.

Clearly, the dynamic model of climate change is conceptually similar to the flow model in the preceding chapters. In particular, quasi-linearity of the utility functions and concavity of the production and convexity of the damage functions are common. This model is also closely related to the previous dynamic models in Dockner, Long, and Sorger (1996) and Dutta and Radner (2009). However, unlike Dockner et al. (1996), the marginal abatement costs are not assumed to be identical and constant, as there is compelling evidence that in reality the marginal abatement costs are decreasing in emission levels and differ significantly across the countries. As will be shown later, decreasing marginal abatement costs, unlike identical and constant marginal abatement costs, imply a *unique* subgame-perfect Nash equilibrium (SPNE) that is also Markov perfect (i.e., the SPNE strategies are functions of the current state variable).[2]

7.1.1. Efficient Emissions and Consumption Time Profiles

Given an initial GHG stock $z_0 \geq 0$, a time profile of consumptions $(x_{1t}, \ldots, x_{nt}, z_t)_{t=1}^T$ is *feasible* if there exists a time profile of emissions $(e_{1t}, \ldots, e_{nt})_{t=1}^T$ such that

$$\sum_{i=1}^n x_{it} = \sum_{i=1}^n g_i(e_{it}) = \sum_{i=1}^n y_{it} \tag{7.1}$$

and

$$z_t = (1 - \delta)z_{t-1} + \sum_{i=1}^n e_{it}, \quad t = 1, \ldots, T. \tag{7.2}$$

---

2. As will be seen below, identical and constant marginal abatement cost functions can give rise to multiple subgame-perfect Nash equilibria and, thus, multiple Markov-perfect equilibria. They also rule out the possibility of any gains from international trade in emissions in contradiction to the Kyoto Protocol, which proposed international trade in emissions to significantly reduce the costs of meeting the obligations under the Protocol.

The parameter, $0 \le \delta < 1$, here is the natural rate of decay of the GHG stock. To minimize notation, I assume henceforth that the initial GHG stock is fixed at $z_0 = 0$.

Because (7.1) does *not* require $x_{it} = g_i(e_{it}) = y_{it}$ for each i, it permits transfers of the composite private good between the countries in each period t, but not across the periods. Given the quasi-linearity of the utility functions $u_i(x_{it}, z_t)$, the latter is not really an assumption, as there is no gain from postponing consumption and there is no possibility of borrowing against future consumption for the world as a whole.

In the optimal control literature, the GHG emissions $(e_{it})_{t=1}^T$, $i = 1, \ldots, n$, are called *control variables*, and the resulting GHG stocks $z_{t-1}, t = 1, \ldots, T$, are called the *state variables*. While the latter are not strategies in the dynamic game introduced below, they are generated by the former and appear in the payoff function of each country. In fact, they have the same role as decision nodes in a dynamic game.

A special category of feasible emissions time profiles are *steady-state* emissions profiles where the emissions remain constant over time. In this connection, we may note already the following points. On the one hand, it is easily seen from the stock accumulation equation (7.2) that

$$z_t = z_{t-1} \Leftrightarrow \sum_{i \in N} e_{it} = \delta z_{t-1}, \tag{7.3}$$

that is, the stock remains constant over two consecutive periods if and only if emissions during the second period are equal to the amount of natural assimilation of the pollutant stock (or "decay" of it) as inherited from the first period. On the other hand, if the time horizon is finite, constant emissions do not necessarily imply a constant stock unless they are related to the stock as in (7.3). A third point to note is that the flow model of chapter 2 corresponds to $\delta = 1$.

Discounting

Because utilities are defined separately for each period t, a feasible consumption time profile generates a stream of utilities $(u_i(x_{it}, z_t))_{t=1}^T$. However, for the purpose of comparing alternative consumption time profiles, it might be more useful to define the payoff of each country over the entire consumption time profile. An additive form of utilities would do. Accordingly, we may define each country i's payoff from a feasible consumption time profile $(x_{1t}, \ldots, x_{nt}, z_t)_{t=1}^T$ as $\sum_{t=1}^T u_i(x_{it}, z_t)$. But doing so implicitly makes an assumption regarding individual preferences with respect to time, namely that goods obtained in different periods of time are of equal value, as for every pair of time periods $(t, t+m)$, $x_{it} = x_{it+m}, z_t = z_{t+m}$ implies $u_i(x_{it}, z_t) = u_i(x_{it+m}, z_{t+m})$. But a vast economic literature rejects this assumption on the ground that goods and commodities obtained in the distant future are of lesser value than if they were available today. This can be formalized by introducing an exponentially declining discount factor $0 < \beta \leq 1, t = 1, \ldots, T$, which is multiplied to each term in the sum that constitutes the above additive function, thus leading to

$$W_i \equiv \sum_{t=1}^T \beta^{t-1} u_i(x_{it}, z_t) = \sum_{t=1}^T \beta^{t-1} [x_{it} - v_i(z_t)]. \qquad (7.4)$$

Notice that under this formulation, it is never optimal to postpone consumption of the composite private good. As will be seen later, choice of the value of the discount factor β has implications for intergenerational equity and can serve as a proxy for bargaining between non-overlapping generations. I now define and characterize efficient consumption and emissions time profiles.

Definition 7.1

Given an initial stock $z_0 \geq 0$, a feasible consumption time profile $(x_{1t}^*, \ldots, x_{nt}^*, z_t^*)_{t=1}^T$ is efficient if there is no other feasible consumption time profile $(x_{1t}', \ldots, x_{nt}', z_t')_{t=1}^T$ such that $\sum_{t=1}^T \beta^{t-1} u_i(x_{it}', z_t') \geq \sum_{t=1}^T \beta^{t-1} u_i(x_{it}^*, z_t^*)$ for each $i \in N$ with strict inequality for at least one i.

Because $u_i(x_{it}, z_t) = x_{it} - v_i(z_t)$, that is, utility is transferable (unit for unit) between the countries, Definition 7.1 implies that a consumption time profile $(x_{1t}^*, \ldots, x_{nt}^*, z_t^*)_{t=1}^T$ is efficient *if and only if* it is a solution of the optimization problem $max_{(x_{1t}, \ldots, x_{nt}, z_t)_{t=1}^T} \sum_{i=1}^n \sum_{t=1}^T \beta^{t-1} [x_{it} - v_i(z_t)]$ subject to (7.1) and (7.2). After switching the summation signs and substituting from (7.1), this optimization problem is equivalent to

$$max_{(e_{1t},\ldots,e_{nt},z_t)_{t=1}^T} W = \sum_{t=1}^T \beta^{t-1} \sum_{i \in N} [g_i(e_{it}) - v_i(z_t)]$$

subject to (7.2). The Lagrangian for this optimization problem is

$$L = \sum_{t=1}^T \beta^{t-1} \sum_{i \in N} [g_i(e_{it}) - v_i(z_t)] + \sum_{t=1}^T \lambda_t [z_t - (1-\delta)z_{t-1} - \sum_{i \in N} e_{it}],$$

where the variables λ_t are the Lagrange multipliers associated with the T constraints. Assuming an interior solution for the reasons stated above and differentiating the Lagrangian, the first-order conditions for optimality imply that $(e_{1t}^*, \ldots, e_{nt}^*, z_t^*)_{t=1}^T$ is a solution of the optimization problem if

$$\frac{\partial L}{\partial e_{it}} = \beta^{t-1} g_i'(e_{it}^*) - \lambda_t^* = 0, t = 1, \ldots, T, i \in N,$$

$$\frac{\partial L}{\partial z_t} = -\beta^{t-1} \sum_{i \in N} v_i'(z_t^*) + \lambda_t^* - \lambda_{t+1}^*(1-\delta) = 0, t = 1, \ldots, T-1,$$

$$\frac{\partial L}{\partial z_T} = -\beta^{T-1} \sum_{i \in N} v_i'(z_T^*) + \lambda_T^* = 0,$$

$$z_t = (1-\delta)z_{t-1} + \sum_{i \in N} e_{it}, t = 1, \ldots, T,$$

$z_0 \geq 0$, given. These conditions imply

$$g_i'(e_{it}^*) - \beta(1-\delta)g_i'(e_{it+1}^*) = \sum_{j \in N} v_j'(z_t^*), i \in N, t = 1, \ldots, T,$$

which (called Euler equations in the optimal control literature) show a link between the marginal abatement costs of each country i in any two consecutive periods and the *sum* of the marginal damage of all countries in the first of those two periods. These equations can be rewritten as

$$g_i'(e_{it}^*) = \sum_{\tau=t}^T [\beta(1-\delta)]^{\tau-t} \sum_{j \in N} v_j'(z_\tau^*), i = 1, \ldots, n, t = 1, \ldots, T. \quad (7.5)$$

Equalities (7.5) embody two distinct notions of efficiency: (a) the marginal abatement costs of all countries must be equal in each period t, and (b) the marginal abatement cost of each country i in each period t is equal to the sum of the discounted marginal damage of all countries that will be avoided over the remaining time horizon $T - t + 1$, if the emissions of country i were reduced by one unit in period t. While (a) is the *equimarginal cost* principle or cost efficiency, (b) is the dynamic version of the well-known Lindahl-Samuelson condition for efficient provision of a public good. For this reason, I shall refer to (7.5) as the efficiency conditions.

It was noted above that each consumption time profile is generated by a unique emissions time profile. Because an efficient consumption time profile must satisfy equalities (7.5) in addition to equations (7.1) and (7.2), I can actually prove a stronger property for *efficient* consumption time profiles in that all efficient consumption time profiles are generated by the *same* emissions time profile $e^* = (e_{1t}^*, \ldots, e_{nt}^*)_{t=1}^{T}$. The proof follows from the fact that the stock accumulation equation (7.2) and the efficiency conditions (7.5) admit a unique solution.

Lemma 7.1

All efficient consumption time profiles $(x_{1t}^*, \ldots, x_{nt}^*, z_t^*)_{t=1}^{T}$ are generated by the same emissions time profile $(e_{1t}^* \ldots, e_{nt}^*)_{t=1}^{T}$.

Proof of Lemma 7.1

Equalities (7.2) and (7.5) form a system of $(n + 1)T$ equations in $(n + 1)T$ variables and, therefore, admit a solution. Let $(e_{1t}^* \ldots, e_{nt}^*)_{t=1}^{T}, (z_t^*)_{t=1}^{T}$ be a solution. I claim that it is unique. Suppose contrary to the assertion that the equalities admit two solutions. Then, because the objective function W is strictly concave and constraints (7.2) are linear, a convex combination of the two solutions is also feasible but implies a higher value of W, contradicting the supposition that both solutions maximize W. Hence, $(e_{1t}^*, \ldots, e_{nt}^*)_{t=1}^{T}$, is unique. ∎

In view of this lemma, I shall refer to $(e_{1t}^*, \ldots, e_{nt}^*)_{t=1}^{T}$ as the unique efficient emissions time profile. This has important consequences as it implies that efficiency cannot be achieved unless all countries emit according to the unique efficient emissions time profile $(e_{1t}^*, \ldots, e_{nt}^*)_{t=1}^T$ and every efficient consumption time profile $(x_{1t}^*, \ldots, x_{nt}^*, z_t^*)_{t=1}^T$ is defined by the equalities $\sum_{i \in N} x_{it}^* = \sum_{i \in N} g_i(e_{it}^*), t = 1, \ldots, T$. Thus, the efficient consumption time profiles differ from each other *only* in terms of the individual consumptions of the composite private good: the time profile of emissions and, thus, the GHG stock is the same for *all* efficient consump tion time profiles. All efficient consumption time profiles, except one, namely, $(g_1(e_{1t}^*), \ldots, g_n(e_{nt}^*), z_t^*)_{t=1}^T$, require transfers of the private good between the countries.

For the purpose of interpreting subgame-perfect cooperative agree ments later on, it may be noted that each efficient consumption time profile $(x_{1t}^*, \ldots, x_{nt}^*, z_t^*)_{t=1}^T$ is equivalent to a time profile $(e_{1t}^0, \ldots, e_{nt}^0)_{t=1}^T$ of emissions rights that are tradable on a competitive international market and such that $\sum_{i=1}^n e_{it}^0 = \sum_{i=1}^n e_{it}^*$, where $(e_{1t}^*, \ldots, e_{nt}^*)_{t=1}^T$ is the unique efficient emissions time profile, and each country i will choose to buy/sell $e_{it}^0 - e_{it}^*$ of rights at the competitive price in period t and emit e_{it}^*.[3] It is easily seen that the market for emissions rights clears in each period t if the international price is $p_t^* = \sum_{\tau=t}^T [\beta(1-\delta)]^{\tau-t} \sum_{j \in N} v_j'(z_\tau^*)$ and each country chooses its emissions to maximize its payoff $g_i(e_{it}) + p_t^*(e_{it}^0 - e_{it}).$[4]

Determinants of the Efficient Emissions Time Profile
and the Terminal Stock

It is easily seen from (7.2) and (7.5) that uniformly higher marginal dam age functions $v_i'(z)$ imply a lower terminal stock z_T^*. This follows from the fact that each g_i is concave, each v_i is convex, and $z_t = (1-\delta)z_{t-1} + \sum_{i \in N} e_{it}$. Therefore, if, contrary to the assertion, the terminal stock is higher, then it should also be higher in the preceding period, because the necessary conditions (7.5) then imply that emissions in the last period must be lower and so on all the way back to period 1, leading to a contradiction, as the initial stock is the same.

---

3. It is worth noting that the right e_{it}^0 can be negative meaning that country i must buy emissions rights that are more than the amount e_{it}^* it can emit under the unique efficient emissions time-profile.

4. It is implicitly assumed here that emission rights for period t can be traded only in period t.

Similarly, a higher discount factor[5] β or a lower decay rate δ implies a lower terminal stock z_T^*. However, while the decay rate δ is determined by natural phenomenon over which humans have no control, the discount factor β is a policy variable. But not much can be said about the time pattern of change in the efficient emissions time profile in general except that for higher β, the efficient emissions must be lower at least over some period(s). But if there are only two periods, present and future (i.e., $T = 2$), then, by the strict concavity of the production functions g_i and the necessary conditions (7.5), a higher discount factor β implies a lower terminal stock and higher emissions for every country in the future period but lower emissions for every country in the present period. If we identify the present and future generations with the present and future periods, respectively, then a higher discount factor is more favorable to the future generation. For this reason, choosing an appropriate value of the discount factor β is an important issue both in theory and in numerical simulation models of climate change. Choosing the value of the discount factor β can also serve as a proxy for bargaining between the present and future generations, which can be facilitated by the national governments acting on behalf of the future generations of their countries. But the situation is loaded against the future generations because the national governments— whether democratically elected or not—are likely to care more about the welfare of the present generation.

7.1.2. Infinite Time Horizon and Efficiency

If the time horizon $T = \infty$ and $0 < \beta(1-\delta) < 1$, then the stock accumulation equation (7.2) and the efficiency conditions (7.5) still hold, but the question is whether the sum on the right-hand side of (7.5), which now will have an infinite number of terms, converges. Given that the multiple $[\beta(1-\delta)]^{\tau-t}$ gets smaller with $\tau \to \infty$, boundedness of the marginal damage function $v_i'(z_t)$ is sufficient for convergence of the sum on the right-hand side of (7.5). Then, $(e_{1t}^*, \ldots, e_{nt}^*)_{t=1}^T$ is the unique efficient emissions time profile if it satisfies equalities (7.2) and (7.5) for $T = \infty$.

---

5. That is, a lower discount rate.

Linear Damage Functions

If the damage functions are linear, that is, $v_i(z) = \pi_i z$, $i = 1, \ldots, n$, and thus $v_i'(z) = \pi_i$, $i = 1, \ldots, n$, the efficiency conditions (7.5) take an especially simple form for $T = \infty$. Substituting in (7.5) and taking $T = \infty$, the efficient emissions time profile is given by the equalities

$$g_i'(e_{it}^*) = \frac{\sum_{j=1}^{n} \pi_j}{1 - \beta(1 - \delta)}, \quad i = 1, \ldots, n, \, t = 1, \ldots. \tag{7.6}$$

Because the right-hand side of (7.6) is independent of time, the efficient emissions time profile requires each country to emit at a constant rate in each period independently of the GHG stock z_t, $t = 1, \ldots$. Thus, I may drop the time subscript and denote the unique solution of (7.6) simply by $(e_1^*, \ldots, e_n^*)$.

Dutta and Radner (2009) assume linear damage functions and introduce the notion of a "global Pareto optimal" (GPO) consumption time profile. Given a (exogenous) profile of welfare weights $(\alpha_1, \ldots, \alpha_n)$ with $\alpha_i > 0$, $i = 1, \ldots, n$ and $\sum_{i=1}^{n} \alpha_i = 1$, the consumption time profile $(\hat{x}_{1t}, \ldots, \hat{x}_{nt}, \hat{z}_t)_{t=1}^{\infty}$ is a GPO consumption time profile if $\hat{x}_{it} = g_i(\hat{e}_{it})$, $i = 1, \ldots, n$, and $(\hat{e}_{1t}, \ldots, \hat{e}_{nt})_{t=1}^{\infty}$ is the solution of the optimization problem $max_{(e_{1t}, \ldots, e_{nt}, z_t)_{t=1}^{\infty}} W^{\alpha} = \sum_{i=1}^{n} \alpha_i \sum_{t=1}^{\infty} \beta^{t-1} [g_i(e_{it}) - v_i(z_t)]$ subject to (7.2). Dutta and Radner (2009, theorem 1) show that the solution to this optimization problem is a unique emissions time profile $\hat{e} = (\hat{e}_{1t}, \ldots, \hat{e}_{nt})_{t=1}^{\infty}$ characterized by the equalities

$$\alpha_i g_i'(\hat{e}_{it}) = \frac{\sum_{j=1}^{n} \alpha_j \pi_j}{1 - \beta(1 - \delta)}, \quad i = 1, \ldots, n, \, t = 1, \ldots, \tag{7.7}$$

when $v_i(z) = \pi_i z$, $i = 1, \ldots, n$, are the (linear) damage functions. The emissions time profile characterized by equalities (7.7) also requires each country to emit at a constant rate, but, as seen by comparing it with (7.6), it is not efficient unless all welfare weights are equal, that is, $\alpha_i = \alpha_j$, $i, j = 1, \ldots, n$.[6] This means that the entire set of GPO consumption time profiles includes just one time profile that is efficient in terms of Definition 7.1.

---

6. In fact, it does not satisfy even cost efficiency or equi-marginal cost principle if the welfare weights are not equal.

This characterization of the set of GPO consumption time profiles is important, as they have been shown to be implementable as the outcomes of history-dependent subgame-perfect equilibria (Dutta and Radner 2009). Accordingly, the next section characterizes them further. To that end, note that the total payoff of each country i in the *unique* efficient GPO consumption time profile without transfers is given by

$$W_i^* = \frac{1}{1-\beta}\left[g_i(e_i^*) - \frac{\pi_i \sum_{j=1}^n e_j^*}{1-\beta(1-\delta)}\right] - \frac{\pi_i z_0}{1-\beta(1-\delta)}, \quad (7.8)$$

where $(e_1^*,\ldots,e_n^*)$ is the unique solution of (7.6) and, therefore, of (7.7) with equal welfare weights $\alpha_i, i=1,\ldots,n$.

7.2. THE DYNAMIC ENVIRONMENTAL GAME

Because different types of strategies and equilibria are possible in dynamic games, unlike strategic games, it is convenient to define the dynamic game of climate change in terms of terminal histories.[7]

Definition 7.2

Given an initial stock $z_0 \geq 0$ and time periods $T > 1$, Γ_{z_0} denotes the dynamic game in which

- $N = \{i = 1, 2, \ldots, n\}$ is the player set;
- $E = E_1 \times E_2 \times \cdots \times E_n$, where $E_i = \{e_i \equiv (e_{it})_{t=1}^T : 0 \leq e_{it} \leq e^0\}$ is the set of all terminal histories;
- $u = (u_1,\ldots,u_n)$ is the profile of payoff functions such that for each terminal history $e \equiv (e_1,\ldots,e_n) = ((e_{1t})_{t=1}^T,\ldots,(e_{nt})_{t=1}^T) \in E$, we have $u_i(e) = \sum_{t=1}^T \beta^{t-1}[g_i(e_{it}) - v_i(z_t)]$, where $z_t = (1-\delta)z_{t-1} + \sum_{j \in N} e_{jt}$, $t = 1,\ldots,T$.

Let $\Gamma_{z_{t-1}}$, $z_{t-1} \geq 0, t \geq 1$, denote the dynamic game in which also the player set is N, the set of all terminal histories is $E_{1t} \times E_{2t} \times \cdots \times E_{nt}$,

---

7. See Osborne and Rubinstein (1994) for a formal definition of a terminal history of a dynamic game.

where $E_{it} = \{(e_{i\tau})_{\tau=t}^T : 0 \le e_{i\tau} \le e^0\}$, and the payoff of player i for each terminal history $e_t \equiv ((e_{1\tau})_{\tau=t}^T, \ldots, (e_{n\tau})_{\tau=t}^T) \in E_{1t} \times E_{2t} \times \cdots \times E_{nt}$ is $u_i(e_t) = \sum_{\tau=t}^T \beta^{\tau-1}[g_i(e_{i\tau}) - v_i(z_\tau)]$, where $z_\tau = (1-\delta)z_{\tau-1} + \sum_{j \in N} e_{j\tau}, t = 1, \ldots, T$. The game $\Gamma_{z_{t-1}}$ has exactly the same structure as the original game Γ_{z_0} except that its origin is at z_{t-1} and the time horizon is shorter. The dynamic game $\Gamma_{z_{t-1}}, t \in \{1, \ldots, T\}$, is a subgame of Γ_{z_0} if there exists a (nonterminal) history $((e_{1\tau})_{\tau=1}^{t-1}, \ldots, (e_{n\tau})_{\tau=1}^{t-1})$ such that $z_{\tau-1} = (1-\delta)z_{\tau-2} + \sum_{j \in N} e_{j\tau-2}, \tau = 2, \ldots, t$. However, notice that each subgame $\Gamma_{z_{t-1}}$ depends only on z_{t-1} and not on the history before the game reaches the state z_{t-1}. Thus, each subgame $\Gamma_{z_{t-1}}, t = 1, \ldots T$, has essentially the same mathematical structure as the original game Γ_{z_0}.

A *strategy* of player i is a function $e_i(z_{t-1}), 0 \le e_i(z_{t-1}) \le e^0, t = 1, \ldots, T$, that is, the emissions of player i in each period depend only on the GHG stock in the beginning of the period. This type of strategy is called Markovian. More conventionally, a Markov strategy is a function $e_i(z, t)$, but to save on notation I simply define it by $e_i(z_{t-1})$. This should not cause any confusion.

7.3. THE NASH EQUILIBRIA OF THE DYNAMIC ENVIRONMENTAL GAME

To motivate the concept of a subgame-perfect Nash equilibrium and its application below, I introduce first the dynamic analog of a Nash equilibrium of a strategic game and note its limitations. This type of equilibrium is called an open-loop Nash equilibrium.

7.3.1. The Open-Loop Nash Equilibrium

The strategic form of the dynamic game Γ_{z_0} is the triple (N, E, u), where N, E, and u are as in Definition 7.2. In the strategic form, the set of terminal histories also represents the set of joint strategies of the players.

Definition 7.3

A strategy profile $\bar{e} = (\bar{e}_{1t}, \ldots, \bar{e}_{nt})_{t=1}^T$ is a Nash equilibrium of the strategic game (N, E, u) if for each $i \in N, (\bar{e}_{it})_{t=1}^T = \text{argmax}_{(e_{it})_{t=1}^T} \sum_{t=1}^T \beta^{t-1}(g_i(e_{it}) - v_i(z_t))$, where $z_t = (1-\delta)z_{t-1} + e_{it} + \sum_{j \in N\backslash i} \bar{e}_{jt}, t = 1, \ldots, T$.

Proposition 7.2

Given $z_0 \geq 0$ and $T > 1$, the game (N, E, u) admits a unique Nash equilibrium.

Proof of Proposition 7.2

(First part: existence) The arguments made to prove Proposition 5.1 for the strategic environmental game can be used presently because the conditions are identical: each player's strategy space is a compact and convex set of a Euclidean space, and each player's payoff function is continuous and concave in the joint strategies $e \in E$. Hence, by Proposition 4.1, there exists a Nash equilibrium.

(Second part: uniqueness) The Lagrangian associated with each player's optimization problem is

$$L_i = \sum_{t=1}^{T} \beta^{t-1} \left(g_i(e_{it}) - v_i(z_t) \right) + \sum_{t=1}^{T} \lambda_{it} \left[z_t - (1-\delta)z_{t-1} - e_{it} - \sum_{j \in N \setminus i} \bar{e}_{jt} \right],$$

where the variables λ_{it} are the nonnegative Lagrange multipliers associated with the T constraints. Limiting ourselves to an interior optimum, the first-order conditions for $(\bar{e}_{it})_{t=1}^{T}$, $(\bar{z}_t)_{t=1}^{T}$, and $(\bar{\lambda}_{it})_{t=1}^{T}$ to be an optimum are

$$\frac{\partial L_i}{\partial e_{it}} = \beta^{t-1} g_i'(\bar{e}_{it}) - \bar{\lambda}_{it} = 0, t = 1, \ldots, T$$

$$\frac{\partial L_i}{\partial z_t} = -\beta^{t-1} v_i'(\bar{z}_t) + \bar{\lambda}_{it} - \bar{\lambda}_{it+1}(1-\delta) = 0, t = 1, \ldots, T-1$$

$$\frac{\partial L_i}{\partial z_T} = -\beta^{T-1} v_i'(\bar{z}_T) + \bar{\lambda}_{iT} = 0$$

$$\bar{z}_t = (1-\delta)\bar{z}_{t-1} + \sum_{j \in N} \bar{e}_{jt}, t = 1, \ldots, T,$$

$z_0 > 0$ given. Substituting for $\bar{\lambda}_{it}$ and $\bar{\lambda}_{it+1}$ and dividing by β^{t-1} yields the Euler equations

$$g_i'(\bar{e}_{it}) - \beta(1-\delta)g_i'(\bar{e}_{it+1}) = v_i'(\bar{z}_t), t = 1, \ldots, T-1.$$

Multiplying each subsequent equation in this sequence of equations by $\beta(1-\delta)$ and then adding them yields the necessary conditions for a Nash equilibrium:

$$g_i'(\overline{e}_{it}) = \sum_{\tau=t}^{T} [\beta(1-\delta)]^{T-\tau} \, v_i'(\overline{z}_\tau), t=1,\ldots,T, i=1,\ldots,n. \quad (7.9)$$

These conditions together with the T constraints characterize a Nash equilibrium $\overline{e} = (\overline{e}_{1t},\ldots,\overline{e}_{nt})_{t=1}^{T}$ of the game (N, E, u).

Given these necessary conditions, the proof for uniqueness of the Nash equilibrium is as follows. Suppose contrary to the assertion that for some initial stock z_0, there exist two Nash equilibria $\overline{e} = (\overline{e}_{1t},\ldots,\overline{e}_{nt})_{t=1}^{T}$ and $\hat{e} = (\hat{e}_{1t},\ldots,\hat{e}_{nt})_{t=1}^{T}$, and let $(\overline{z}_t)_{t=1}^{T}$ and $(\hat{z}_t)_{t=1}^{T}$ be the associated sequences of the stock. If $\overline{z}_T = \hat{z}_T$, then by the necessary conditions, $\overline{e}_{iT} = \hat{e}_{iT}$, and, therefore, $\overline{z}_{T-1} = \hat{z}_{T-1}$. Repeating this reasoning, one obtains $\overline{z}_t = \hat{z}_t$ and $\overline{e}_{it} = \hat{e}_{it}$ for all $t=1,\ldots,T$. This means that the game can have more than one Nash equilibrium only if $\overline{z}_T \neq \hat{z}_T$.

Suppose without loss of generality that $\overline{z}_T > \hat{z}_T$. Because each v_i is an increasing and convex function, $v_i'(\overline{z}_T) \geq v_i'(\hat{z}_T)$ for all i. In view of the necessary conditions for a Nash equilibrium, $g_i'(\overline{e}_{iT}) = v_i'(\overline{z}_T)$ and $g_i'(\hat{e}_{iT}) = v_i'(\hat{z}_T)$. Therefore, $g_i'(\overline{e}_{iT}) \geq g_i'(\hat{e}_{iT})$ for all i. As a consequence, $\overline{e}_{iT} \leq \hat{e}_{iT}$ for all i because the function g_i is increasing and concave. Thus, $\sum_{i \in N} \overline{e}_{iT} \leq \sum_{i \in N} \hat{e}_{iT}$. Combining this last property with the fact that $\overline{z}_T = (1-\delta)\overline{z}_{T-1} + \sum_{i \in N} \overline{e}_{iT} > \hat{z}_T = (1-\delta)\hat{z}_{T-1} + \sum_{i \in N} \hat{e}_{iT}$, one obtains $\overline{z}_{T-1} > \hat{z}_{T-1}$. However, by the Euler equations and the necessary conditions, (7.9), for a Nash equilibrium, we have $g_i'(\overline{e}_{iT-1}) = v_i'(\overline{z}_{T-1}) + \beta(1-\delta)v_i'(\overline{z}_T)$ and $g_i'(\hat{e}_{iT-1}) = v_i'(\hat{z}_{T-1}) + \beta(1-\delta)v_i'(\hat{z}_T)$. Together, $\overline{z}_T > \hat{z}_T$ and $\overline{z}_{T-1} > \hat{z}_{T-1}$ just established imply $g_i'(\overline{e}_{iT-1}) > g_i'(\hat{e}_{iT-1})$. Thus, $\overline{e}_{iT-1} < \hat{e}_{iT-1}$ for all i. Repeating this reasoning for all $t < T-1$, we eventually have $\sum_{i \in N} \overline{e}_{it} < \sum_{i \in N} \hat{e}_{it}$ and $\overline{z}_t > \hat{z}_t$ for all $t=1,\ldots,T$. But this is a contradiction, as it implies that $\overline{z}_t > \hat{z}_t$ for $t=1$, but $\overline{z}_1 = (1-\delta)z_0 + \sum_{i \in N} \overline{e}_{i1}$, $\hat{z}_1 = (1-\delta)z_0 + \sum_{i \in N} \hat{e}_{i1}$, and $\sum_{i \in N} \overline{e}_{i1} < \sum_{i \in N} \hat{e}_{i1}$. Thus, we must have $\overline{z}_T = \hat{z}_T$, and hence the Nash equilibrium is unique. ∎

It may be noted that if $T = \infty$, then the necessary conditions (7.9) for a Nash equilibrium still hold, as the marginal damage functions, by assumption, are bounded above. Furthermore, because the Nash equilibrium is unique for every T however large, it is also unique for $T = \infty$.

I now critically examine other aspects of the open-loop Nash equilibrium of a dynamic environmental game and note why it may not be a suitable concept for defining a subgame-perfect cooperative agreement. Though the open-loop Nash equilibrium of a dynamic game seems to be a natural and simple extension of the concept of Nash equilibrium of a strategic game, it suffers from an important limitation because of the way it is constructed. While it aptly formulates the equilibrium strategy of each player in terms of well-defined actions/emissions in each period, all these actions are determined essentially by the initial state z_0. Such modeling of strategic behavior is often interpreted in the dynamic games literature as one of commitment on the part of players.[8] More specifically, an open-loop Nash equilibrium does not guarantee that the actions prescribed by it will remain the best for the players irrespective of the state that may occur in a future period: while they are the best in the beginning of the game in period 1 and initial state z_0, they may no longer be so if the state z_t in a future period t happens to be different from the state $\bar{z}_t$ implied by the open-loop Nash equilibrium. But if this is not the case (i.e., if the state in each period is exactly the same as that implied by the open-loop Nash equilibrium), then the prescribed actions for the players remain the best for them, and there is no reason for the commitment not to be respected. The open-loop Nash equilibrium indeed has this property of *time consistency*. However, if the actual state of the system happens not to belong to the time path of the open-loop Nash equilibrium (for some reason exogenous to the model, such as a random shock for instance, or more endogenously, due to an action of the other players that was not anticipated), then the action prescribed by the open-loop Nash equilibrium is likely not to be the best for the players from that point of time onward. The open-loop Nash equilibrium strategy is not an equilibrium strategy anymore, and the commitment it represents loses its credibility.

Formally, (7.2) and the necessary conditions (7.9) for a Nash equilibrium show that if $(\bar{e}_{1\tau}, \ldots, \bar{e}_{n\tau})_{\tau=1}^{T}$ is a Nash equilibrium of the dynamic game Γ_{z_0}, then $(\bar{e}_{1\tau}, \ldots, \bar{e}_{n\tau})_{\tau=t}^{T}, t = 1, \ldots, T$, is a Nash equilibrium of the subgame $\Gamma_{\bar{z}_{t-1}}, t = 1, \ldots, T$. But conditions (7.2) and (7.9) do not imply

---

8. A view originally presented in Reinganum and Stokey (1985).

that $(\overline{e}_{1\tau}, \ldots, \overline{e}_{n\tau})_{\tau=t}^{T}$ is also a Nash equilibrium of every subgame $\Gamma_{z_{t-1}}$ with $z_{t-1} \neq \overline{z}_{t-1}$. In other words, an open-loop Nash equilibrium satisfies a *time consistency* property but not subgame perfection.

Because of its rigidity over time as explained above, the open-loop Nash equilibrium appears not to be an appropriate solution concept for describing the equilibrium payoffs of the players in all possible states that may occur in a future period. From a decision-making perspective in general, and interpretation of coalitional payoffs from noncooperation as a reference for international negotiations in particular, this is a serious drawback of the concept. Fixed (i.e., "committed") strategies irrespective of the state that may occur can hardly be useful in negotiations. The players should be able to evaluate their prospectus not only in particular states but also in all conceivable states that may arise in future periods as a result of cooperation or noncooperation. What is needed, therefore, is a solution concept that would have the equilibrium property not only relative to the initial state z_0 but also relative to all conceivable states z_t in any future period t. Only such a noncooperative Nash equilibrium can serve as a reference position in players' dilemma of choosing between noncooperation and cooperation in every state z_t that may possibly occur rather than only in states z_0 and $\overline{z}_t$, $t = 1, \ldots, T-1$, on the time path of the open-loop Nash equilibrium.

Another reason for the inappropriateness of the open-loop Nash equilibrium is directly linked with the central message of this book, namely the feasibility of achieving stable cooperative agreements among the countries (i.e., agreements that are efficient, at least Pareto superior to the Nash equilibrium), and such that no player will have incentive to withdraw from the agreement in any conceivable state that may occur. Chapter 5 revealed that such agreements should include the possibility for countries to make voluntary transfers of resources among themselves. These transfers should now be specified in terms that rest in an essential way on the state that may occur.

For that purpose, one might propose in each period t the payoffs corresponding to the open-loop Nash equilibrium defined at period 1 and state z_0 as the payoffs from noncooperation in period t.[9] I shall not do so,

---

9. This has been done, for example, in Jorgensen, Martin-Herran, and Zaccour (2003, 2005).

however, as that would amount to assuming that no cooperation has at all occurred before period t. This is so because if the players cooperated in all or some periods before t, then the state in period t will not be $\bar{z}_t$ (as implied by the open-loop Nash equilibrium), but something else depending on what cooperation or deviations from the open-loop Nash equilibrium strategies have occurred before period t.

To make precise our argument, which essentially bears on the issue of cooperation versus noncooperation, it is important to state explicitly what payoffs are possible for each player both from cooperation and non-cooperation in every conceivable state. There is, however, one special case in which the open-loop Nash equilibrium of the dynamic environmental game is independent of the state that may occur, and, therefore, it is also subgame perfect.[10] Thus, in this special case, the problem of choosing between alternative types of the Nash equilibria can be avoided altogether. I now consider this special case in some detail.

Linear Damage Functions

In the particular case of *linear damage functions*, that is, $v_i'(z) = \pi_i > 0$ for all z, the necessary conditions (7.9) reduce to:

$$g_i'(\bar{e}_{it}) = \pi_i \left(1 + \beta(1 - \delta) + (\beta(1 - \delta))^2 + \cdots + (\beta(1 - \delta))^{T-t}\right). \quad (7.10)$$

This condition allows us to state sharper properties of the open-loop Nash equilibrium, two of which bear on the *time profiles* of the emissions and the stock while the others are qualitative.

(i) As to the time profiles of the emissions and stock, as t increases, the emissions $\bar{e}_{it}$, $i \in N$, also *increase*. This is implied by the facts that the summation on the right of (7.10) always decreases as t increases, and that g_i is an increasing and strictly concave function. The rationale for this is quickly understood: the justification for keeping emissions low diminishes as the system approaches the terminal period, simply because the

---

10. See Hoel (1992) for a more thorough discussion of an open-loop versus a subgame-perfect equilibrium in the context of a dynamic game of climate change, and Reinganum (1982) for a class of games in which open-loop Nash equilibria are also subgame perfect.

remaining periods over which future damage will be incurred are getting fewer. By the same argument, (7.10) also implies that the longer the time horizon T, the lower the emissions $\bar{e}_{it}$, $t=1,\ldots,T$, $i \in N$.

(ii) As to the time profile of stock $\bar{z}_t$, depending on the level of the initial stock z_0, it might be decreasing in early periods in spite of increasing emissions if z_0 is such that $\sum_{i \in N} \bar{e}_{i1} < \delta z_0$. With linear damage functions this can occur only until some period τ when $\sum_{i \in N} \bar{e}_{i\tau} \geq \delta \bar{z}_{\tau-1}$, after which the stock rises in every period until the terminal period. Clearly, the time period τ depends on both the level of the initial stock z_0 and the rate of decay δ.

(iii) Decisions by countries other than country i itself do not play any role in (7.10). Therefore, for each country $i \in N$, the Nash equilibrium strategy is a dominant strategy. This property was also shown to hold for the Nash equilibrium of the environmental game in chapter 5 with linear damage functions.

(iv) In contrast to strictly convex damage functions, the equilibrium condition (7.10) does not depend on the stock $\bar{z}_t$, $t=1,\ldots,T$, either. Thus, the stock $\bar{z}_t$ plays no role in the determination of the Nash equilibrium emissions at time t. To put it another way: the equilibrium strategies are independent of the states. Therefore, the open-loop Nash equilibrium remains a Nash equilibrium irrespective of the state that may occur and is actually a subgame-perfect Nash equilibrium.

(v) Because $\sum_{j \in N} \pi_j > \pi_i$ for each $i \in N$, comparing (7.10) and (7.5) for linear damage functions, the strict concavity of g_i implies $e_{it}^* < \bar{e}_{it}$ for all $t=1,\ldots,T$ and $i \in N$. Thus, the open-loop Nash equilibrium emissions are uniformly higher than the efficient emissions.[11]

Infinite Time Horizon and the Open-Loop Nash Equilibrium

Any environmentally inclined reader must be surprised by property (i) of the open-loop Nash equilibrium. Most climate scientists would probably dismiss the linear damage function model for this reason alone. These

---

11. No such general assertion can be made if the damage functions are nonlinear. But it is often observed in numerical simulation models (e.g., Nordhaus and Yang 1996) that the Nash equilibrium emissions are uniformly higher than the efficient emissions.

understandable concerns can be alleviated by adding to the model an exogenous upper bound, say $z^0 > 0$, as an additional constraint on the variables z_t, $t = 1, \ldots, T$. Such a method is mathematically quite tractable and has been used in many simulation models[12] where z^0 is picked, for instance, equal to the 550 ppm GHGs concentration target mentioned in many scenarios of the Intergovernmental Panel on Climate Change (IPCC) or equal to the concentration of GHGs that may limit the temperature rise to 2°C as agreed upon by 196 countries at Paris in 2015. But such an upper bound on the terminal stock is exogenously imposed on the model, with no economic justification for the chosen level z^0.

Yet the comment following property (i) points to a most likely reason for the ever rising equilibrium emissions; namely, that the time horizon T is *finite*. Would an infinite horizon change the solution? If $T = \infty$, (7.10) becomes

$$g_i'(\bar{e}_{it}) = \pi_i \frac{1}{1 - \beta(1 - \delta)}, \, t = 1, \ldots, \infty, \, i \in N,$$

implying constant emissions from the start to infinity. The emissions are also uniformly lower than in the finite time horizon model. The stock z_t will either always increase or always decrease depending on the initial stock z_0. The stock z_t will continue to rise (resp. fall) if $\sum_{i \in N} \bar{e}_{i1} > (\text{resp.} <) \delta z_0$ until it stabilizes at the unique level z_∞ such that $\delta z_\infty = \sum_{i \in N} \bar{e}_{i1}$.

7.4. THE SUBGAME-PERFECT NASH EQUILIBRIUM

To decide whether to withdraw from an agreement, each country should be able to evaluate its payoffs not only if it abides by the agreement, but also if it withdraws from it in any period and state. Thus, I show that the dynamic game admits a subgame-perfect Nash equilibrium and interpret it as the situation that will prevail in the absence of an agreement or withdrawal by a player from an agreement in some period and state.

---

12. Many simulation models use the ad hoc method of choosing a very long time horizon T and then consider only the projections for the early periods of the trajectories, in the hope that they are least influenced by the assumed terminal stock.

To be precise, I show that the dynamic game Γ_{z_0} admits a unique SPNE if the production functions are strictly concave and quadratic and the damage functions are linear or convex and quadratic.[13] It will be clear to the reader that the game admits a unique SPNE also if the production functions are linear and the damage functions are strictly convex and quadratic. I do not consider both linear production and damage functions, as they can lead to corner solutions.

Theorem 7.3

The dynamic game Γ_{z_0} admits a unique subgame-perfect Nash equilibrium if the production functions $g_i, i = 1, \ldots n$, are strictly concave, the damage functions $v_i, i = 1, \ldots, n$, are strictly convex or linear, and the third derivatives $g_i''' = v_i''' = 0$, $i = 1, \ldots n$ (i.e., all production and damage functions are quadratic).

The proof of this theorem uses the method of backward induction and consists of the following steps: First prove that each subgame $\Gamma_{z_{T-1}}, z_{T-1} \geq 0$, in the last period T admits a unique SPNE $(e_{1T}(z_{T-1}), \ldots, e_{nT}(z_{T-1}))$, which results in payoffs $q_i(z_{T-1}) \equiv g_i(e_{iT}(z_{T-1})) - v_i((1 - \delta)z_{T-1} + \Sigma_{j \in N}\, e_{jT}(z_{T-1}))$, $i = 1, \ldots, n$. Then, using the assumptions regarding the functions g_i and v_i, prove that each $q_i(z_{T-1})$, $z_{T-1} \geq 0$, is a concave function and $q_i'''(z_{T-1}) = 0$, and note that the players' payoff functions in the reduced form of the subgame $\Gamma_{z_{T-2}}, z_{T-2} \geq 0$, in period $T - 1$ are given by $g_i(e_{iT-1}) - [v_i((1 - \delta)z_{T-2} + \Sigma_{j \in N}\, e_{jT-1}) - q_i((1 - \delta)z_{T-2} + \Sigma_{j \in N}\, e_{jT-1})]$, $i = 1, \ldots, n$, and satisfy concavity in exactly the same way as do the payoff functions in the subgame $\Gamma_{z_{T-1}}$, $z_{T-1} \geq 0$, in period T. Thus, the reduced form of the subgame $\Gamma_{z_{T-2}}$ has exactly the same mathematical structure as the game $\Gamma_{z_{T-1}}$, as $q_i(z_{T-1}), i = 1, \ldots, n$, is a nonincreasing concave and quadratic function of z_{T-1}. Therefore, $\Gamma_{z_{T-2}}$ admits a unique SPNE, and the SPNE payoffs $q_i(z_{T-2}), i = 1, \ldots, n$, are similarly nonincreasing concave and quadratic functions of z_{T-2}.

---

13. Germain et al. (2003, last line of p. 82) claim existence and uniqueness of an SPNE in a dynamic game under more general conditions. But their claim is false (as also pointed out in Chander 2017a) and is based on the erroneous belief that the conditions for the existence of an SPNE of a dynamic game are the same as the conditions for the existence of a Nash equilibrium of a strategic game.

Continuing in this manner, the backward induction leads to a unique SPNE of the dynamic game Γ_{z_0}.

Proof of Theorem 7.3

To keep the algebra simple, I prove the theorem for $\beta=1$ and $\delta=0$. The proof for the more general case $\beta \leq 1$ and $\delta \geq 0$ is analogous.

I show that backward induction leads to a unique SPNE. Begin with a subgame in the last period T. A strategy profile $(e_{1T},\ldots,e_{nT})$ is an SPNE of a last-period subgame $\Gamma_{z_{T-1}}$ if each e_{iT} maximizes $g_i(e_{iT}) - v_i(z_{T-1} + \sum_{j\in N} e_{jT})$, given $e_{jT}, j \neq i$. Therefore, the first-order conditions for payoff maximization imply

$$g_i'(e_{iT}) = v_i'(z_{T-1} + \sum_{j\in N} e_{jT}), i=1,\ldots,n. \tag{7.11}$$

I claim that these equations have a unique solution. Suppose not, and let $(\overline{e}_{1T},\ldots,\overline{e}_{nT})$ and $(\overline{\overline{e}}_{1T},\ldots\overline{\overline{e}}_{nT})$ be two different solutions such that $\sum_{i\in N} \overline{e}_{iT} = (>) \sum_{i\in N} \overline{\overline{e}}_{iT}$. Then, because each v_i is convex and g_i is strictly concave, (7.11) implies $\overline{e}_{iT} = (<)\overline{\overline{e}}_{iT}$ for $i=1,\ldots,n$, which contradicts our supposition. Hence, the last-period subgame $\Gamma_{z_{T-1}}$ admits a unique SPNE for $z_{T-1} \geq 0$. Let $(e_{1T}(z_{T-1}),\ldots,e_{nT}(z_{T-1}))$ denote the unique SPNE of $\Gamma_{z_{T-1}}$. By differentiating (7.11), we have

$$g_i''(e_{iT}(z_{T-1}))e_{iT}'(z_{T-1}) = v_i''\left(z_{T-1} + \sum_{j\in N} e_{jT}(z_{T-1})\right)$$
$$\times \left(1 + \sum_{j\in N} e_{jT}'(z_{T-1})\right), i \in N. \tag{7.12}$$

Because $g_i'' < 0$ and $v_i'' \geq 0$, equations (7.12) imply $e_{iT}'(z_{T-1}) \leq 0$ and $(1 + \sum_{j\in N} e_{jT}'(z_{T-1})) \geq 0$. By differentiating (7.12), $g_i'''(e_{iT}(z_{T-1}))(e_{iT}'(z_{T-1}))^2 + g_i''(e_{iT}(z_{T-1}))e_{iT}''(z_{T-1}) = v_i'''(z_{T-1} + \sum_{j\in N} e_{jT}(z_{T-1}))(1 + \sum_{j\in N} e_{jT}')^2 + v_i''(z_{T-1} + \sum_{j\in N} e_{jT}(z_{T-1}))\sum_{j\in N} e_{jT}''(z_{T-1}), i=1,\ldots,n$. Therefore,

$$g_i''(e_{iT}(z_{T-1}))e_{iT}''(z_{T-1}) = v_i''\left(z_{T-1} + \sum_{j\in N} e_{jT}(z_{T-1})\right)$$
$$\times \sum_{j\in N} e_{jT}''(z_{T-1}), i \in N, \tag{7.13}$$

as $g_i''' = v_i''' = 0, i=1,\ldots n$. Because $g_i'' < 0$ and $v_i'' \geq 0$, equations (7.13) imply $e_{iT}''(z_{T-1}) = 0$. Let $q_i(z_{T-1}) \equiv g_i(e_{iT}(z_{T-1})) - v_i(z_{T-1} + \sum_{j\in N} e_{jT}(z_{T-1}))$, $i=1,\ldots,n$. Then, $q_i'(z_{T-1}) = g_i'(e_{iT}(z_{T-1}))e_{iT}'(z_{T-1}) - v_i'(z_{T-1} + \sum_{j\in N} e_{jT}(z_{T-1}))$

$(1+\sum_{j\in N}e'_{jT}(z_{T-1}))\leq 0$, as $g'_i>0, v'_i>0$, and, as shown, $e'_{jT}(z_{T-1})\leq 0$ and $(1+\sum_{j\in N}e'_{jT}(z_{T-1}))\geq 0$. Furthermore, $q''_i(z_{T-1})=g''_i(e_{iT}(z_{T-1}))(e'_{iT}(z_{T-1}))^2 - v''_i(z_{T-1}+\sum_{j\in N}e_{jT}(z_{T-1}))(1+\sum_{j\in N}e'_{jT}(z_{T-1}))^2+g'_i(e_{iT}(z_{T-1}))e''_{iT}(z_{T-1})-v'_i(z_{T-1}+ \sum_{j\in N}e_{jT}(z_{T-1}))\sum_{j\in N}e''_{jT}(z_{T-1})=g''_i(e_{iT}(z_{T-1}))(e'_{iT}(z_{T-1}))^2 - v''_i(z_{T-1}+\sum_{j\in N}e_{jT}(z_{T-1}))(1+\sum_{j\in N}e'_{jT}(z_{T-1}))^2\leq 0$, as $g''_i<0, v''_i\geq 0$, and, as shown, $e''_{iT}(z_{T-1})=0, i=1,\ldots,n$.

This proves that each $q_i(z_{T-1}), i=1,\ldots,n$, is a nonincreasing concave function of z_{T-1}. In fact, by differentiating the above expression and using $g'''_i=v'''_i=0$ for each $i=1,\ldots n$, it is seen that $q'''_i(z_{T-1})=0$. Thus, the reduced form of the subgame $\Gamma_{z_{T-2}}$ with payoff functions $g_i(e_{iT-1})-[v_i(z_{T-2}+\sum_{j\in N}e_{jT-1})-q_i(z_{T-2}+\sum_{j\in N}e_{jT-1})]$ has exactly the same mathematical structure as the game $\Gamma_{z_{T-1}}$, as $q_i(z_{T-1}), i=1,\ldots,n$, is a nonincreasing concave function of z_{T-1}. Therefore, $\Gamma_{z_{T-2}}$ admits a unique SPNE, and the SPNE payoffs $q_i(z_{T-2}), i=1,\ldots,n$, are similarly nonincreasing and concave functions of z_{T-2}. Continuing in this manner, the backward induction leads to a unique SPNE of the dynamic game Γ_{z_0}. ∎

Because the SPNE is a noncooperative solution concept in which the countries maximize their own individual payoffs without any consideration whatsoever regarding the damage their actions would inflict on the other countries, I shall refer to it as the *status quo*; that is, the situation that may prevail in the absence of a cooperative agreement.[14]

I take note of some properties of the unique SPNE. First, the SPNE strategy of each player is linear and nonincreasing in the state variable $z_{t-1}, t=1,\ldots,T$. This follows from the proof of Theorem 7.3, which shows that the subgame-perfect equilibrium strategy $e_{it}(z_{t-1})$ of each player i satisfies $e'_{it}(z_{t-1})\leq 0$ and $e''_{it}(z_{t-1})=0$ [see equations (7.12) and (7.13), respectively]. Thus, the SPNE is also a Markov-perfect equilibrium (MPE). Second, the SPNE is not efficient. Indeed, if $\beta=1$ and $\delta=0$, then the efficiency conditions (7.5) imply $g'_i(e_{iT})=\sum_{j\in N}v'_j(z_T), i=1,\ldots,n$. Comparing them with (7.11) implies that the SPNE outcome is inefficient. Clearly,

---

14. Mäler and de Zeeuw (1998) also prove existence of an SPNE, but for damage functions of the form $v_i(z)=c_iz^2, i=1,\ldots,n$, and apply dynamic programming. In contrast, Theorem 7.3 holds for general quadratic damage functions; for example, $v_i(z)=a_iz+c_iz^2$, where a_i or c_i may be equal to zero for some countries but positive for others.

this is also true for $\beta \leq 1$ and $1 \geq \delta \geq 0$. Third, if the production functions are identical and linear and the damage functions are strictly convex, the game admits infinitely many subgame-perfect Nash equilibria. Indeed, if the production functions are linear and identical, equations (7.11) admit infinitely many solutions such that only the sum total of emissions is uniquely determined, but not the individual emissions. In fact, in this case, equations (7.12) and (7.13) do not reveal much about the characteristics of the individual equilibrium strategies except in the case of symmetric countries.

7.4.1. Infinite Time Horizon and the SPNE

If the time horizon $T = \infty$, the existence of an SPNE or equivalently an MPE of the dynamic game Γ_{z_0} can be proved by showing that the functional equations $q_i(z) = g_i(e_{i1}(z)) - [v_i((1-\delta)z + \sum_{j \in N} e_{j1}(z)) - \beta q_i((1-\delta)z + \sum_{j \in N} e_{j1}(z))], i \in N$, with $g_i''' = v_i''' = 0, i \in N$, admit a solution $q_i(z), i \in N$, such that each $q_i(z)$ is a nonincreasing and concave function with $q_i''' = 0$, where $(e_{11}(z), \ldots, e_{n1}(z))$ is the Nash equilibrium of the strategic game with payoff functions given by $g_i(e_{i1}) - [v_i((1-\delta)z + \sum_{j \in N} e_{j1}) - \beta q_i((1-\delta)z + \sum_{j \in N} e_{j1})], i \in N$. One can guess that the solution of the above functional equations is of the form $q_i(z) = (a_i z + b_i)^2$ and then find the values of the parameters a_i and b_i such that the conditions for the Nash equilibrium of the strategic game as well as the functional equations $q_i(z) = g_i(e_{i1}(z)) - [v_i((1-\delta)z + \sum_{j \in N} e_{j1}(z)) - \beta q_i((1-\delta)z + \sum_{j \in N} e_{j1}(z))], i \in N$ are satisfied for all values of z.

This line of proof, however, does not prove that the MPE is unique. In fact, for a model with two symmetric countries and infinite time horizon, Dockner and Long (1993) show that the dynamic game may also admit a Markov-perfect equilibrium in nonlinear strategies that is better for both the countries and can even sustain an efficient consumption time profile. However, this line of research still has to be developed for the asymmetric case with more than two countries. As Dockner and Long (1993, p. 24) note in the concluding section of their paper, "If countries are highly asymmetrical, it would be difficult to agree on the selection of a given pair of strategies."

7.4.2. Linear Damage Functions and Efficiency

I now return to the infinite horizon linear damage functions model discussed in the preceding sections. As in the case of the efficient emissions time profile or the efficient GPO emissions time profile, the SPNE equilibrium strategies also require each country to emit at a constant rate independent of the GHG stock and are characterized by[15]

$$g_i'(e_{it}) = \frac{\pi_i}{1-\beta(1-\delta)}, i = 1, \ldots, n; t = 1, \ldots \tag{7.14}$$

7.4.3. Necessity of Transfers

To further reinforce my claim that transfers between countries that are not sufficiently symmetric are necessary, I now return to the well-known model by Dutta and Radner (2009) and show that if the countries are not sufficiently symmetric, then the unique efficient GPO consumption time profile is not a Pareto improvement over the SPNE; that is, the business-as-usual (BAU) equilibrium.

Proposition 7.4

If the damage functions are linear and the countries are not sufficiently symmetric, the unique efficient GPO consumption time profile is not a Pareto improvement over the SPNE.

Proof of Proposition 7.4

For linear damage functions, the SPNE strategies are independent of the GHG stock and characterized by (7.14). Let $(\overline{e}_1, \ldots, \overline{e}_n)$ denote the solution

---

15. There are several ways to prove this. The simplest is perhaps to first show that the SPNE for T finite is characterized by the equations $g_i'(e_{it}) = \frac{\pi_i(1-(\beta(1-\delta))^{T-t+1})}{1-\beta(1-\delta)}, i = 1, \ldots, n; t = 1, \ldots, T$, and then take the limit $T \to \infty$.

of (7.14). Then, comparing (7.6) and (7.14), the strict concavity of g_i implies $\bar{e}_i > e_i^*, i = 1, \ldots, n$, and the SPNE payoff of country i is

$$\bar{W}_i = \frac{1}{1-\beta} \left[g_i(\bar{e}_i) - \frac{\pi_i \sum_{j=1}^{n} \bar{e}_j}{1-\beta(1-\delta)} \right] - \frac{\pi_i z_0}{1-\beta(1-\delta)}, i = 1, \ldots, n. \quad (7.15)$$

From (7.15) and (7.8), we obtain

$$\bar{W}_i - W_i^* = \frac{1}{1-\beta} \left[g_i(\bar{e}_i) - \frac{\pi_i \sum_{j=1}^{n} \bar{e}_j}{1-\beta(1-\delta)} \right] - \frac{1}{1-\beta} \left[g_i(e_i^*) - \frac{\pi_i \sum_{j=1}^{n} e_j^*}{1-\beta(1-\delta)} \right]$$

$$= \frac{1}{1-\beta} \left[g_i(\bar{e}_i) - g_i(e_i^*) - \frac{\pi_i}{1-\beta(1-\delta)} \left(\sum_{j=1}^{n} \bar{e}_j - \sum_{j=1}^{n} e_j^* \right) \right], i = 1, \ldots, n.$$

In this expression, the terms $g_i(\bar{e}_i) - g_i(e_i^*) > 0$ and $\frac{\pi_i}{1-\beta(1-\delta)} \times (\sum_{j=1}^{n} \bar{e}_j - \sum_{j=1}^{n} e_j^*) > 0$, because, as shown, $\bar{e}_i > e_i^*$. As seen from (7.6) and (7.14), e_i^* is a decreasing function of $\sum_{j=1}^{n} \pi_j$, whereas $\bar{e}_i$ is a decreasing function of π_i alone. Thus, for π_i sufficiently small but $\sum_{j=1}^{n} \pi_j$ sufficiently large with some $\pi_j > \pi_i, j \neq i$, we have $\bar{W}_i > W_i^*$. ∎

Lemma 7.1 together with the observation that no GPO consumption time profile except one with equal welfare weights (i.e. $\alpha_i = \alpha_j, i, j = 1, \ldots, n$) is efficient show that if the countries are not sufficiently symmetric, the GPO consumption time profiles that can be supported as the outcomes of history-dependent subgame-perfect equilibria through the use of trigger strategies may be either inefficient or not a Pareto improvement over the SPNE. Typically, countries with relatively low marginal damage will be worse off if efficiency requires them to reduce their emissions by large amounts. This stands to reason because such countries benefit little from climate change mitigation but have to bear costs of reducing their emissions that are higher than their benefits. Example 1 below illustrates this fact. Because in reality countries are sovereign and highly asymmetric, it follows that some countries may not be willing to participate in games whose outcomes can be supported as history-dependent subgame-perfect equilibria through the use of trigger strategies. In contrast, subgame-perfect cooperative agreements use transfers to balance the costs and benefits of controlling climate change and induce the countries to voluntarily participate in the agreement. Example 1 illustrates this fact.

Example 1

Let $T=2$, $N=\{1, 2\}$, $g_i(e_{it})=2e_{it}^{\frac{1}{2}}$, $i=1, 2$, $v_1(z_t)=\frac{1}{2}z_t$, and $v_2(z_t)=z_t, z_0=0$, $\beta=1$, and $\delta=0$. Notice that the marginal damage of country 1 is only half that of country 2.

In view of (7.2) and (7.5), the unique efficient emissions time profile is given by $e_{i1}^*=\frac{1}{9}, e_{i2}^*=\frac{4}{9}, i=1, 2$. Thus, $z_1^*=e_{11}^*+e_{21}^*=\frac{2}{9}$, $z_2^*=z_1^*+e_{12}^*+e_{22}^*=\frac{2}{9}+\frac{8}{9}=\frac{10}{9}$. Using backward induction, the unique SPNE strategies are given by $\bar{e}_{11}=1, \bar{e}_{21}=\frac{1}{4}, \bar{e}_{12}=4$, and $\bar{e}_{22}=1$. Therefore, $\bar{z}_1=\bar{e}_{11}+\bar{e}_{21}=1+\frac{1}{4}=\frac{5}{4}$, $\bar{z}_2=\bar{z}_1+\bar{e}_{12}+\bar{e}_{22}=\frac{5}{4}+4+1=\frac{25}{4}$. Thus, compared to the SPNE emissions, efficiency requires country 1 to reduce its emissions in both periods by higher amounts than country 2: $\bar{e}_{11}-e_{11}^*=1-\frac{1}{9}>\bar{e}_{21}-e_{21}^*=\frac{1}{4}-\frac{1}{9}$ and $\bar{e}_{12}-e_{12}^*=4-\frac{4}{9}>\bar{e}_{22}-e_{22}^*=1-\frac{4}{9}$, though its benefits from efficient control of climate change are only half as much as those of country 2. The question is: Will both countries be better off if they indeed reduce their emissions to efficient levels?

Using the computations above, the payoffs of countries 1 and 2 if they reduce their emissions to efficient levels are $W_1^*=2(e_{11}^*)^{\frac{1}{2}}-\frac{1}{2}z_1^*+2(e_{12}^*)^{\frac{1}{2}}-\frac{1}{2}z_2^*=\frac{2}{3}-\frac{1}{9}+\frac{4}{3}-\frac{5}{9}=\frac{4}{3}$ and $W_2^*=2(e_{21}^*)^{\frac{1}{2}}-z_1^*+2(e_{22}^*)^{\frac{1}{2}}-z_2^*=\frac{2}{3}-\frac{2}{9}+\frac{4}{3}-\frac{10}{9}=\frac{2}{3}$, respectively, whereas their SPNE payoffs are $\bar{W}_1=2(\bar{e}_{11})^{\frac{1}{2}}-\frac{1}{2}\bar{z}_1+2(\bar{e}_{12})^{\frac{1}{2}}-\frac{1}{2}\bar{z}_2=2-\frac{5}{8}+4-\frac{25}{8}=\frac{9}{4}$ and $\bar{W}_2=2(\bar{e}_{21})^{\frac{1}{2}}-\bar{z}_1+2(\bar{e}_{22})^{\frac{1}{2}}-\bar{z}_2=1-\frac{5}{4}+2-\frac{25}{4}=-\frac{9}{2}$, respectively This shows that $W_1^*<\bar{W}_1$. Thus, country 1 will not participate in any agreement for efficient control of climate change unless it is given transfers to compensate it for the resulting loss in its welfare. But is it possible to make transfers such that, after transfers, both countries will be better off?

Because $W_1^*+W_2^*=2>\bar{W}_1+\bar{W}_2=-\frac{9}{4}$, there indeed exist transfers s_1 and s_2 such that $s_1+s_2=0$ and $W_1^*+s_1\geq\bar{W}_1$ and $W_2^*+s_2\geq\bar{W}_2$. For example, if $s_1=\frac{7}{3}$ and $s_2=-\frac{7}{3}$, then $W_1^*+s_1=\frac{4}{3}+\frac{7}{3}\geq\bar{W}_1=\frac{9}{4}$ and $W_2^*+s_2=\frac{2}{3}-\frac{7}{3}\geq\bar{W}_2=-\frac{9}{2}$. A further question is: What should be the time profile of these transfers? Should the transfers be such that they make both countries better off in each period rather than just over the entire duration of the agreement? I pursue this question in the next section.

Finally, it may be noted that if the countries were symmetric, then $W_1^* = W_2^* > \overline{W}_1 = \overline{W}_2$; that is, no transfers are necessary for the countries to be better off if they both reduce their emissions to efficient levels.

7.5. THE SUBGAME-PERFECT COOPERATIVE AGREEMENTS

I interpret the SPNE payoff of each country as the payoff that the country can ensure for itself without cooperation of the other countries. However, because I also consider coalitional behavior for reasons mentioned above, I need to specify the payoff that a non-singleton coalition can similarly ensure for itself without cooperation of the other countries. To that end, given the dynamic game Γ_{z_0}, for each coalition $S \subset N$ let $\Gamma_{z_0}^S$ denote the induced dynamic game in which coalition S acts as one single player; that is, within the coalition the individual strategies are selected so as to maximize the sum of the payoffs of its members, given the strategies of the nonmembers. Similarly, let $\Gamma_{z_{t-1}}^S$ denote an induced game of the subgame $\Gamma_{z_{t-1}}$, to be called an induced subgame. The induced games $\Gamma_{z_0}^S$ and $\Gamma_{z_{t-1}}^S$ may seem to have the same structures as the original games Γ_{z_0} and $\Gamma_{z_{t-1}}$, respectively, except that the number of players is $n - |S| + 1$ instead of n. But there is an important difference in that, unlike the original game Γ_{z_0}, the payoff of one of the players, namely coalition S, is a function of as many variables as the number of members of S. Therefore, existence and characterization of an SPNE for the induced games $\Gamma_{z_0}^S, S \subset N$, do not follow simply from Theorem 7.3 and have to be established, except in two cases of coalition structures in that the unique SPNE of Γ_{z_0} is also a unique SPNE of each induced game $\Gamma_{z_0}^{\{i\}}, i = 1, \ldots, n$, and the efficient emission time profile $(e_{1t}^*, \ldots, e_{nt}^*)_{t=1}^T$ is the unique SPNE of the induced game $\Gamma_{z_0}^N$.

Theorem 7.5

For each coalition $S \subset N$, each induced subgame $\Gamma_{z_{t-1}}^S, z_{t-1} \geq 0, t = 1, \ldots, T$, admits a unique subgame-perfect Nash equilibrium if the production functions $g_i, i = 1, \ldots n$, are strictly concave, the damage functions $v_i, i = 1, \ldots, n$, are strictly convex or linear, and the third derivatives $g_i''' = v_i''' = 0, i = 1, \ldots n$.

The proof of this theorem is also by backward induction, though, unlike in the proof for the existence of a unique SPNE, the payoff of one of the players, namely coalition S, in each induced subgame $\Gamma^S_{z_{t-1}}$, $z_{t-1} \geq 0, t = 1, \ldots, T$, is a function of as many variables as the number of members of S. The method of backward induction works in this case also because the damage functions of all countries (though not necessarily identical) are functions of the same variables, and, as seen from a comparison of equations (7.11) and (7.15)–(7.16) below, the conditions characterizing the Nash equilibrium in both cases have the same mathematical structure.

Proof of Theorem 7.5

As in the case of Theorem 7.3, we need to prove the theorem only for $\beta = 1$ and $\delta = 0$. Thus, an emission profile $(e_{1T}, \ldots, e_{nT})$ is an SPNE of an induced subgame $\Gamma^S_{z_{T-1}}$ in the last period T if $(e_{iT})_{i \in S}$ maximizes $\sum_{i \in S} g_i(e_{iT}) - \sum_{i \in S} v_i(z_{T-1} + \sum_{j \in N} e_{jT})$ and each e_{jT}, $j \in N \backslash S$, maximizes $g_j(e_{jT}) - v_j(z_{T-1} + \sum_{i \in N} e_{iT})$. Therefore, by first-order conditions (FOCs) for payoff maximization, $(e_{1T}, \ldots, e_{nT})$ must be a solution of the equations

$$g_i'(e_{iT}) = \sum_{j \in S} v_j'(z_{T-1} + \sum_{k \in N} e_{kT}), i \in S, \tag{7.15}$$

$$g_j'(e_{jT}) = v_j'(z_{T-1} + \sum_{i \in N} e_{iT}), j \in N \backslash S. \tag{7.16}$$

As in the proof of Theorem 7.3, these equations admit a unique solution $(e_{1T}(z_{T-1}), \ldots, e_{nT}(z_{T-1}))$. Differentiating (7.15) and (7.16), I obtain

$$g_i''(e_{iT}(z_{T-1}))e_{iT}'(z_{T-1}) = \sum_{k \in S} v_k''\left(z_{T-1} + \sum_{j \in N} e_{jT}(z_{T-1})\right)$$
$$\times \left(1 + \sum_{j \in N} e_{jT}'(z_{T-1})\right), i \in S.$$

$$g_j''(e_{jT}(z_{T-1}))e_{jT}'(z_{T-1}) = v_j''\left(z_{T-1} + \sum_{j \in N} e_{jT}(z_{T-1})\right)$$
$$\times \left(1 + \sum_{j \in N} e_{jT}'(z_{T-1})\right), j \in N \backslash S.$$

Because each g_i is strictly concave and each v_i is convex, these equations imply $(1 + \sum_{j \in N} e_{jT}'(z_{T-1})) \geq 0$ and $e_{iT}'(z_{T-1}) \leq 0$. Differentiating these equations once more and using $g_i''' = v_i''' = 0$ for each i implies $e_{iT}''(z_{T-1}) = 0$, $i = 1, \ldots, n$, and the SPNE payoffs are $w(S, z_{T-1}) = \sum_{i \in S} g_i(e_{iT}(z_{T-1})) - \sum_{i \in S} v_i(z_{T-1} + \sum_{j \in N} e_{jT}(z_{T-1}))$ for coalition S and $m_j(z_{T-1}) = g_j(e_{jT}(z_{T-1})) - v_i(z_{T-1} + \sum_{j \in N} e_{jT}(z_{T-1}))$ for each $j \in N \backslash S$. Differentiating $w(S, z_{T-1})$ and

$m_j(z_{T-1})$ twice shows that each $w(S, z_{T-1}), S \subset N$, and $m_j(z_{T-1}), j=1,\ldots,n,$ are nonincreasing concave function of z_{T-1}. Furthermore, as in the proof of Theorem 7.3, $w'''(S, z_{T-1}) = m'''_j(z_{T-1}) = 0$. Thus, the reduced form of the induced subgame $\Gamma^S_{z_{T-2}}$ has exactly the same mathematical structure as the game $\Gamma^S_{z_{T-1}}$. Continuing in this manner, the backward induction leads to a unique SPNE of the induced subgame $\Gamma^S_{z_{t-1}}, t=1,\ldots,T$. ∎

The proof of Theorem 7.5 also shows that for each coalition S and the induced subgame $\Gamma^S_{z_{t-1}}$, the subgame-perfect Nash equilibrium strategy of each player, including each individual player in coalition S, is linear and nonincreasing in the state variables $z_{t-1}, t=1,\ldots,T$; that is, $e'_{it}(z_{t-1}) \leq 0$ and $e''_{it}(z_{T-1}) = 0$ irrespective of whether player i is a member of coalition S or not. This means that the equilibrium strategy of each individual player, including each *individual* player in coalition S, is a Markov strategy. This extends applications of Markov strategies even to a concept of cooperation in a dynamic game. Even more important, the proof of Theorem 7.5 also shows that the unique subgame-perfect Nash equilibrium payoff $w(S, z_{t-1})$ of coalition S is a nonincreasing and concave function of z_{t-1}.

The definition of $w(S, z_{t-1})$ involves two assumptions. First, it does not require coalition S to cooperate with the countries outside S in any future period. Because the countries are sovereign and coalition S is free to not cooperate with the outside countries now or in the future, $w(S, z_{t-1})$ is among the possible payoffs that coalition S can achieve for itself without cooperation of the other countries. This means a cooperative agreement cannot be stable unless it promises each coalition S a payoff of at least $w(S, z_{t-1})$ in the subgame $\Gamma_{z_{t-1}}$. Second, as in the definition of the γ-core of a strategic game, the definition of $w(S, z_{t-1})$ assumes that the countries outside the deviating coalition S choose their *individually* best reply strategies rather than their *joint* best reply strategies. This assumption is also implicit in Dutta and Radner (2009), who, however, restrict deviations to singleton coalitions only and assume that a deviation by a single country will result in reversal to the MPE (i.e., the SPNE). Thus, in their formulation as well as here, if a single country deviates, then its payoff is equal to its MPE/SPNE payoff; that is, $w(\{i\}, z_{t-1}) = q_i(z_{t-1})$. But if I assume instead that the remaining countries form a coalition of their own and choose their joint best reply strategies—which is as arbitrary as assuming that the remaining countries form singletons and choose their individually best reply strategies—then the payoff of a deviating singleton country will not be equal to its MPE/SPNE payoff, and my approach would no longer be consistent with the concept of an MPE/SPNE. In fact, it would lead to a different concept in which a deviation by a single country will not result— except in the case of only two countries—in reversion to the SPNE/MPE but to an equilibrium in which the deviating country plays its best reply strategy against the joint best reply strategies of the other countries.

Responding to a deviation by a coalition with individually best reply strategies can also be interpreted as imposing a mild punishment on the deviating coalition. It is mild because the remaining countries can impose instead a harsher punishment on the deviating coalition by choosing their highest feasible emissions in the current and all future periods (i.e., $e_{i\tau} = e^0$ for each $i \notin S$ and $\tau \geq t$), where t is the period in which the deviation occurs.[16]

Definition 7.4

A subgame-perfect cooperative agreement in the dynamic game Γ_{z_0} is a feasible consumption time profile $(x_{1t}, \ldots, x_{nt}, z_t)_{t=1}^T$ such that for each coalition $S \subset N, w(S, z_{t-1}) \leq \sum_{i \in S} \sum_{\tau=t}^T \beta^{\tau-t}(x_{i\tau} - v_i(z_\tau))$ for each, $t = 1, \ldots T$.

The definition requires that the transfers between countries, implicit in the variables $x_{it}, i = 1, \ldots, n; t = 1, \ldots, T$, should be such that no country or coalition of countries is worse off in *any* subgame. Because the payoffs $w(S, z_{t-1}), S \subset N$, depend on the state variable z_{t-1}, each x_{it} also depends on the state variable z_{t-1}, though, to save on notation, this is not explicitly indicated in the definition. Furthermore, as will also become clear later, the transfers in any subgame $\Gamma_{z_{t-1}}$ do not depend on transfers in other subgames. This is because if a coalition deviates in some subgame, then, by definition, it deviates forever, and therefore its payoffs—before or after transfers—in other subgames are not relevant.

Indeed, because the unique efficient emissions time profile $(e_{1t}^*, \ldots, e_{nt}^*)_{t=1}^T$ is also the unique SPNE of the "one-player" induced game $\Gamma_{z_0}^N$, we have $w(N; z_0) = \sum_{t=1}^T \beta^{t-1} \sum_{i \in N}[g_i(e_{it}^*) - v_i(z_t^*)]$, where $(z_t^*)_{t=1}^T$ is generated by

---

16. This harsher punishment is conceptually equivalent to the punishment in Marx and Matthews (2000) that requires the remaining players to contribute nothing to the public good once a player deviates because that is the worst the remaining players can do in their model.

the unique efficient emissions time profile $(e_{1t}^*,\ldots,e_{nt}^*)_{t=1}^T$ according to equation (7.2). This means that only an efficient consumption time profile $(x_{1t}^*,\ldots,x_{nt}^*;z_t^*)_{t=1}^T$, where $\sum_{i\in N}x_{it}^*=\sum_{i\in N}g_i(e_{it}^*),t=1,\ldots,T$, can be a subgame-perfect cooperative agreement. Thus, the search for a subgame-perfect cooperative agreement is restricted to the set of efficient consumption time profiles, and the set of subgame-perfect cooperative agreements is a refinement of the set of efficient consumption time pro-files. Furthermore, an efficient consumption time profile $(x_{1t}^*,\ldots,x_{nt}^*;z_t^*)_{t=1}^T$ is a subgame-perfect cooperative agreement if and only if each restriction $(x_{1\tau}^*,\ldots,x_{n\tau}^*;z_\tau^*)_{\tau=t}^T$, $t=1,\ldots,T$, of the efficient consumption time profile belongs to the core of the characteristic function game $(N,w(.,z_{\tau-1}^*))$, where $w(.,z_{\tau-1}^*)$ denotes the characteristic function $w(S,z_{\tau-1}^*),S\subset N$. Because $w(N,z_{t-1}^*) = \sum_{\tau=t}^T \beta^{\tau-t}\sum_{i\in N}[g_i(e_{i\tau}^*)-v_i(z_\tau^*)]$, the dynamic game Γ_{z_0} admits a subgame-perfect cooperative agreement if and only if each of the T characteristic function games $(N,w(.,z_{\tau-1}^*)),\tau=1,\ldots,T$, is bal-anced and appropriate restrictions of the *same* efficient consumption time profile belong to the core of each of these T characteristic function games.

Definition 7.4 implies that an efficient consumption time profile $(x_{1t}^*,\ldots,x_{nt}^*;z_t^*)_{t=1}^T$ satisfies subgame perfection if each restricted consump-tion time profile $(x_{1t}^*,\ldots,x_{nt}^*;z_t^*)_{t=\tau}^T$ satisfies subgame perfection in the sub-game $\Gamma_{z_{\tau-1}^*}$, $\tau=1,\ldots,T$.[17] To understand why an agreement should have this property, consider an efficient consumption time profile $(x_{1t}',\ldots,x_{nt}';z_t^*)_{t=1}^T$ such that $\sum_{i\in S}\sum_{t=1}^T \beta^{t-1}(x_{it}'-v_i(z_t^*))\geq w(S,z_0)$ for all $S\subset N$, but for some $S\subset N$ and period $\tau>1$, $\sum_{i\in S}\sum_{t=\tau}^T \beta^{t-\tau}(x_{it}'-v_i(z_t^*)<w(S,z_{\tau-1}^*))$. In words, consumption time profile $(x_{1t}',\ldots,x_{nt}';z_t^*)_{t=1}^T$ as an agreement is such that no coalition has incentive to withdraw from it in period 1, but some coali-tion S has incentive to withdraw from it at a future date τ and force rene-gotiations of the terms of the agreement—in contradiction to the beliefs of the other countries in period 1 that S will do its part when the time comes.[18] We confirm this possibility by building on Example 1.

---

17. It may be noted in this connection that if a consumption time-profile $(x_{1t}^*,\ldots,x_{nt}^*;z_t^*)_{t=1}^T$ is efficient in the game Γ_{z_0}, then the restricted consumption time-profile $(x_{1\tau}^*,\ldots,x_{n\tau}^*;z_\tau^*)_{\tau=t}^T$ is efficient in the subgame $\Gamma_{z_{\tau-1}^*}$, $\tau=1,2,\ldots,T$.

18. Becker and Chakrabarti (1995) show that the issue of subgame perfection of an agreement can also arise in dynamic games without externalities and, thus, propose the recursive core as the set of allocations such that no coalition can improve upon its consumption stream at any time.

Example 2

In continuation of Example 1 with $\beta=1$ and $\delta=0$ and using the computations for the unique efficient emissions time profile, the set of all efficient consumption time profiles $((x_{11}, x_{21}; z_1), (x_{12}, x_{22}; z_2))$ is described by the equalities $x_{11} + x_{21} = \frac{4}{3}$, $x_{12} + x_{22} = \frac{8}{3}$, $z_1^* - \frac{2}{9}$, and $z_2^* = \frac{10}{9}$. Thus, $w(\{1,2\}, z_0) = x_{11} + x_{21} - \frac{1}{2} z_1^* - z_1^* + x_{12} + x_{22} - \frac{1}{2} z_2^* - z_2^* = \frac{4}{3} - \frac{1}{9} - \frac{2}{9} + \frac{8}{3} - \frac{5}{9} - \frac{10}{9} = 2$, and $w(\{1,2\}, z_1^*) = x_{12} + x_{22} - \frac{1}{2} z_2^* - z_2^* = \frac{8}{3} - \frac{5}{9} - \frac{10}{9} = 1$. Next, using the computations for the unique SPNE in Example 1, we have $w(\{1\}, z_0) = 2(\overline{e}_{11})^{\frac{1}{2}} - \frac{1}{2}\overline{z}_1 + 2(\overline{e}_{12})^{\frac{1}{2}} - \frac{1}{2}\overline{z}_2 = 2 - \frac{5}{8} + 4 - \frac{25}{4} = \frac{9}{4}$, $w(\{2\}, z_0) = 2(\overline{e}_{21})^{\frac{1}{2}} - \overline{z}_1 + 2(\overline{e}_{22})^{\frac{1}{2}} - \overline{z}_2 = 1 - \frac{5}{4} + 2 - \frac{25}{4} = -\frac{9}{2}$, $w(\{1\}, z_1^*) = 2(\overline{e}_{12})^{\frac{1}{2}} - \frac{1}{2}(z_1^* + \overline{e}_{12} + \overline{e}_{22}) = 4 - \frac{1}{2}(\frac{2}{9} + 5) = \frac{25}{18}$, and $w(\{2\}, z_1^*) = 2(\overline{e}_{22})^{\frac{1}{2}} - (z_1^* + \overline{e}_{12} + \overline{e}_{22}) = 2 - (\frac{2}{9} + 5) = -\frac{29}{9}$.

Clearly, $((x_{11}, x_{21}; z_1), (x_{12}, x_{22}; z_2)) = ((-\frac{5}{3}, 3, \frac{2}{9}), (6, -\frac{10}{3}, \frac{10}{9}))$ is an efficient consumption time profile and requires country 1 to transfer $\frac{7}{3}$ units of the private good to country 2 in period 1 and country 2 to transfer $\frac{14}{3}$ units to country 1 in period 2 and generates total payoffs of $r_{11} = x_{11} - \frac{1}{2} z_1^* + x_{12} - \frac{1}{2} z_2^* = -\frac{5}{3} - \frac{1}{9} + 6 - \frac{5}{9} = \frac{11}{3} > \frac{9}{4} = w(\{1\}, z_0)$ for country 1 and $r_{21} = x_{12} - z_1^* + x_{22} - z_2^* = 3 - \frac{2}{9} - \frac{10}{3} - \frac{10}{9} = -\frac{5}{3} > -\frac{9}{2} = w(\{2\}, z_0)$ for country 2. Furthermore, $r_{11} + r_{21} = 2 = w(\{1,2\}, z_0)$. Thus, no country or coalition of countries can be better off by withdrawing in period 1 if the agreement is the efficient consumption time profile $((-\frac{5}{3}, 3, \frac{2}{9}), (6, -\frac{10}{3}, \frac{10}{9}))$. But for the restricted consumption time profile $(x_{12}, x_{22}; z_2) = (6, -\frac{10}{3}, \frac{10}{9})$ in period 2, $w(\{2\}, z_1^*) = -\frac{29}{9} > -\frac{10}{3} - \frac{10}{9} = -\frac{40}{9}$, and therefore country 2 will be better off if it withdraws from the agreement in period 2, but not if it withdraws in period 1.

Country 1 would adhere to the agreement in period 1 because it expects to receive a large enough transfer in period 2, though it has to make a transfer in period 1, and country 2 would adhere to it in period 1 because it is to receive a large enough transfer in period 1, though it has to make a transfer in period 2. But come period 2 and state z_1^* (after country 1 has reduced its emissions in period 1 and made transfers), country 2 realizes that it would be better off if it now leaves the agreement rather than complies with it. In fact, country 2 can force renegotiation of the terms of the agreement after the game reaches period 2, but not before period 2. For instance, it can abandon the current agreement and then propose instead the consumption profile $(4, -\frac{4}{3}, \frac{10}{9})$ as the new agreement in period 2, which can make both countries better off but offers only 4 units of the private good to country 1 compared to 6 units in the original agreement.

The example explains why Definition 7.4 requires that no country or coalition of countries should have incentive to leave the agreement in any period and force renegotiations of the terms of the agreement. It can easily be verified that if the agreed-upon efficient consumption time profile were instead $((\frac{1}{3}, 1, \frac{2}{9}), (4, -\frac{4}{3}, \frac{10}{9}))$, then no country will have incentive to withdraw from the agreement in any period.

7.5.1. Existence and Further Characterization

I now identify sufficient conditions for the existence of a subgame-perfect cooperative agreement for the dynamic game Γ_{z_0}.

Theorem 7.6

The dynamic game Γ_{z_0} admits a subgame-perfect cooperative agreement $(x_{1t}^*, \ldots, x_{nt}^*; z_t^*)_{t=1}^T$ if the benefit functions g_i are strictly concave with $g_i''' = 0$, $i = 1, \ldots, n$, and the damage functions $v_i, i = 1, \ldots, n$ are linear.

The proof of this theorem shows that if the damage functions are linear, then the transfers required by a subgame-perfect cooperative agreement, like the SPNE/MPE strategies, do not depend on the state variable. In fact, the dynamic game can be decomposed into a sequence of strategic games that do not depend on the state variable and that can be solved independently of each other. But to meet the requirement of subgame perfection, their γ-core solutions must be combined such that no coalition will have incentive to deviate in any period. For instance, if some coalition is given "too high" transfers in period $T-1$, then it will have to make "too high" transfers in period T, and thus it will have incentive to withdraw from the agreement in period T, though not in period $T-1$.

Proof of Theorem 7.6

To keep the algebra simple, I prove the theorem for $\beta = 1$ and $\delta = 0$. The proof for the more general case $\beta \leq 1$ and $\delta \geq 0$ is analogous. Let $v_i(z) = \pi_i z, i = 1, \ldots, n$, be the damage functions.

I first prove the theorem for $T=2$ and then extend the proof to any finite T. Because $\beta=1$ and $\delta=0$, the payoff of player i in period $T-1$ is given by $g_i(e_{iT-1}) - \pi_i(z^*_{T-2} + \sum_{j\in N} e_{jT-1}) + g_i(e_{iT}) - \pi_i(z^*_{T-2} + \sum_{j\in N} e_{jT-1} + \sum_{j\in N} e_{jT}) = g_i(e_{iT-1}) - 2\pi_i \sum_{j\in N} e_{jT-1} + g_i(e_{iT}) - \pi_i \sum_{j\in N} e_{jT} - 2\pi_i z^*_{T-2}$. Because $= 2$, $z^*_{T-2} = z_0 = 0$. Now consider two strategic form games, say Ω_{T-1} and Ω_T, in which the strategy sets are $E_{1t} \times E_{2t}$, where $E_{it} = \{(e_{i\tau})^T_{\tau=t} : 0 \le e_{it} \le e^0\}, i=1,2; t=T-1, T$, respectively, and the payoff function of player $i \in \{1,2\}$ is $g_i(e_{iT-1}) - 2\pi_i \sum_{j\in N} e_{jT-1}$ in Ω_{T-1} and $g_i(e_{iT}) - \pi_i \sum_{j\in N} e_{jT}$ in Ω_T. The two games can be solved independently of each other. Because each g_i is strictly concave, let $w_{T-1}(S), S \subset N$, and $w_T(S), S \subset N$, denote the unique Nash equilibrium payoff of coalition S in the induced games of Ω_{T-1} and Ω_T, respectively, in which coalition S acts as one player and the remaining players act as singletons. By Theorem 5.5, the core (i.e., the γ-core) of each characteristic function game (N, w_{T-1}) and (N, w_T) is nonempty. By definition of these strategic games, $w_{T-1}(N) = \sum_{i\in N} [g_i(e^*_{iT-1}) - 2\pi_i \sum_{j\in N} e^*_{jT-1}]$ and $w_T(N) = \sum_{i\in N} [g_i(e^*_{iT}) - \pi_i \sum_{j\in N} e^*_{jT}]$, where the emission time profile $((e^*_{1T-1}, \ldots, e^*_{nT-1}), (e^*_{1T}, \ldots, e^*_{nT}))$ is the unique efficient emission time profile in the (two-period) dynamic game and $z^*_{T-1} = z^*_{T-2} + \sum_{i\in N} e^*_{iT-1}$. Thus, there exist $(x^*_{1T-1}, \ldots, x^*_{nT-1})$ and $(x^*_{1T}, \ldots, x^*_{nT})$ such that $\sum_{i\in N} x^*_{iT-1} = \sum_{i\in N} g_i(e^*_{iT-1})$, $\sum_{i\in N} x^*_{iT-2} = \sum_{i\in N} g_i(e^*_{iT-2})$, and for each $S \subset N$, $w_{T-1}(S) \le \sum_{i\in S} [x^*_{iT-1} - 2\pi_i \sum_{i\in S} e^*_{iT-1}]$ and $w_T(S) \le \sum_{i\in S} [x^*_{iT} - \pi_i \sum_{i\in S} e^*_{iT}]$. By definition, $w(S, z^*_{T-2}) = w_{T-1}(S) + w_T(S) - 2\sum_{i\in S} \pi_i z^*_{T-2}$ and $w(S, z^*_{T-1}) = w_T(S) - \pi_i z^*_{T-1}$. Because $z^*_{T-1} = z^*_{T-2} + \sum_{i\in N} e^*_{iT-1}$, this implies that there exists a feasible consumption time profile $(x^*_{1t}, \ldots, x^*_{nt}; z^*_t)^T_{t=1}$ such that for each $S \subset N$, we have $w(S, z^*_{T-2}) \le \sum_{i\in S} [x^*_{iT-1} - \pi_i(z^*_{T-2} + \sum_{j\in N} e^*_{jT-1})] + \sum_{i\in S} [x^*_{iT} - \pi_i(z^*_{T-1} + \sum_{j\in N} e^*_{jT})]$ and $w(S, z^*_{T-1}) \le \sum_{i\in S} [x^*_{iT} - \pi_i(z^*_{T-1} + \sum_{j\in N} e^*_{jT})]$.

The proof for $T \ge 3$ is analogous and follows from the fact that if the damage functions are linear, then every T period dynamic game can be decomposed into T strategic games and a constant term. A consumption time profile that satisfies subgame perfection can then be constructed by combining a γ-core imputation of each of the strategic games. ∎

Subgame-perfect cooperative agreements also exist if the damage functions are strictly convex but the countries are symmetric. In this case, no transfers are necessary and $(x^*_{1t}, \ldots, x^*_{nt}; z^*_t)^T_{t=1}$, where $x^*_{it} = g_i(e^*_{it}), i=1, \ldots, n; t=1, \ldots, T$, is a subgame-perfect cooperative agreement. But if the damage functions are strictly convex and the countries are not symmetric, proving existence of a subgame-perfect cooperative agreement is technically far more challenging, as the dynamic game then cannot be decomposed into a sequence of strategic games that are independent of the state variable and thus can be solved independently of each other. However, proving existence in this more general case will not lead to additional insights regarding subgame-perfect cooperative agreements and is at best only of technical interest.

7.5.2. Infinite Time Horizon

Theorem 7.6 also holds for $T=\infty$ and $0<\beta(1-\delta)<1$. In fact, if the time horizon $T=\infty$, a subgame-perfect cooperative agreement is constructed as follows. As in the proof of Theorem 7.6, the dynamic game can be decomposed into (infinitely many) strategic games $\Omega_{z^*_{t-1}}, t=1,2,\ldots$ with $g_i(e_{it}) - \pi_i \sum_{\tau=t}^{\infty} [\beta(1-\delta)]^{\tau-t} (\sum_{j \in N} e_{jt} + (1-\delta)z^*_{t-1}) = g_i(e_{it}) - \frac{\pi_i}{1-\beta(1-\delta)} (\sum_{j \in N} e_{jt} + (1-\delta)z^*_{t-1})$ (because $0<\beta(1-\delta)<1$) as the payoff function of player i in the strategic game $\Omega_{z^*_{t-1}}$. This payoff function is strictly concave in e_{it}, as g_i is strictly concave. Thus, each strategic game $\Omega_{z^*_{t-1}}$ admits a unique Nash equilibrium. This Nash equilibrium is independent of the state variable z^*_{t-1} and characterized by equalities (7.14). This implies that the dynamic game with infinite time horizon and linear damage functions admits a unique SPNE in stationary strategies. Let $(\overline{e}_1,\ldots,\overline{e}_n)$ denote the stationary SPNE strategies. Similarly, let $(e^*_1,\ldots,e^*_n)$ denote the stationary efficient emissions as characterized by (7.6) above.

Let $w_t(S), S \subset N, t=1,2,\ldots$ denote the unique Nash equilibrium payoff of coalition S in the induced game of the strategic game $\Omega_{z^*_{t-1}}$ in which coalition S acts as one player and the remaining players act as singletons. Because the payoff functions are strictly concave and the damage functions are linear, by Theorem 5.5 the core of each characteristic function game $(N, w_t), t=1,2,\ldots$ is nonempty, and the imputation

$$r_{it} = x^*_i - \frac{\pi_i}{1-\beta(1-\delta)} \left(\sum_{j \in N} e^*_j + (1-\delta)z^*_{t-1} \right), i=1,\ldots,n, \qquad (7.17)$$

where $x^*_i = g_i(\overline{e}_i) - \frac{\pi_i}{\sum_{j \in N} \pi_j} \sum_{j \in N} [g_j(\overline{e}_j) - g_j(e^*_j)]$ belongs to the core; that is, $w_t(S) \leq \sum_{i \in S} r_{it}$.

Because the Nash equilibrium of each induced game of each strategic game $\Omega_{z_{t-1}^*}$ is independent of the stock z_{t-1}^*, we have $w(S, z_{t-1}^*) = \sum_{t=1}^{\infty} \beta^{t-1} w_t(S)$. Because $w_t(S) \leq \sum_{i \in S} r_{it}$, we have $w(S, z_{t-1}^*) \leq \sum_{t=1}^{\infty} \beta^{t-1} w_t(S) = \sum_{t=1}^{\infty} \beta^{t-1} \sum_{i \in S} [x_i^* - \frac{\pi_i}{1-\beta(1-\delta)}(\sum_{j \in N} e_j^* + (1-\delta)z_{t-1}^*)]$, by (7.17) above. This means the efficient consumption time profile $(x_1^*, \ldots, x_n^*; z_t^*)_{t=1}^{\infty}$ with $z_t^* = (1-\delta)z_{t-1}^* + \sum_{j \in N} e_j^*$ is a subgame-perfect cooperative agreement.

Notice that the above arguments are analogous to those in the proof of Theorem 7.6 in that they also exploit the fact that the damage functions are linear and, therefore, both the SPNE strategies and the efficient emission time profile in each subgame are independent of the GHG stock. However, because each subgame of the dynamic game with infinite horizon has exactly the same structure, unlike the dynamic game with finite horizon, the SPNE strategies and the efficient emissions are stationary, and exactly the same transfers in each period ensure a subgame-perfect cooperative agreement.

Subgame-Perfect Cooperative Agreements as Pollution Rights

Analogous to cooperative agreements in the strategic environmental game, every subgame-perfect cooperative agreement $(x_{1t}^*, \ldots, x_{nt}^*; z_t^*)_{t=1}^T$ can be interpreted as an intertemporal assignment of pollution rights that can be traded on a competitive international market. These pollution rights, denoted by $(e_{1t}^0, \ldots, e_{nt}^0)$, are defined, in view of equations (7.5), by the equations

$$(e_{it}^0 - e_{it}^*)p_t^* + g_i(e_{it}^*) = x_{it}^*, \quad i = 1, \ldots, n; \, t = 1, \ldots, T,$$

where $p_t^* = \sum_{\tau=t}^T [\beta(1-\delta)]^{\tau-t} \sum_{j \in N} v_j'(z_\tau^*)$. The proof for this is analogous to that of Proposition 5.10.

Additional Interpretations

The subgame-perfect cooperative agreements introduced and studied in this chapter are to be viewed as proposals that if put on the table will not be rejected in any period by any *coalition* of countries. This last requirement is important because the countries often react to proposed agreements on climate change as a group. For example, developing and developed countries often react to proposals for controlling climate change as separate groups. Thus, the approach in this chapter allows withdrawals from an agreement not only by single countries but also by coalitions of many countries. This leads to a more exclusive notion of a subgame-perfect cooperative agreement and more stringent sufficient conditions for its existence. The analysis also holds if withdrawals are restricted to single countries or if coalitions cannot write binding agreements. In particular, a subgame-perfect cooperative agreement would still be a subgame-perfect agreement. The chapter focused mostly on the finite time horizon version of the dynamic game but also discussed extensions to the infinite time horizon version of the model.

The analysis in this chapter, as in chapter 3, shows that if an agreement has to be efficient and acceptable to sovereign governments, then transfers between countries are necessary unless the countries are sufficiently similar, if not identical. But "front loading" transfers to some countries as a compensation for agreeing to reduce their emissions in the future may induce them to comply with the agreement in the beginning but provide them incentives to withdraw from it in a future period—when it is their turn to reduce emissions—and force renegotiations of the terms of the agreement. Thus, the time profile of transfers needs to be structured appropriately to guard against withdrawals or attempts to renegotiate the terms of the agreement in a future period.

By their appearances, the subgame-perfect cooperative agreements may look quite different from actual climate agreements, as neither the Kyoto Protocol nor the Paris Agreement requires significant direct transfers between the countries. But this is not really so. Because the Kyoto Protocol made provision for trade in emissions, it was equivalent to an agreement in terms of tradable emissions quotas that after (competitive) trade in emissions would have resulted in transfers (see, e.g., Chander 2003) that are similar to those in a cooperative agreement. Similarly, as shown earlier, a subgame-perfect cooperative agreement is generally equivalent to an agreement in terms of tradable emissions quotas that does not require direct transfers between the countries, but would nevertheless result in after-trade transfers that are exactly the same. Thus, a subgame-perfect cooperative agreement defined equivalently in terms of tradable quotas means that the time profile of tradable quotas should not be such that some country or coalition of countries will have incentive to comply with the agreement in the early periods but withdraw from it in a future period and force renegotiations of the terms of the agreement. Because the Paris Agreement also provides for trade in emissions, the intended nationally determined contributions (INDCs) chosen voluntarily by the signatories to the agreement can be interpreted as tradable emissions quotas. It is an open question whether some countries that have currently signed the Paris Agreement will have incentives to withdraw from it in a later period when it would be their turn to do deeper cuts in emissions.

Future research should address a number of simplifying assumptions that were made in order to highlight the strategic aspects of the climate change problem. We assumed no capital accumulation and no technological progress.[19]

7.5.3. Other Approaches to Cooperative Agreements in Dynamic Games

Rubio and Ulph (2007) study the impact of changes in the GHG stock on the internal-external stability of a coalition in a dynamic game.[20] As in the γ-core, they also assume that when a coalition forms, the remaining players follow their individually best reply strategies. But unlike the approach in this book, they assume symmetric countries. There is also a fairly small but important literature on dynamic games for efficient public good provision including the papers by Marx and Matthews (2000) and Harstad (2012).[21] Like Dockner et al. (1996), Marx and Matthews (2000) make assumptions that are equivalent to assuming constant and identical marginal abatement costs, and thus they too face the problem of choosing among alternative MPEs.[22]

---

19. See Hong and Karp (2012), Dutta and Radner (2004), and Harstad (2012) for static and dynamic models of technical progress without transfers between countries.

20. Recall from chapter 6 that a coalition is internally stable if no member of the coalition can be better off by leaving the coalition (assuming that the coalition left behind will remain stable) and externally stable if no nonmember can be better off by joining the coalition (assuming that the expanded coalition will be stable).

21. Also, see Chander (1993) for a differential game model of public good provision.

22. Marx and Matthew (2000, p. 348) propose extending their analysis to the case in which the players are more heterogeneous in future research.

Marx and Matthews (2000) prove existence of efficient Bayesian equilibria sustained by trigger strategies that impose maximal possible punishment. This punishment, as noted earlier, is conceptually harsher than adopting individually best reply strategies assumed in this chapter. In contrast, Harstad (2012) studies incentives to invest in green technologies that aid in reducing emissions. The countries face no threat of punishment and can write contracts that commit them to a time profile of emissions. Though innovative and insightful, Harstad for reasons of tractability abstracts from the fact of highly asymmetric countries. As a result, both international trade in emissions and transfers are missing from his analysis. He also does not consider coalitional behavior and the possibility of a country opting out of an agreement.

In contrast to the dynamic model in this chapter, Dockner et al. (1996) assume linear production and nonlinear damage functions and two countries. As noted in the context of Theorem 7.6 above, they also find that the degree of nonlinearity of both production and damage functions introduce significant complexities: (1) Unlike linear damage functions, marginal damage from emissions is not independent of the GHG stock if the damage functions are nonlinear, and therefore each country's emissions cannot be treated in isolation; and (2) if the production functions are nonlinear, the marginal abatement costs, unlike linear production functions, are not constant and independent of the time profile and amount of emissions, and therefore the costs and benefits of emitting according to the efficient emissions time profile cannot be balanced simply by reallocating the emissions.

Noncooperative as well as efficient solutions in dynamic games have been studied previously in de Zeeuw and van der Ploeg (1992), Nordhaus and Yang (1996), Dutta and Radner (2004, 2009), and Dockner and Long (1993), among others. These contributions characterize the Nash equilibrium of the dynamic game as well as efficient solutions. The paper by Dockner and Long (1993) considers a model with two (identical) agents. The paper by Dutta and Radner (2009) finds conditions under which an efficient solution can be supported as a history-dependent subgame-perfect equilibrium through the use of a trigger strategy. They assume that once a country cheats on the agreement by over-emitting, punishment begins and continues forever. Apart from the usual concern that trigger strategies are not renegotiation proof, they allow deviations by only single players. In fact, none of above-mentioned studies considers robustness of the efficient solution against coalitional deviations or transfers that have the property of subgame perfection.

Germain et al. (2003) study a dynamic model of climate change and discuss a cooperative solution for which they claim to prove analytically a "core property," but the solution discussed by them may not even exist let alone satisfy subgame perfection despite the fact that it is the outcome of what they call backward induction. More specifically, it may not be stable against the following strategy of a coalition: the coalition decides to not cooperate with the remaining players in *any* future period and thereby guarantees itself a payoff that is equal to its subgame-perfect Nash equilibrium payoff in the game induced by the coalition. Moreover, Germain et al. (2003) do not offer a formal definition of their solution concept but propose an algorithm to find it.

Though the method of backward induction is known to lead to a subgame-perfect equilibrium outcome in noncooperative dynamic games, the same has not been demonstrated for solutions in which players may cooperate and form the grand coalition and share the resulting surplus at each stage of the game. Thus, just applying a kind of backward induction does not imply that the resulting outcome is subgame perfect. In fact, the method of backward induction, applied in Germain et al. (2003), amounts to modifying the payoff *functions* of the players at each stage of the dynamic game, and the solution of the so-defined game may not be a meaningful solution of the original game. In terms of the model in this chapter, this is seen as follows.

Begin with the γ-coalitional function $w(S, z_{T-1}), S \subset N, z_{T-1} > 0$, of the subgame $\Gamma_{z_{T-1}}$ in the last period T. This coalitional function is well defined and admits a nonempty core, as every subgame in the last period satisfies the usual assumptions of a strategic environmental game. Thus, there exist nonnegative distribution parameters $\theta_{iT}(z_{T-1}), i = 1, \ldots, n$, such that $\sum_{i \in N} \theta_{iT}(z_{T-1}) = 1$ and $\sum_{i \in S} \theta_{iT}(z_{T-1}) w(N, z_{T-1}) \geq w(S, z_{T-1})$, $S \subset N$. Now consider a reduced form game in period $T-1$ in which the payoff *function* of each player i is defined as

$$g_i(e_{iT-1}) - v_i(z_{T-2} + \sum_{i=1}^{n} e_{iT-1}) + \theta_{iT}(z_{T-2} + \sum_{i=1}^{n} e_{iT-1}) \times w(N, z_{T-2} + \sum_{i=1}^{n} e_{iT-1}). \tag{7.18}$$

(Here, $z_{T-1} = z_{T-2} + \sum_{i \in N} e_{iT-1}$.) However, the strategic game with these payoff functions is not a reduced form of the subgame in period $T-1$ in the sense in which the term is used in noncooperative game theory. This definition of players' payoff functions suffers from two serious problems: First, these payoff functions may not be concave, and thus the γ-core of the reduced game may be empty.[23] Germain et al. (2003) seem to be aware of this and thus comment (in section 5 of their paper), "An original algorithm is needed because in the backward induction with the complex transfer parameters, the problem is not analytically tractable." Second, the reduced-form game with these modified payoff functions, called the "rational expectation" game in Germain et al. (2003), is not a reduced form of the subgame $\Gamma_{z_{T-2}}$ in the sense in which the term is used in noncooperative game theory, as it is defined by modifying the *original payoff functions* of the players. The payoff function of player i in the reduced form of the subgame $\Gamma_{z_{T-2}}$, as in the proof of Theorem 7.3, is instead given by

$$g_i(e_{iT-1}) - v_i(z_{T-2} + \textstyle\sum_{i \in N} e_{iT-1}) + q_i(z_{T-2} + \textstyle\sum_{i \in N} e_{iT-1}),$$

where $q_i(z_{T-2} + \sum_{i \in N} e_{iT-1})$ is the SPNE payoff of player i in the subgame $\Gamma_{z_{T-1}}$, $z_{T-1} = z_{t-2} + \sum_{i \in N} e_{iT-1}$ in period T and, as shown, is concave with $q_i''' = 0$.[24] To conclude, the backward solution in Germain et al. (2003) may not lead to a solution of the dynamic game because the reduced form as per (7.16) may not admit a nonempty γ-core, and even if it does it may not belong to the core of the coalition game $w(S, z_{T-2})$, $S \subset N$. To put it differently, the method of backward induction in Germain et al. (2003) may not converge to a solution of the dynamic game, and even if it does the so-computed solution may not be stable against coalitional deviations, especially against deviations that may last forever. Because the distortion of the players' payoff functions of the original game gets compounded as the backward induction works its way backwards, one cannot even be sure how the so-computed solution is related to the original dynamic game.

---

23. Though $w(N, z_{T-2} + \sum_{i \in N} e_{iT-1})$, as Theorem 7.5 shows, is a concave function of $\sum_{i \in N} e_{iT-1}$, the distribution parameters $\theta_{iT}(z_{T-2} + \sum_{i \in N} e_{iT-1})$ are typically not concave.

24. Recall from footnote 17 that $q_i(z_{T-2} + \sum_{i \in N} e_{iT-1}) = w(\{i\}, z_{T-2} + \sum_{i \in N} e_{iT-1})$.

In contrast to the approach in this chapter, Filar and Petrosjan (2000) consider a sequence of superadditive coalitional games as the primitive concept.[25] Thus, they abstract away from interesting strategic interactions that are behind the definition of coalitional payoffs in this chapter. Their main concern is how the relative power of the coalitions may change over time and not with subgame perfection of the solution. The dynamics in their models comes from their assumption that the coalitional game in each period depends on an *exogenously* given rule and on the solution of the coalitional game in the previous period, whereas in my formulation the coalitional game in each period depends only on an endogenously determined state of the system, and its solution does not depend on the coalitional game in the preceding period. Similarly, Kranich, Perea, and Peters (2005) consider a sequence of coalitional games as the primitive concept. In their framework, a coalitional game in each period is given *exogenously* and does not depend on the state of the system or on the solution of the game in the previous period. Despite these differences, Filar and Petrosjan (2000), Kranich et al. (2005), Becker and Chakrabarti (1995), and this chapter have one thing in common in that they all assume that if a coalition deviates in some period, then it deviates forever.

---

25. Recall from chapter 5 that the γ-coalitional function may not be even superadditive.

8

LIMITS TO CLIMATE CHANGE

I n the preceding chapters, we studied both static and dynamic models of climate change in detail. But two important issues were deliberately left out: first, the role of technology, which has become increasingly important in the international negotiations and efforts to tackle climate change; second, a long-term view of climate change, which has also become important since nearly 200 countries committed—via the Paris Agreement in 2015—to restrict temperature rise due to climate change to "well below" 2°C above pre-industrial levels by the end of the century. Though a more ambitious goal of limiting global warming to 1.5°C, proposed by the small island states and the least developed small countries, is also mentioned in the Paris Agreement, but either target gives climate change negotiations a long-run perspective. This chapter focuses on both of these important issues.

It was assumed in the preceding chapter that the production functions g_i and the damage functions v_i do not change over time. As far as the damage functions are concerned, this does not seem to be an unreasonable assumption, as damage from climate change mainly depends on such invariants as location, geography, and size of a country. Thus, in what follows, I will continue to assume invariance of the damage functions. But production involves inputs other than fossil fuels, such as capital and labor. These and production technology were implicitly assumed to be fixed in the models in the previous chapters. Such an assumption makes sense in the short-run models but does not in a long-run model because as climate change becomes stronger, each country may adopt less polluting production technologies. Furthermore, each country may over time become sufficiently abundant in capital and labor, and these may no longer be the limiting factors. Therefore, a model in which all countries have an identical production technology with constant returns to scale seems to be more appropriate for a long-run analysis of climate change.

I begin the long-run analysis of climate change by first considering whether the countries will have incentives to accept some long-run pollution rights according to some normative criterion. However, different normative criteria to allocate pollution rights have different welfare implications. For example, it has been proposed to assign pollution rights on a per capita basis (favoring countries with large populations such as India and China) in contrast to grandfathering the current emissions (favoring the industrialized countries such as the United States, Japan, and those of the European Union). Because the welfare implications of these normative criteria are different for different countries, there may be no consensus among them on the normative criterion to be followed for assigning pollution rights.

Besides the fact that there may be no agreement on the normative criterion to allocate pollution rights in general, an additional issue is time consistency. For instance, if the allocation of pollution rights is to be made on a per capita basis, then it is to be relative to which population? Is it to be on the basis of population that each country had in the year 1950 or in 2000 or what it will be in 2100? Because the population of a country may change over time, pollution rights on a per capita basis will have to be revised every few years—even for countries whose population may not have changed—if they are to be time consistent. Similarly, if allocation of rights is to be on the basis of grandfathering of emissions, then should that be on the basis of emissions of the year 1992 when some countries were far less industrialized than others or of the year 2030 when some developing countries are likely to be significantly more industrialized in terms of production and emissions? Developing countries, especially India and China, have indeed argued against the principle of grandfathering of current emissions on the ground that they are not yet as industrialized as the developed countries and therefore they should not be required to limit their emissions until they have reached comparable levels of industrialization. Though the polluter-pays or the pollutee-pays principles—which were discussed in chapter 3—do not suffer from time inconstancy, there is no agreement among the countries about which one of the two principles to apply in the case of climate change, as they too have different welfare implications.

In this chapter, I propose long-run pollution rights that are not only time consistent but also free from the above-discussed objections. Specifically, the proposed rights are equal to the emissions that the countries themselves will choose in a steady-state equilibrium if they did not have binding constraints on their capacities to produce and pollute. But a country may still choose to produce only limited quantities of the consumption goods, as producing them generates emissions that add to the already high greenhouse gas (GHG) stock and, thus, cause additional damage not only to others but also to the country itself. To be precise, I consider an idealized world in which all countries have industrialized so much that their capacities to produce and pollute are no longer the binding constraints, and climate change is the only limiting factor. However, the countries differ in terms of their damage functions, which, as noted above, are taken to be fixed even in the long run.[1] If the countries choose to produce and pollute differently, it is not because of constraints on their capacities to produce and pollute, but because of how much they are affected by climate change.

Such an idealized world corresponds to a long-run model in which all countries would have industrialized so much that climate change is the only binding constraint on production in the sense that individual benefits to a country from additional production, which leads to higher emissions and contribution to the GHG stock, would be less than the additional damage the country itself will suffer. Such a long-run model roughly represents a dismal end-state in which the GHG stock will be so high that no country can be better off by increasing production that generates emissions and raises the already high GHG stock. Though such a dismal end-state may never be reached,[2] the corresponding payoffs and emissions can be strategically used by the countries as their reference

---

1. In contrast, we studied, in section 5.4.2, a model in which the damage functions are identical, but not necessarily the production functions.

2. In fact, climate change policy must aim at preventing the occurrence of such a dismal end-state.

emissions for negotiating the terms of cooperation to control climate change, as that is what their emissions and payoffs would eventually be in the absence of any cooperation between them.

Incidentally, such a long-run view of climate change seems to reconcile the debate between the two fields of environmental economics and ecological economics, a reference to which was made in chapter 1. According to environmental economics, the value of the environment to society derives from the individual values held by human members of society. However, ecological economics takes a more biophysical view of value and often measures it in terms of embodied energy content. According to this view, consumer preferences have no role, and a similar good produced by using less energy is more valuable. Thus, public policy should aim at minimizing the energy content of goods and services produced. The economists' argument against this energy theory of value is that besides energy, there are many other resources in short supply, including capital and labor. Reducing the value of a good to the embodied content of any single factor is an oversimplification.[3] However, if climate change becomes the only limiting factor, then the two theories converge, as then it is indeed optimal to minimize the fossil energy (which causes pollution and climate change) content of goods produced and consumed.

As in chapter 7, I consider a dynamic game formulation of climate change, but in continuous time and with an infinite time horizon.[4] As before, the players (i.e., the countries) in this game choose their strategies so as to maximize their discounted payoffs over the entire time horizon. I define their steady-state Nash equilibrium emissions as their long-run pollution rights.[5] The so-defined pollution rights are both time consistent and free from the conventional normative considerations. They are also self-enforcing, as they are equal to the emissions that the countries

---

3. The same as in the labor theory of value.
4. Similar continuous time models have been considered by Hoel (1992) and Dockner and Long (1993). Related models appear in Tahvonen (1994) and Mäler and de Zeeuw (1998). A classic reference on differential games is Basar and Olsder (1995).
5. Thus, two different types of equilibriums are involved at the same time: one is game theoretic in the sense of a Nash equilibrium, and the other is dynamic in the sense of a steady-state equilibrium.

themselves will choose in a long-run equilibrium if the constraints on the capacity to produce and pollute were nonbinding.[6] Such rights therefore may be used by the countries as their reference emissions in negotiating the terms of cooperation for controlling climate change.

However, assigning pollution rights is not an end in itself. A further question is whether assigning such rights would lead to an efficient out-come through Coasian bargaining. That is, whether the assignment of rights and Coasian bargaining among the countries will lead to coop-eration and socially optimal control of climate change. In chapter 5, this issue was resolved by explicitly stating the payoffs that are possible for the players from both cooperation and noncooperation. In this chapter, I assume that the countries take a long-term view of climate change and are concerned with their long-run payoffs from cooperation vis-à-vis nonco-operation. In theoretical terms, this translates into the countries compar-ing their payoffs in the cooperative and noncooperative steady states.

8.1. THE MODEL

Time is continuous. Each country i produces a composite consumption good y_i with a constant returns-to-scale technology. Production of a unit of the consumption good results in an amount of pollution e_i given by the emission-production relationship

$$y_i = \sigma e_i. \tag{8.1}$$

The "pollution factor" $\sigma > 0$ indicates how much pollution is produced if country i produces an amount y_i. It is the inverse of "pollution intensity" of production and depends on both energy efficiency of production using fos-sil fuels and the proportion of green energy in the country's energy mix. A higher σ means a cleaner technology overall. The linearity of the emission-production relationship and absence of any other constraints on production reflects the assumption that in the long run, other factors of production can be changed if required and are not a binding constraint. Though I assume

---

6. See Barrett (1994a) for the notion of a self-enforcing international agreement.

all countries to have the same production technology, the analysis below also holds if their technologies differ or become cleaner over time.

The current emissions of the countries $(e_1(t), \ldots, e_n(t))$ add to the current GHG stock $z(t)$ according to the following differential equation:

$$\dot{z}(t) = \sum_{i \in N} e_i(t) - \delta z(t), 1 > \delta > 0 \text{ and } z(0) = z_0 > 0 \text{ given.}^{[7]} \quad (8.2)$$

This equation is exactly the same as the stock accumulation equation in the preceding chapter except that time is continuous. It implies that the steady-state emissions and GHG stock are related by the equation

$$\sum_{i \in N} e_i = \delta z. \quad (8.3)$$

Thus, in a steady state, the total emissions of all countries are equal to the amount of natural decay of the GHG stock. In other words, a higher steady-state GHG stock can sustain higher total steady-state emissions and, conversely, higher total steady-state emissions imply a higher steady-state GHG stock. In fact, whatever is the initial GHG stock z_0, if each country i emits at a constant rate e_i, the GHG stock will eventually become equal to z such that $\delta z = \sum_{i \in N} e_i$. In particular, if each country i emits at a constant rate e_i and the current GHG stock z_0 is such that $\delta z_0 < \sum_{i \in N} e_i$, then the GHG stock will continue to rise and eventually become stable at level z such that $\delta z = \sum_{i \in N} e_i$.

Each country i suffers damage both from its emissions and from climate change due to the GHG stock. Damage from the *emissions* of a country are local and given by $h_i(e_i)$ (i.e., the damage depends only on country i's own emissions).[8] I assume that $h_i'(e_i) > 0, h_i''(e_i) > 0$, and $h_i'(0) = 0$ (i.e., the "local" damage functions are increasing and strictly convex). As will be seen, the strict convexity of the local damage functions implies a unique solution for the system of equations below even if damage from climate change is linear.

---

7. Several studies use this specification for the GHG stock transition (Dutta and Radner 2004, 2009; Karp and Zhang 2005; Nordhaus and Yang 1996).

8. For example, burning coal to produce electricity not only adds to the stock of CO_2 but also causes air pollution and damage locally.

Let $g_i(e_i) \equiv \sigma e_i - h_i(e_i)$. Then, $g_i(e_i)$ is country i's production net of local damage from its own emissions. It is strictly concave, as $h_i(e_i)$ is strictly convex. In contrast, damage from *climate change* is global and given for each country i by $v_i(z), i \in N$, with $v_i'(z) > 0$, $v_i''(z) \geq 0$, and $v_i'(0) = 0$. In words, the "national" damage functions are increasing and convex. The countries may differ in terms of their national damage functions, as they may be affected differently by climate change because of differences in their location, area, and geography. As already noted, the fact that the countries may be affected differently by climate change is central to the problem.

8.1.1. The Socially Optimal Steady State

Because global warming is a function of the GHG stock z, the goal of restricting climate change, say to +2°C for the sake of concreteness, is effectively equivalent to the goal of stabilizing the GHG stock at a specific level, say $\hat{z}$, which is determined by the ecological phenomenon independently not only of the current stock z_0 but also of how, when, and who may contribute to it.[9] The implicit goal of stabilizing the GHG stock at $\hat{z}$ has been proposed by climate scientists and others because damage from climate change of more than +2°C would be devastating for many countries. But from an economics perspective, we must ask whether limiting global warming to +2°C is socially optimal. I now identify conditions under which it is. For simplicity, I assume that all countries have the same constant discount rate r.[10] A socially optimal, stable GHG stock is then a steady-state solution of the following optimal control problem:

$$\max_{e_i(t), i \in N} \int_0^\infty e^{-rt} \sum_{i \in N} \left[\sigma e_i(t) - h_i(e_i(t)) - v_i(z(t)) \right] dt,$$

subject to

$$\dot{z}(t) = \sum_{i \in N} e_i(t) - \delta z(t) \qquad (z(0) = z_0).$$

9. It has been estimated that the GHG concentration in the atmosphere of no more than 550 ppm can limit global warming to 2°C.

10. Extension to hyperbolic discounting, as in Karp (2005), would be interesting but is beyond the scope of this chapter.

Because this is a single-state variable optimal control problem, the current value Hamiltonian (after suppressing the time argument) implies that the optimum is characterized by the equations

$$\sigma - h_i'(e_i) + \lambda = 0 \qquad (i = 1, 2, \ldots, n)$$

$$\dot{\lambda} = (r + \delta)\lambda + \sum_{i \in N} v_i'(z),$$

and the boundary condition $\lim_{t \to \infty} \lambda(t)z(t)e^{-rt} = 0$, where λ is the current-value adjoint variable. After differentiation and substitution, I obtain

$$\dot{e}_i = \frac{(r + \delta)\left(h_i'(e_i) - \sigma\right) + \sum_{i \in N} v_i'(z)}{h_i''(e_i)} \qquad (i = 1, 2, \ldots, n).$$

Thus, the socially optimal steady-state emissions $(e_1^*, \ldots, e_n^*)$ and GHG stock z^* must satisfy

$$(r+\delta)\left(\sigma - h_i'(e_i^*)\right) = \sum_{i \in N} v_i'(z^*) \quad \text{and} \quad \sum_{i \in N} e_i^* = \delta z^*, \, i = 1, 2, \ldots, n. \quad (8.4)$$

Conditions (8.4) characterize the socially optimal steady-state GHG stock and emissions and are the dynamic counterpart of the familiar Lindahl-Samuelson condition for optimal public good provision.[11] The steady-state emissions $(e_1^*, \ldots, e_n^*)$ are what Geden (2016) calls the actionable climate targets. Because each h_i is strictly convex and each v_i is convex, these equations admit a unique solution, which depends on the magnitude of the pollution factor σ. In fact, they imply that the higher the pollution factor σ, the higher the socially optimal steady-state emissions $(e_1^*, \ldots, e_n^*)$ and the GHG stock z^*. This is because a higher σ implies a higher abatement cost for every country with a nonbinding constraint on its production capacity. In other words, the GHG stock $\hat{z}$ for stabilizing global warming at +2°C is socially optimal (i.e., $\hat{z} = z^*$) only for a specific value of the technology parameter σ. But because (as many scientific studies show) the *world* damage from climate change rises very steeply once the GHG stock surpasses

---

11. Thus, $\frac{1}{r+\delta}\sum_{i \in N} v_i'(z^*)$ is the long-run "social value" of the GHG stock.

the implicitly defined threshold $\hat{z}$, it follows that $\hat{z}$ is also "nearly" socially optimal for all higher values of the pollution factor σ. Put another way, once the GHG stock crosses the threshold $\hat{z}$, it will no longer be possible for the world to achieve a higher social welfare unless it adopts a cleaner technology that can produce higher amounts of the consumption good without increasing the emissions and the GHG stock. Thus, stabilizing climate change at +2°C, if socially optimal, implies that a higher output of the consumption good may not lead to a higher social welfare unless the higher output is produced with a cleaner technology such that the emissions and the GHG stock are not higher. In other words, the world output of the consumption good, once the threshold $\hat{z}$ is reached, may stagnate unless there is continuous innovation and adoption of cleaner technologies leading eventually to innovation and adoption of "carbon-free" technologies.

The difference between the base-year GHG stock z_0 and $\hat{z}$ determines what has come to be known as the "carbon budget": it is the amount of GHGs that can be emitted such that the GHG stock will not exceed $\hat{z}$ by a specified date. As per the Fifth Assessment Report (AR5) of the IPCC (2013), for an even chance of limiting temperature rise to 2°C by the end of the century, the total amount of GHGs that can be emitted between the base year of 2011 until the intended year 2100 is the equivalent of 900 gigatons (Gt) of CO_2. The *Synthesis Report on the Aggregate Effect of Intended Nationally Determined Contributions (INDCs)* by the United Nations Framework Convention on Climate Change (UNFCCC) estimates that the CO_2 equivalent emissions for 2011–2025 will aggregate to 542 Gt and for 2011–2030 to 748 Gt, which leaves a meager carbon budget of 122 Gt of CO_2 equivalent for the world for the remaining 70 years from 2030 to 2100. Therefore, unless there are great scientific discoveries and adoption of significantly cleaner technologies, the world faces the prospectus of an economic collapse and hardship either because of devastation due to climate change or because of lower production of the consumption goods to avoid the devastation.[12]

---

12. However, it may be noted that the scientific evidence regarding the impact of climate change of more than 2°C is not exactly known except that it is likely to be severe and in a range that is best avoided. Similarly, the cost of adopting cleaner technologies is falling and may turn out to be not as high as appears to be the case at present.

However, the negotiating countries in Paris did not even try to agree on how the so-defined carbon budget of 900 Gt may be shared by them. This is not surprising because the problem of distributing the carbon budget is similar to that of splitting a pie of fixed size among agents. It was noted earlier why no normative criterion for distributing a fixed amount of a valuable resource may be acceptable to all countries. The same can be said about noncooperative game forms or mechanisms designed to obtain a split of a pie of fixed size as their equilibrium outcomes. Because the outcomes of different game forms or mechanisms may differ, no game form or mechanism may be acceptable to *all* agents, which in this case are sovereign governments. Thus, there does not seem to be much difference between normative and noncooperative game theory approaches as far as applications to the problem of climate change are concerned.

It may be noted that besides the size of the carbon budget, the time profile of emissions also matters. To see why, figure 8.1 illustrates two alternative trajectories for the GHG stock, both of which lead to the threshold level $\hat{z}$ in the year 2100.

The upper trajectory implies implicitly that immediately before the stock rises to $\hat{z}$, the world may emit at a rate that is just a little above the rate of natural decay of the GHG stock, that is, $\sum_{i \in N} e_i(t) \cong \delta \hat{z}$ for t close to 2100, and then become equal to it. But that is not so in the case of the lower trajectory. In both cases the stock rises to $\hat{z}$ in the same year, but in the case of the lower trajectory, unless there is a sudden reduction in the rate of world emissions, the stock may continue to rise beyond $\hat{z}$. The point is that if the GHG stock is to be stabilized at $\hat{z}$ without risking a sudden economic collapse, then the countries need to negotiate not only their shares of the carbon budget but also the time profiles of their emissions. At the minimum, they need to agree on the years in which each country's emission rates will peak, then decline, and eventually stabilize at the steady-state emission rates $(e_1^*, \ldots, e_n^*)$.[13] Thus, implementation of the Paris Agreement

---

13. Equations (8.4) show that the steady-state emissions $(e_1^*, \ldots, e_n^*)$ (i.e., the actionable emission targets) depend on the implicit goal for the level at which the GHG stock is to be stabilized. In fact, the more ambitious the goal for limiting temperature rise and thus the GHG stock, the more ambitious the goal for the aggregate actionable emission target.

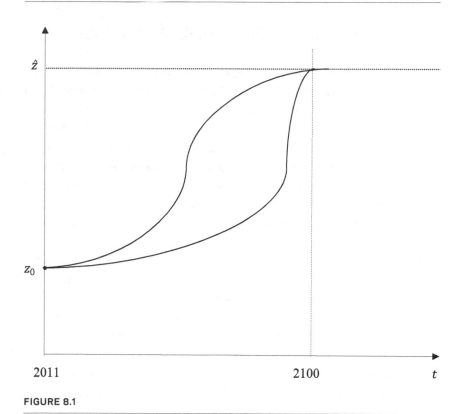

FIGURE 8.1

Alternative trajectories for stabilizing the GHG stock at the targeted level $\hat{z}$ by the end of the century

would require the countries to also agree on clearly defined time profiles of their emissions that are consistent with the stated goal of limiting climate change to +2°C without risking a sudden economic collapse.

However, the carbon budget and its distribution are not the only important implications of stabilizing climate change at +2°C. An equally important implication, which has not been given sufficient attention, is the rate at which the countries may emit GHGs as the stock nears or rises to $\hat{z}$. In fact, if $\hat{z}$ is socially optimal (i.e., $\hat{z} = z^*$), then, as seen from the stock accumulation equation (8.3) and conditions (8.4), each country i must emit at a constant rate e_i^* once the GHG stock nears the threshold $\hat{z}$, and the threshold is not to be surpassed. Will the countries have incentives to emit at these constant rates?

8.2. POLLUTION RIGHTS AND CLIMATE CHANGE

Though all 196 countries present at the Conference of the Parties 21 (COP 21) in Paris agreed to the long-term goal of limiting global warming to 2°C or less, it has been left to each country to voluntarily set its own emissions and mitigation targets, which are to be revised and reviewed every 5 to 10 years. To date, 148 countries have made voluntary pledges (i.e., submitted INDCs) to the UNFCCC for cutting GHG emissions and undertaking mitigation actions. However, the actual pledges made have come under criticism because, as noted earlier, they are not sufficient for restricting climate change to +2°C. Figure 8.2 shows the temperature rise under various scenarios.

However, it is not surprising that the voluntarily submitted INDCs are not sufficient. Otherwise, it would have contradicted both theoretical and experimental evidence concerning voluntary provision of public goods. There is a vast literature (see, e.g., Bergstrom, Blume, and Varian 1986; Palfrey and Prisbrey 1997) that shows that voluntary contributions to a public good may fall well short of the optimal provision.

This raises the question: As the GHG stock rises and nears the socially optimal stock z^*, will the countries with a long-run perspective have incentives to agree to emit at the rates $(e_1^*, \ldots, e_n^*)$ that are necessary for stabilizing the stock at z^*? I first characterize the INDCs that the countries with

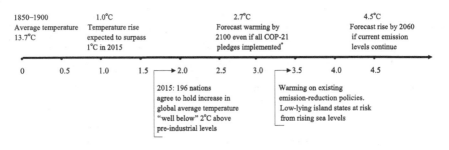

FIGURE 8.2

Temperature rise implied by alternative climate policies

Source: UK Metrology Office Hadley Centre, Climate Action Tracker

Source: United Nations *Synthesis Report on the Aggregate Effect of Intended Nationally Determined Contributions (INDCs)*

a long-run perspective of climate change may voluntarily choose, given that, by then, they all would have developed so much that the capacity to produce and pollute would no longer be a binding constraint for any one of them. I show that bargaining among the countries in which each country may trade off reductions in its own INDCs against the reductions in INDCs of other countries may indeed induce them to emit at the rates $(e_1^*,\ldots,e_n^*)$.

8.2.1. The Steady-State Nash Equilibrium

The above optimistic view comes from the fact that INDCs are in effect self-enforcing pollution rights, as they are voluntarily chosen by each country after much deliberation to serve its own national interests, keeping in view the INDCs of the other countries.[14] I show that if so interpreted, then, as seen in section 5.4.1, the INDCs can—after bargaining and mutually beneficial trades among the countries—induce the countries with a long-run perspective to indeed emit at the socially optimal steady-state rates $(e_1^*,\ldots,e_n^*)$.

I interpret the steady-state Nash equilibrium emissions in an appropriately defined differential game as the INDCs that each country with a long-run perspective will choose voluntarily. In this differential game, the strategy of each country i is a time function $e_i(t), t \geq 0$, that maximizes its individual payoff given the strategies of the other countries and evolution of the GHG stock. Thus, in the framework of the dynamic model stated above, each country i may choose a function $e_i(t), t \geq 0$, that maximizes

$$\int_0^\infty e^{-rt}\big[y_i(t)-h_i\big(e_i(t)\big)-v_i\big(z(t)\big)\big]dt = \int_0^\infty e^{-rt}\big[\sigma e_i(t)-h_i\big(e_i(t)\big)-v_i\big(z(t)\big)\big]dt$$

subject to

$$\dot{z}(t) = \sum_{i \in N} e_i(t) - \delta z(t), z(0) \text{ given.} \qquad (8.5)$$

---

14. The then Indian Minister for Environment and Forests, Prakash Javadekar, was reported to have commented during the course of negotiations in Paris that the INDCs were an innovative idea and were very helpful in facilitating the negotiations.

The motivation for describing the open-loop rather than closed-loop Nash equilibrium strategies as the INDCs chosen by the countries comes from the fact that the Paris Agreement requires each country to commit to its INDCs for a period of at least 5 years independently of how the GHG stock may evolve during the 5-year period. This will be required of the INDCs also for the last five years of the twenty-first century when the GHG stock is to be stabilized at a predetermined level. Thus, nearer to a steady state, it is reasonable to interpret the open-loop Nash equilibrium emissions $e_i(t)$ as the pollution right or INDCs claimed by country i at time t.

With this in mind, I note that because the maximization problem of each country as described above is also a one-state variable optimal control problem, the current-value Hamiltonian (after suppressing the time argument) is

$$H(e_1, e_2, \ldots e_n, z, \lambda_i) = \sigma e_i - h_i(e_i) - v_i(z) + \lambda_i \left(\sum_{j \in N} e_j - \delta z \right),$$

where λ_i is the current-value adjoint variable. Thus, the solution to this differential game is characterized by constraint (8.5), the equations

$$\frac{\partial H}{\partial e_i} = 0 = \sigma - h_i'(e_i) + \lambda_i \qquad (8.6)$$

$$\dot{\lambda}_i = r\lambda_i - \frac{\partial H}{\partial z} = (r + \delta)\lambda_i + v_i'(z), \qquad (8.7)$$

and the boundary condition $\lim_{t \to \infty} \lambda_i(t)z(t)e^{-rt} = 0, i = 1, 2, \ldots, n$. Differentiating (8.6) and substituting,

$$\dot{e}_i = \frac{\dot{\lambda}_i}{h_i''(e_i)} = \frac{(r + \delta)(h_i'(e_i) - \sigma) + v_i'(z)}{h_i''(e_i)}, i = 1, 2, \ldots, n. \qquad (8.8)$$

Equations (8.5) and (8.8) are differential equations in z and $e_1, \ldots, e_n$. The steady state of the above system of differential equations and the boundary condition describe the emissions that each country acting individually will choose in the long run. Because a steady state is described by $\dot{z} = \dot{e}_i = 0$, the steady-state Nash equilibrium emissions of country i, to be denoted by $\bar{e}_i$, and the steady-state GHG stock, to be denoted by $\bar{z}$, in view of (8.5) and (8.8), are given by the solution to the equations

$$(r+\delta)(\sigma - h'_i(\bar{e}_i)) = v'_i(\bar{z}), \; i = 1,\ldots,n, \text{ and } \sum_{i \in N} \bar{e}_i = \delta\bar{z}. \qquad (8.9)$$

By convexity of the damage functions $v_i(z)$ and strict concavity of the net production functions $\sigma e_i - h_i(e_i)$, $i \in N$, it is easily verified that (8.9) admit a unique solution $\bar{z}, \bar{e}_i, i \in N$, which depends on the pollution factor σ.

8.2.2. The Pollution Rights

Because the emissions of each country i would eventually converge to $\bar{e}_i$, it cannot be prevented from emitting in the long run at the constant rate $\bar{e}_i$. Thus, I interpret $\bar{e}_i$ as the long-run pollution right of country i that it can eventually assert without the agreement of others.

The steady-state Nash equilibrium emissions interpreted as pollution rights are self-enforcing and free from any normative considerations such as per capita allowances proposed by developing countries or grandfathering of current emissions proposed by the industrialized countries. Because each country is assumed to have access to the same technology and faces no capacity constraints to produce and pollute, the so-defined pollution rights are also independent of any differences in the countries' stages of industrialization or development. However, as seen from (8.9), they could be different for different countries, unless both local and national damage functions are the same for the countries. In fact, if a country is more affected by its own emissions and climate change, then it will be able to claim fewer pollution rights.

To be precise, a country i is *more affected* by its own emissions and climate change than country j if $h'_i(e_i) \geq h'_j(e_j)$ for all $e_i = e_j \geq 0$ and $v'_i(z) \geq v'_j(z)$ for all $z \geq 0$ (i.e., if both local and national marginal damages of country i are uniformly higher). Because both damage functions of each country are convex, if country i is more affected by its own emissions and climate change than country j, then its damage functions h_i and v_i are uniformly higher than the damage functions h_j and v_j of country j. The notion of "more affected" introduces a partial ordering on the set of countries. Equations (8.9) imply $\bar{e}_i \leq \bar{e}_j$ if country i is more affected by its own emissions and climate change than country j. This implies that a country that is more affected by its own emissions and climate change will be able to claim fewer rights, even if it faces no binding capacity constraint to produce and pollute. This seems to suggest that the so-defined pollution rights, though self-enforcing, are more in line with the victim-pays principle. This is not surprising because in the context of climate change, no country really has the right not to be polluted.

The steady-state Nash equilibrium emissions as pollution rights reflect the bargaining power of each country. They can be interpreted as a measure of each country's ultimate capacity to exploit Earth's atmosphere. This capacity is not absolute, as it depends on the capacity of other countries to similarly exploit Earth's atmosphere, and it is lower for a country that is affected relatively more by its own emissions and climate change. In reality, these rights may never be exercised. In fact, the aim of international climate agreements is precisely to prevent the situation in which it becomes optimal for each country to exercise them.

Stability of the Steady State

Interpretation of the steady-state Nash equilibrium emissions as pollution rights becomes even more compelling if the steady state is stable. I identify two sets of conditions under which it is indeed stable. If there are only two countries (i.e., $n = 2$), we have three differential equations [i.e., (8.5) and (8.8)]. Thus, as shown in Long (1992), the Jacobian matrix, after ignoring the terms involving third-order derivatives, is given by

$$
\begin{bmatrix}
-\delta & 1 & 1 \\
\dfrac{v_1''(z)}{h_1''(e_1)} & (r+\delta) & 0 \\
\dfrac{v_2''(z)}{h_2''(e_2)} & 0 & (r+\delta)
\end{bmatrix}.
$$

Because, under my assumptions, the last two terms in the first column of this matrix are nonnegative, the trace of this matrix is positive and the determinant is negative. Hence, for $n = 2$, there exists a negative eigenvalue that implies saddle-point stability of the steady state. If $n > 2$, the Jacobian matrix satisfies the same properties if the damage functions $v_i, i = 1, \ldots, n$, are linear, as then each $v_i''(z) = 0$. Hence, the steady-state Nash equilibrium is saddle-point stable if the damage functions are linear and $n \geq 2$. However, linearity of the damage functions is a sufficient but not a necessary condition for the stability of the steady state. It could also be stable for specific nonlinear damage functions, but an analytical proof does not seem possible except for $n = 2$.

8.3. A ROAD MAP FOR STABILIZING THE CLIMATE

It is seen from a comparison of (8.4) and (8.9) that the steady-state Nash equilibrium emissions $(\bar{e}_1, \ldots, \bar{e}_n)$ and GHG stock $\bar{z}$ are not socially optimal. In fact, as can easily be verified, the sum total of steady-state Nash equilibrium emissions of all countries is higher than the sum total of steady-state socially optimal emissions. Despite that, each country with a long-run perspective will be willing to emit at the socially optimal rates $(e_1^*, \ldots, e_n^*)$ if the GHG stock is socially optimal and $\sigma e_i^* - h_i(e_i^*) - v_i(z^*) > \sigma \bar{e}_i - h_i(\bar{e}_i) - v_i(\bar{z})$ for each i. Though by definition of the steady-state Nash equilibrium and socially optimal emissions and GHG stock, we do have $\sum_{i \in N}(\sigma e_i^* - h_i(e_i^*) - v_i(z^*)) > \sum_{i \in N}(\sigma \bar{e}_i - h_i(\bar{e}_i) - v_i(\bar{z})),$[15] this does not imply $\sigma e_i^* - h_i(e_i^*) - v_i(z^*) > \sigma \bar{e}_i - h_i(\bar{e}_i) - v_i(\bar{z})$ for each i unless the damage functions of the countries are identical or sufficiently similar, which they are not. However, if the rights are tradable on a competitive international market, then there exists an allocation of rights $(e_1^0, \ldots, e_n^0)$ such that $\sum_{i \in N} e_i^0 = \sum_{i \in N} e_i^*$, each country i chooses to emit at the socially optimal rate e_i^* and buy/sell $e_i^0 - e_i^*$ of rights at competitive prices, and after the trades each country is better off than in the steady-state Nash

---

15. The inequality is strict because the objective of the optimal control problem is strictly concave, and therefore the steady state is unique.

equilibrium. Thus, each country will be willing to accept the allocation of rights $(e_1^0, \ldots, e_n^0)$, rather than emit according to the steady-state Nash equilibrium emissions $(\bar{e}_1, \ldots, \bar{e}_n)$, provided the rights are tradable on an international market. To be specific, note that for *any* allocation of tradable rights $(e_1^0, \ldots, e_n^0)$ such that $\sum_{i \in N} e_i^0 = \sum_{i \in N} e_i^*$, the market for buying and selling the rights clears if the competitive price is $\left(\frac{1}{r+\delta}\right) \sum_{i \in N} v_i'(z^*)$. That is so because if each country takes this price as given and maximizes its after-trade payoff $\sigma e_i - h_i(e_i) + \left(\frac{1}{r+\delta}\right) \sum_{i \in N} v_i'(z^*)(e_i^0 - e_i)$, then it will choose

$e_i = e_i^*$, as $\sigma - h_i'(e_i^*) = \left(\frac{1}{r+\delta}\right) \sum_{i \in N} v_i'(z^*)$ by (8.4). I claim that there exists an

allocation of rights $(e_1^0, \ldots, e_n^0)$ such that $\sum_{i \in N} e_i^0 = \sum_{i \in N} e_i^*$ for which not only the market clears but also

$$(e_i^0 - e_i^*) \left(\frac{1}{r+\delta}\right) \sum_{i \in N} v_i'(z^*) + \sigma e_i^* - h_i(e_i^*) - v_i(z^*)$$
$$\geq \sigma \bar{e}_i - h_i(\bar{e}_i) - v_i(\bar{z}), \tag{8.10}$$

for each $i = 1, 2, \ldots, n$. That is, the after-trade payoff of each country is higher than its steady-state Nash equilibrium payoff. I prove this claim by showing that the allocation $(e_1^0, \ldots, e_n^0)$ defined by the equations

$$(e_i^0 - e_i^*) \left(\frac{1}{r+\delta}\right) \sum_{j \in N} v_j'(z^*) = \sigma \bar{e}_i - h_i(\bar{e}_i) - \sigma e_i^* + h_i(e_i^*)$$
$$- \frac{v_i'(z^*)}{\sum_{j \in N} v_j'(z^*)} \left[\sum_{j \in N} \left(\sigma \bar{e}_j - h_j(\bar{e}_j) \right) - \sum_{j \in N} \left(\sigma e_j^* - h_j(e_j^*) \right) \right], \tag{8.11}$$
$$i = 1, 2, \ldots, n,$$

indeed leads to after-trade payoffs that satisfy inequality (8.10). Clearly, these equations imply $\sum_{i \in N} e_i^0 = \sum_{i \in N} e_i^*$. Furthermore, because each h_j is strictly convex,

$$\frac{v_i'(z^*)}{\sum_{j \in N} v_j'(z^*)} \left[\sum_{j \in N} \left(\sigma \bar{e}_j - h_j(\bar{e}_j) \right) - \sum_{j \in N} \left(\sigma e_j^* - h_j(e_j^*) \right) \right]$$
$$\leq \frac{v_i'(z^*)}{\sum_{j \in N} v_j'(z^*)} \left(\sigma - h_j'(e_j^*) \right) \left(\sum_{j \in N} \bar{e}_j - \sum_{j \in N} e_j^* \right) = \left(\frac{\delta}{r+\delta} \right) v_i'(z^*)(\bar{z} - z^*), \tag{8.12}$$

by using (8.4) and (8.9). Because the local damage functions h_i are strictly convex and the global damage functions v_i are convex, a comparison of (8.9) and (8.4) implies that the sum total of socially optimal steady-state emissions $\sum_{i \in N} e_i^*$ and the GHG stock z^* are both lower than the total steady-state Nash equilibrium emissions $\sum_{i \in N} \bar{e}_i$ and stock $\bar{z}$. Because $\bar{z} > z^*$ and v_i is convex,

$$v_i(\bar{z}) - v_i(z^*) \geq v_i'(z^*)(\bar{z} - z^*) > \left(\frac{\delta}{r+\delta}\right) v_i'(z^*)(\bar{z} - z^*),$$

as $r > 0$. Therefore, substituting in (8.12), it follows that

$$\frac{v_i'(z^*)}{\sum_{j \in N} v_j'(z^*)} \left[\sum_{j \in N} \left(\sigma \bar{e}_j - h_j(\bar{e}_j) \right) - \sum_{j \in N} \left(\sigma e_j^* - h_j(e_j^*) \right) \right] < v_i(\bar{z}) - v_i(z^*).$$

This inequality and (8.11) imply (8.10).

This shows that there exist tradable rights $(e_1^0, \ldots, e_n^0)$ such that the countries would be willing to emit $e_1^*, \ldots, e_n^*$, as required for the stability of the GHG stock at the socially optimal level, if they can exchange their pollution rights $(\bar{e}_1, \ldots, \bar{e}_n)$ for $(e_1^0, \ldots, e_n^0)$ and trade them on a competitive international market. The complete process involves two forms of exchange: a non-tatonnement process that starts from the initial allocation $(\bar{e}_1, \ldots, \bar{e}_n)$ of rights and converges to $(e_1^0, \ldots, e_n^0)$, followed by a tatonnement process that starts from $(e_1^0, \ldots, e_n^0)$ and converges to $(e_1^*, \ldots, e_n^*)$.[16] In the non-tatonnement process, each country reduces its pollution right in exchange for reductions in pollution rights by other countries. This is a barter exchange that requires the countries to offer and accept barter trades. In the tatonnement process, the countries or their entities need to respond to market prices, and the prices may continue to adjust until demand equals supply—no barter trade is required.

---

16. See Arrow and Hahn (1971) for the meaning and distinction between non-tatonnement and tatonnement processes. See Chander (1993) for an application of a non-tatonnement process in the context of a public good economy.

8.3.1. The Constrained Core

In the actual negotiations on climate change, many countries often react to proposals as members of a group with common interests. For example, the developing and developed countries often form and negotiate as separate blocs. I list below the various blocs that actively participated in the negotiations in Paris, as reported in *Time* magazine on December 14, 2015.[17]

Major Negotiating Groups

1. OPEC nations: Oil exporters that have a history of blocking meaningful climate action
2. European Union (EU): The EU's 28 nations negotiate in a bloc and push for tougher climate action
3. Umbrella Group: A coalition of non-EU developed countries that have been foot draggers on climate change in past summits
4. Group of 77 and China: This influential group of developing countries now includes 134 countries
5. Environmental Integrity Group: This mix of developed and developing countries tries to find common ground on climate change
6. Alliance of Small Island States: A coalition of 44 low-lying and small island countries that pushes for ambitious carbon cuts

Other Groups in the Climate Negotiation Process

1. African nations: The group aims to raise influence of Africa, which is very vulnerable to climate change
2. League of Arab States: These nations could face terrible heat—but depend on oil revenue

---

17. Formation of blocs for the purpose of negotiations has no impact on the tatonnement process because the countries take the market prices as given and, thus, it is immaterial whether they react to prices individually or as a bloc. But, as we know from the general equilibrium theory, formation of blocs can have an impact on the outcome of the non-tatonnement process.

3. Coalition of Rainforest Nations: This group advocates reforestation to mitigate climate change
4. Less-Developed Countries: All are very poor and need help for adapting to climate change
5. Like-Minded Group of Developing Countries: This group represents more than 50 percent of world's population
6. Agence intergouvernementale de la Francophonie: This alliance is composed of French-speaking nations
7. BASIC Countries: The major developing nations: Brazil, South Africa, India, and China

This means that the allocation of tradable rights $(e_1^0, \dots, e_n^0)$ must be such that not only each individual country but also every possible coalition of countries is better off. In other words, the allocation of tradable rights $(e_1^0, \dots, e_n^0)$ must satisfy a core property.

Proving the existence of an allocation of tradable rights $(e_1^0, \dots, e_n^0)$ with a core property requires a modification of the γ-core concept, to be called the constrained core. The modification is necessary because, by definition, no country either individually or as a member of a coalition can emit more than its right. Recall that the definition of the γ-core does not impose any such restriction on the choice of emissions of a country. In fact, the coalitional function in Example 1 of chapter 5 is not superadditive precisely because country 1 outside the coalition $\{2, 3\}$ could emit even more than its Nash equilibrium emissions.

We first need to specify the payoffs that a coalition can achieve on its own without the support of the outside countries. For each $S \subset N, S \neq N$, these payoffs are defined by the steady-state solution of the following optimal control problem:

$$\max_{e_i(t), i \in S} \int_0^\infty e^{-rt} \sum_{i \in S} \left[\sigma e_i(t) - h_i\left(e_i(t)\right) - v_i\left(z(t)\right) \right]$$

subject to

$$\dot{z}(t) = \sum_{i \in N} e_i(t) - \delta z(t), \quad e_i(t) \leq \overline{e}_i \text{ for all } t \geq 0 \text{ and } i \in S, \quad z(0) = z_0,$$

and for each $j \in N \backslash S$,

$$\max_{e_j(t)} \int_0^\infty e^{-rt} \left[\sigma e_j(t) - h_j \left(e_j(t) \right) - v_j \left(z(t) \right) \right]$$

subject to

$$\dot{z}(t) = \sum_{i \in N} e_i(t) - \delta z(t), \quad e_j(t) \leq \overline{e}_j \text{ for all } t \geq 0, \quad z(0) = z_0.$$

Notice the constraint $e_i(t) \leq \overline{e}_i$ for each i irrespective of whether i is a member of S or not. This is necessitated by the fact that the pollution rights are binding. Let $(e_1^S, \ldots, e_n^S)$ and $z^S = \sum_{i \in N} e^S$ denote the steady-state solution of the above optimal control problem. Then, $(e_1^S, \ldots, e_n^S)$ is a steady-state Nash equilibrium of the "constrained" game induced by coalition S. I claim that $(e_1^S, \ldots, e_n^S)$ and z^S must be such that

$$(r + \delta)\left(\sigma - h_i'(e_i^S) \right) = \sum_{i \in S} v_i'(z^S), i \in S. \tag{8.13}$$

To that end, I show that equality (8.13) holds if and only if the constraints $e_i^S \leq \overline{e}_i$ for all $i \in S$ are not binding. The proof for the "if" part follows from the first-order-conditions for a Nash equilibrium—analogous to (8.4) and (8.9). I prove, by induction, the "only if" part.

If $|S| = 1$, then $e_i^S = \overline{e}_i, i \in S$, and as seen from (8.9), the claim is true. I next show that if $|S| > 1$, then $z^S < \overline{z}$. Suppose contrary to the assertion that $z^S \geq \overline{z}$. Then, by comparing (8.9) and (8.13), $e_i^S < \overline{e}_i$ for all $i \in S$, as the functions h_i are increasing and strictly convex, functions v_i are increasing and convex, and $|S| > 1$. Because, by definition, $e_i^S \leq \overline{e}_i$ for all $i \in N \backslash S$, $\sum_{i \in N} e_i^S < \sum_{i \in N} \overline{e}_i$. This implies $z^S < \overline{z}$, as by definition $z^S = \delta \sum_{i \in N} e_i^S$ and $\overline{z} = \delta \sum_{i \in N} \overline{e}_i$. This contradicts the supposition that $z^S \geq \overline{z}$. Hence, $z^S < \overline{z}$. Given $z^S < \overline{z}$, the constraints on emissions must be binding for each $j \in N \backslash S$ (i.e., $e_j^S = \overline{e}_j$ for $j \in N \backslash S$). That is because if $e_j^S < \overline{e}_j$ for some $j \in N \backslash S$, then in the steady state, $(r + \delta)(\sigma - h_j'(e_j^S)) = v_j'(z^S)$. But as seen from (8.9), this contradicts that $z^S < \overline{z}$ and h_j is strictly convex and v_j is convex. This means $e_j^S = \overline{e}_j$ for all $j \in N \backslash S$. The proof is now completed by noting that inequalities (8.9) and (8.13) imply that either $e_i^S \geq \overline{e}_i$ for all $i \in S$ or $e_i^S < \overline{e}_i$ for all $i \in S$. The former, however, cannot be true, as $z^S < \overline{z}$, $z^S = \delta \sum_{i \in N} e_i^S$, $\overline{z} = \delta \sum_{i \in N} \overline{e}_i$, and

$e_j^S = \bar{e}_j$ for all $j \in N \backslash S$. Hence, $e_i^S < \bar{e}_i$ for all $i \in S$. The following lemma summarizes these results.

Lemma 8.1

For each $S \subset N, S \neq N, |S| > 1$, $z^S < \bar{z}$, $e_i^S < \bar{e}_i$ for each $i \in S$, and $e_j^S = \bar{e}_j$ for each $j \in N \backslash S$.

8.3.2. The Constrained Coalitional Function

For each $S \subset N$, let

$$w(S) = \sum_{i \in S} \left[\sigma e_i^S - h_i(e_i^S) - v_i(z^S) \right], \tag{8.14}$$

where $(e_1^S, e_2^S, \ldots, e_n^S)$ denote the steady-state Nash equilibrium emissions of the constrained game induced by coalition S, and z^S denotes the corresponding steady-state stock. Let $w(N) = \sum_{i \in N} [\sigma e_i^* - h_i(e_i^*) - v_i(z^*)]$ and $w(\{i\}) = \sigma \bar{e}_i - h_i(\bar{e}_i) - v_i(\bar{z})$. The function $w(S)$ specifies the per-period steady-state payoff of coalition S if it leaves the grand coalition. However, unlike the definition of the γ-core, the strategies of the players must satisfy the exogenously given constraints $(\bar{e}_1, \bar{e}_2, \ldots, \bar{e}_n)$, which, as Lemma 8.1 shows, are binding for the countries not in coalition S. These constraints therefore provide coalition S some protection against free riding and thus a higher payoff. This means a smaller and possibly empty core.

8.3.3. Existence of the Constrained Core

Theorem 8.2

The coalitional game (N, w) has a nonempty core; that is, there exists an imputation $(a_1, a_2, \ldots, a_n)$ such that $\sum_{i \in N} a_i = w(N)$ and $\sum_{i \in S} a_i \geq w(S)$ for all $S \subset N, S \neq N$.

I show that the game (N, w) is balanced. Let P denote the set of all coalitions. A collection of weights $(d_S)_{S \in P}$ such that $d_S \in [0, 1]$ for all $S \in P$ is balanced if for all $i \in N$, $\sum_{S \in P(i)} d_S = 1$. I first take note of a mathematical equality, which, by Lemma 8.1, implies a useful inequality. Let $P(i) \subset P$ denote the subset of all coalitions that contain player i and $P(\backslash i)$ of all those coalitions that do not contain player i. Then, $P(i) \cup P(\backslash i) = P$.

Given a balanced collection of weights $(d_S)_{S \in P}$, for any vector $(e_1, e_2, \ldots, e_n)$ and $i \in N$, $\sum_{j \in N} e_j = \sum_{S \in P(i)} d_S \sum_{j \in S} e_j + \sum_{S \in P(\backslash i)} d_S \sum_{j \in S} e_j$. Thus, $\sum_{S \in P(\backslash i)} d_S \sum_{j \in S} e_j =$

$\sum_{j \in N} e_j - \sum_{S \in P(i)} d_S \sum_{j \in S} e_j = \sum_{j \in N} e_j - \sum_{S \in P(i)} d_S (\sum_{j \in N} e_j - \sum_{j \in N \backslash S} e_j) = \sum_{j \in N} e_j - \sum_{S \in P(i)} d_S \sum_{j \in N} e_j +$

$\sum_{S \in P(i)} d_S \sum_{j \in N \backslash S} e_j = \sum_{S \in P(i)} d_S \sum_{j \in N \backslash S} e_j$ (as the collection of weights is balanced,

i.e., $\sum_{S \in P(i)} d_S = 1$). In particular, this equality must be true for the steady-state

Nash equilibrium emissions $(\bar{e}_1, \bar{e}_2, \ldots, \bar{e}_n)$; that is, $\sum_{S \in P(\backslash i)} d_S \sum_{j \in S} \bar{e}_j = \sum_{S \in P(i)} d_S \sum_{j \in N \backslash S} \bar{e}_j$.

Because by Lemma 8.1, $e_j^S < \bar{e}_j$ for all $j \in S$ and $e_j^S = \bar{e}_j$ for all $j \in N \backslash S$,

$$\sum_{S \in P(\backslash i)} d_S \sum_{j \in S} e_j^S < \sum_{S \in P(i)} d_S \sum_{j \in N \backslash S} e_j^S. \tag{8.15}$$

Using convexity of the damage function v_i, we derive next another useful inequality. For each $i \in N$, let $\tilde{e}_i = \sum_{S \in P(i)} d_S e_i^S$ and $\tilde{z} = (1/\delta) \sum_{i \in N} \tilde{e}_i$. Because

$\sum_{S \in P(i)} d_S = 1$, $\tilde{e}_i$ denotes country i's average (steady-state) emissions across

all possible coalitions that contain player i, and $\tilde{z}$ denotes the average (steady-state) stock of pollution. By definition,

$$v_i(\tilde{z}) = v_i \left((1/\delta) \sum_{j \in N} \sum_{S \in P(j)} d_S e_j^S \right) = v_i \left((1/\delta) \sum_{S \in P} d_S \sum_{j \in S} e_j^S \right)$$

$$= v_i \left((1/\delta) \left(\sum_{S \in P(i)} d_S \sum_{j \in S} e_j^S + \sum_{S \in P(\backslash i)} d_S \sum_{j \in S} e_j^S \right) \right)$$

$$< v_i \left((1/\delta) \left(\sum_{S \in P(i)} d_S \sum_{j \in S} e_j^S + \sum_{S \in P(i)} d_S \sum_{j \in N \backslash S} e_j^S \right) \right) \text{ (using (8.15))}$$

$$= v_i \left((1/\delta) \sum_{S \in P(i)} d_S \sum_{j \in N} e_j^S \right) = v_i \left(\sum_{S \in P(i)} d_S z^S \right) \le \sum_{S \in P(i)} d_S v_i (z^S),$$

as the damage function v_i is convex and the collection of weights is balanced. Hence, for each $i \in N$,

$$v_i(\tilde{z}) < \sum_{S \in P(i)} d_S v_i (z^S) \tag{8.16}$$

for any balanced collection of weights $(d_S)_{S \in P}$.

Proof of Theorem 8.2

I need to show that $\sum_{S \in P} d_S w(S) \leq w(N)$ for any balanced collection of weights

$(d_S)_{S \in P}$. For each $i \in N$, let $\tilde{e}_i = \sum_{S \in P(i)} d_S e_i^S$ and $\tilde{z} = (1/\delta) \sum_{i \in N} \tilde{e}_i$. By definition

$$w(N) \geq \sum_{i \in N} [\sigma \tilde{e}_i - h_i(\tilde{c}_i) - v_i(\tilde{z})]$$

$$= \sum_{i \in N} \left[\sigma \sum_{S \in P(i)} d_S e_i^S - h_i \left(\sum_{S \in P(i)} d_S e_i^S \right) \right] - \sum_{i \in N} v_i(\tilde{z})$$

$$\geq \sum_{i \in N} \sum_{S \in P(i)} d_S \sigma e_i^S - \sum_{i \in N} \sum_{S \in P(i)} d_S h_i(e_i^S) - \sum_{i \in N} \sum_{S \in P(i)} d_S v_i(z^S)$$

(using convexity of h_i and inequality (8.16))

$$= \sum_{S \in P} d_S \sum_{i \in S} \sigma e_i^S - \sum_{S \in P} d_S \sum_{i \in S} h_i(e_i^S) - \sum_{S \in P} d_S \sum_{i \in S} v_i(z^S)$$

$$= \sum_{S \in P} d_S \sum_{i \in S} \left[\sigma e_i^S - h_i(e_i^S) - v_i(z^S) \right]$$

$$= \sum_{S \in P} d_S w^v(S) \quad \text{(using (8.14))}. \quad \blacksquare$$

8.3.4. An Overview of the Proposed Solution

The analysis in this chapter suggests that the problem of limiting climate change to +2°C can be divided into two parts: (a) the countries must first agree on a division of the carbon budget, and (b) they must then agree on their long-run emission rights that can be traded on a competitive international market.

While I could propose a solution for part (b) that involves both non-tatonnement and tatonnement processes, part (a) of the problem is equivalent to that of splitting a pie of fixed size among the agents.

Should It Be +2°C or +1.5°C?

The Paris Agreement sets a target of holding global warming to "well below 2 degrees Celsius" and to "pursue efforts to limit the temperature increase to 1.5 degree Celsius above pre-industrial levels" by the end of the century. The inclusion of the 1.5°C target was done at the insistence of the small island states that are most vulnerable to sea-level rise because of climate change and the least developed small countries that do not have enough resources to cope with climate change. But is it really socially optimal to restrict climate change to 1.5°C rather than 2°C? Social optimality requires weighing the damage that the small island states and the least developed small countries will suffer against the economic gains net of additional damage that the rest of the world will suffer if the temperature rise is allowed to be 2°C rather than 1.5°C. If the gains net of additional damage for the rest of the world are more than the damage suffered by the small island states and the least developed small countries, then allowing a temperature rise of 2°C is welfare improving, though not a Pareto improvement.[18] But if it is the opposite, that is, the gains net of additional damage for allowing a temperature rise of 2°C for the rest of the world are positive but less than the damage suffered by the small island states, then restricting temperature rise to 1.5°C, though welfare improving, may not be acceptable to the rest of the world as that would make it worse off. Therefore, the rest of the world may not agree to restrict its emissions such that global warming would be restricted to 1.5°C unless it is adequately compensated for doing so by the small island states and the least developed small countries. But this may not be possible because the small island states and the least developed small countries neither have enough resources to compensate the rest of the world for restricting its emissions so that the temperature rise is only 1.5°C, nor can they compensate the rest of the world by sufficiently reducing their own emissions as they themselves emit little. In this case, only considerations of fairness, if at all, can motivate the rest of the world to restrict temperature rise to 1.5°C and achieve efficiency without receiving adequate compensation for it.

In the next chapter, I further discuss the issue of fairness in climate policy and whether climate change should be restricted to only +1.5°C.

---

18. It may be noted in this regard that the differential game model in this chapter is amenable to numerical computations and simulations. Therefore, it can be used to calculate the welfare gains and losses from alternative targets for limiting global warming.

9

THE JOURNEY FROM KYOTO TO PARIS

This chapter is not intended for game theorists—unless they are interested in learning how game theory, and the game-theoretic approach developed in this book, can be applied to control climate change. The chapter is meant mainly for practitioners and policymakers who have only some knowledge of basic economic theory. It is written in a style that is accessible even to those who may not want to master the theory in the previous chapters. It is self-contained, and references to the material in the previous chapters are intentionally kept to a minimum. Informed readers who are familiar with the previous chapters will be able to make the connections but may sometimes find some material repetitive.

As the title suggests, the chapter interprets the Kyoto Protocol and the more recent Paris Agreement on climate change in light of the game-theoretic approach developed in the previous chapters. Though the Kyoto Protocol is a failed agreement, studying it and the reasons for its failure can help us understand the issues at stake. In contrast, the Paris Agreement is a work in progress. Though it will, on its own, not solve global warming, it provides a floor on which to build ambition and action. The rest of the chapter interprets the Paris Agreement and proposes a road map for making it effective. As will be seen, the Kyoto Protocol and the Paris Agreement differ in many ways but are also similar in an important respect. However, those readers who are more interested in interpretation and analysis of the Paris Agreement may go directly to sections 9.4 and 9.5.

The negotiations on climate change, which have been taking place since the late 1980s within the United Nations institutions, are obviously a worldwide process, judging by the length of the list of countries that have taken part in the successive meetings.[1] But these negotiations, prior to the Kyoto Protocol in 1997, had led only to a "framework convention," signed in 1992 in Rio de Janeiro, that was little more than a declaration of intent.[2] The real issue then was: Will the continuing negotiations eventually lead to a sustainable agreement bearing on effective actions that is also worldwide? Or will they lead to formation of separate independent blocs, each acting or not acting at all to the best of its own interests?

The Kyoto Protocol, signed in December 1997 amid scenes of jubilation in Japan, was hailed as a breakthrough that could set the world on a new low-carbon path. It is a milestone in the post-Rio evolution of the international climate change negotiations. Its importance lies mainly in the fact that it required some countries to take effective actions that were to become binding on them once they ratified the Protocol. After a summary presentation of the relevant features of the Protocol in section 9.1, section 9.2 provides a sketch of the economic model that is appropriate for interpreting and analyzing the Protocol.[3] I then proceed in two steps. First, taking the Protocol as it was signed, I consider in section 9.3 a series of its characteristics, features, and properties such as reference emissions, efficiency and stability, desirability of competitive trade in emissions, and the clean development mechanism as a form of emission trading. Independently of the subsequent problems concerning its implementation, I use it as a benchmark for understanding the various issues

---

1. According to the relevant United Nations Framework Convention on Climate Change (UNFCCC) website, 165 countries were present at the time of signing of the 1992 convention in Rio, 84 at the December 1997 Conference of the Parties 3 (COP-3) to sign the Kyoto Protocol, 192 at the December 2009 COP-15 held in Copenhagen as well as at the December 2011 COP-17 in Durban, and 196 at COP-21 in December 2015 for the Paris Agreement.

2. The "little more" in the sentence is far from being negligible, since for the implementation of any policy, an essential and preliminary component is that the emissions be known. To that effect, the signatories to the Rio Convention committed themselves to submit information regarding inventories of their emissions, annually for the countries listed in Annex 1 of the Convention, less frequently for others. A United Nations (UN) administration has been set up, located in Geneva and Bonn, which is in charge of receiving, reviewing, compiling, and publishing "national communications" containing these inventories (which did not exist before) as well as other reports on actions taken to reduce the emissions. It also organizes the successive Conferences of the Parties.

3. It closely follows Chander (2003) and Chander, Tulkens, van Ypersele, and Willems (1998).

concerning the climate change problem in general—in contrast to the more critical views offered by many other commentators.[4]

In section 9.4, I extend this exercise for an appraisal of the situation that emerged after the non-ratification of the Protocol by the United States, which eventually led to the Paris Agreement. Because the Protocol did not require all countries to commit to achieve quantified reductions or limitations of their emissions, a list of parties that agreed to such reductions or limitations of their emissions appeared in Annex B of the Protocol.[5] The role of other countries in the implementation of the Protocol, although not ignored, was less precisely specified in terms of emissions. Therefore, a central question is whether the Kyoto Protocol was to be considered just an Annex B agreement or could it be seen, after further thought and beyond appearances, as a worldwide agreement? In sections 9.3 and 9.4, I defend the second thesis.

9.1. MAIN FEATURES OF THE KYOTO PROTOCOL

I briefly note the main features of the Kyoto Protocol[6] that are relevant from the perspective of my interpretation and analysis:

(i) The Protocol proposed dated quotas of yearly emissions, expressed in percentages of 1990 emissions, for Annex B countries, to be met on average over the period 2008–2012.

---

4. For instance, Nordhaus and Boyer (1999, p. 125) have argued that "the strategy behind the Kyoto Protocol has no grounding in economics or environmental policy. The approach of freezing emissions at a given level for a group of countries is not related to a particular goal for concentrations, temperature, or damages. Nor does it bear any relation to an economically oriented strategy that would balance the costs and benefits of greenhouse-gas reductions."

5. Annex B of the Kyoto Protocol was distinct from Annex 1 of the Rio Convention, but both essentially listed the Organization of Economic Cooperation and Development (OECD) countries, the former Soviet Union countries, and the Eastern European economies in transition. This group is often referred to as the "developed" or "industrialized" countries.

6. The text of the Kyoto Protocol was adopted unanimously by the delegates of the 84 countries that participated in the negotiations at Kyoto in 1997. Signing of the text by governments and ratification by parliaments was to take place later on. The Protocol was to enter into force only if 55 countries, representing 55 percent of the world total emissions, ratified it. That occurred in February 2005, but ratifications by more countries continued, and by October 2009, 189 countries had ratified the Protocol. In the meantime, the United States, under the Bush administration, decided not to ratify it and, thus, did not submit the Protocol to the U.S. Congress for ratification. A few countries, including Canada and Japan, later withdrew from the Protocol.

(ii) It proposed the principles of (a) emission trading by countries (or their entities) and of (b) joint implementation by Annex B countries.

(iii) It proposed a clean development mechanism (CDM) as a way to involve the non–Annex B countries (especially the developing countries) in some particular form of joint implementation and emission trading.

(iv) It allowed trade in emissions only among those countries that ratified the Protocol. It also proposed to not permit trade in emissions with countries that did not fulfill their obligations under the Protocol.

I also note some of the features the Protocol did not have:

(i)   The Protocol did not set targets in terms of the greenhouse gas (GHG) stock or temperature rise, and its object was not a trajectory for the GHG stock. Rather, it was to control emissions per year over a specific period.

(ii) No explicit emissions ceilings were proposed for non–Annex B countries, and such ceilings, if at all, were to be negotiated in the future.

(iii) The parties to the Protocol were expected to fulfill the commitments made by them. But no sanctions were proposed if a ratifying country did not fulfill its obligations under the Protocol, except for the above provision of being excluded from emissions trading.

9.2. A MODEL FOR INTERPRETING THE KYOTO PROTOCOL

The Kyoto Protocol can be interpreted in the framework of an environmental game introduced in chapter 5. In simplest terms, there are n countries (indexed by $i = 1, \dots, n$) each of which enjoys an aggregate consumption x_i equal to the aggregate value of its production y_i minus damage D_i, which consists of lost production due to global warming.[7] The production activities of country i during a given period are described most simply by an

---

7. Several studies give estimates of these damages (e.g., Fankhauser 1995; Nordhaus and Yang 1996; Stern et al. 2006; and Tol 2009). However, the estimates differ widely.

increasing and strictly concave production function $y_i = g_i(e_i)$, where e_i is the input of fossil fuels during the period.[8] Assume that the units have been so defined that burning a unit of fossil fuel generates a unit of emissions as a by-product. The emissions of country i are thus equal to e_i. Accordingly, $g_i'(e_i)(= dg_i(e_i)/de_i)$ is the marginal product of fossil energy or the marginal cost of abatement, depending on the context. Damage in each country depends on the sum total of emissions of all countries, that is, $\sum_{i=1}^{n} e_i$, and represented by a linear damage cost function $D_i = d_i \sum_{j=1}^{n} e_j$. Each country's output net of damage is thus given by the expression

$$x_i = g_i(e_i) - d_i \sum_{j=1}^{n} e_j, \qquad (9.1)$$

where $d_i > 0$ is country i's damage per unit of emissions or, equivalently, benefit per unit of abatement. Ignoring distributional issues for the time being, the optimal world consumption is equal to the maximum of $\sum_{i=1}^{n} x_i$ with respect to the variables $e_1, \ldots, e_n$. Let $(e_1^*, \ldots, e_n^*)$ denote the profile of emissions that achieves such a *world optimum*. These emissions are a solution to the first-order conditions for a maximum; that is,

$$g_i'(e_i^*) = \sum_{j=1}^{n} d_j, i = 1, \ldots, n. \qquad (9.2)$$

Thus, at the world optimum, the marginal abatement cost of each country must be equal to the sum of the marginal damage of all countries. Notice that the *world efficient emissions* are independent of the actual or current emissions of the countries. They depend only on the total marginal damage $\sum_{j=1}^{n} d_j$ of all countries, assuming that the production functions do not change.[9]

The negotiations on climate change must aim, at least in principle, to induce the countries to choose the world efficient emissions. But economic reasoning tells us that the countries may not have incentives to do

---

8. Despite development of alternative sources of energy, more than 95 percent of world energy still comes from fossil fuels. However, the share of renewable energy, especially solar energy, is fast increasing.

9. However, if the production functions g_i change, the same world emissions may not be optimal even if the damage as a function of total world emissions remains the same.

so, as the cost of doing so for a country may be higher than its benefits. However, the theory in the preceding chapters also shows that appropriately designed transfers between countries can remedy for that. In addition, I show in section 9.3 that a system of tradable permits with properly specified initial allowances can play the same role as such transfers. Therefore, in the sections to follow, I argue that the Kyoto Protocol, thanks to its "cap-and-trade" architecture and with appropriately selected reference emissions, was a step in the direction of an efficient and stable regime for the world climate, and that a sequence of such steps could have led the countries eventually to an efficient and stable trajectory of emissions and consumptions.

9.2.1. Reference Emissions

How does a country decide how much to emit? Higher production means higher emissions, which entail greater damage due to climate change. Following classical economics reasoning, each country can achieve its domestic optimum by maximizing its net consumption x_i with respect to emissions e_i as defined in (9.1), taking as given the variables e_j for $j \neq i$. If all countries adopt such behavior, the result would be a noncooperative equilibrium that consists of a vector of emissions $(\overline{e}_1, \ldots, \overline{e}_n)$ such that[10]

$$g_i'(\overline{e}_i) = d_i, \, i = 1, \ldots, n. \tag{9.3}$$

This noncooperative equilibrium is actually a Nash equilibrium, though we have not formally defined the game here. We note two characteristics of this equilibrium: (i) the equilibrium emissions $(\overline{e}_1, \ldots, \overline{e}_n)$ are clearly not equal to the world efficient emissions $(e_1^*, \ldots, e_n^*)$, as can be seen by comparing (9.2) and (9.3), and (ii) $\overline{e}_i > e_i^*$ for each country i, as g_i is strictly concave and $\sum_{j=1}^{n} d_j > d_i$ for each i. Thus, the world efficient emissions are *lower* than those prevailing at the Nash equilibrium.

---

10. Uniqueness of this vector is ensured under my assumptions of strict concavity of the functions g_i and linearity of the functions D_i.

Fulfillment of conditions (9.3), which characterize the Nash equilibrium, requires domestic policies that involve either an energy tax or appropriately priced pollution permits such that the fossil fuel price including the tax or the permit price is equal to the domestic marginal damage cost d_i. Such domestic policies, which are nationally rational, are often called "no regrets policies."

However, there is little empirical evidence to support that the countries do indeed decide their emissions in this rational manner. For example, if fossil energy consumers and firms in a country have strong lobbying power, they may be able to influence their government to keep energy prices low.[11] Because profit maximization by firms implies equality between the marginal product and the price of energy, this may lead to emissions $\check{e}_i$ that are higher than $\bar{e}_i$ and such that $g_i'(\check{e}_i) < d_i$, thus emitting more than what the rational domestic policy requires. If the governments and firms behave in this manner, a different equilibrium—also noncooperative in nature—results, called the "market solution" by Nordhaus and Yang (1996) or "business-as-usual" by others.

Another reason why a domestic rational policy may not be followed is that the producers in a country may simply not be profit maximizers, as is the case with large public sector firms in some nonmarket economies. In such cases, the emissions of a country are neither "rational" nor of the "business-as-usual" type, and the domestic energy prices may not induce any well-defined emissions policy—except for a generally low concern for efficient use of energy.

In sum, at least three types of country behavior are possible. But whatever is a country's behavior if its producers maximize profits and markets are competitive, its marginal abatement cost must be equal to the (average) domestic fossil fuel price in real terms. Given the strict concavity of the production function g_i, it follows that the higher the domestic fossil fuel price, the higher the marginal abatement cost. As seen from table 9.1, such a relationship more or less holds (except in the case of China, where, as is known, the state-owned firms do not necessarily

---

11. It is not just the firms; some other economic agents may do the same. For example, in some countries, governments subsidize fossil fuels such as kerosene used by the poor.

TABLE 9.1 Fossil Fuel Prices, Marginal Abatement Costs, and the Type of Domestic Equilibrium

COUNTRY OR REGION	HEAVY FUEL OIL FOR INDUSTRY (PER TON; US$)	STEAM COAL FOR INDUSTRY (PER TON; US$)	NATURAL GAS FOR INDUSTRY (PER 10 KCAL; US$)	MARGINAL ABATEMENT COST PER TON FOR FIRST 100 MEGATON REDUCTION (US$)	ANNUAL DAMAGE COST AS A PERCENTAGE OF GDP (%)	TYPE OF DOMESTIC EQUILIBRIUM CONJECTURED
United States	138.00	35.27	136.62	12	1.3	$g_i'(\bar{e}_i) < d_i$
European Union	187.4	76.0	182.0	40	1.4	$g_i'(\bar{e}_i) \geq d_i$
Japan	172.86	49.90	423.12	350	1.4	$g_i'(\bar{e}_i) \geq d_i$
India	191.15	19.36	n.a.	22	n.a.	?
FSU	n.a.	n.a.	n.a.	22	0.7	?
China	150.60	30.12	n.a.	3.5	4.7	?

Note: FSU: Former Soviet Union; kcal: kilo calories; n.a.: not available.

Source: Chander (2003).

maximize profits).[12] In particular, the energy prices in the United States are systematically lower and so is the marginal abatement cost. Moreover, for the three market economies of the United States (US), the European Union (EU), and Japan, the higher the fossil fuel prices, the higher the marginal abatement costs.[13] For the other countries, I cannot say much, not only because of lack of data but also because they are either nonmarket or less developed economies, or both.

The marginal abatement cost of the United States is low compared to that of the European Union or Japan; it is next only to that of China and significantly below that of India. Because the marginal damage cost

---

12. The marginal cost of abatement may seem exceptionally high in the case of Japan, but this is because of its large dependence on natural gas, the price of which is relatively high, and less dependence on coal and oil.

13. Coal in Japan is a noticeable exception, but its use there is considerably lower.

of the United States, which is one of the largest countries, cannot be lower than, say, that of the European Union, this suggests that the U.S. emissions were determined by the "business-as-usual" policy rather than by maximization of its national welfare.[14] In contrast, the domestic oil prices were kept high in India by imposing tariffs to reduce its import of oil, which is one of the largest items in its import basket. The last column of table 9.1 presents an educated guess about the type of domestic equilibrium that was prevailing in each of the countries and regions.

9.3. THE NECESSITY OF TRANSFERS

The actual emissions, that is, the *reference emissions*, of a country could be equal to its Nash equilibrium emissions or business-as-usual emissions or, worse, the outcome of a generally low concern for the efficient use of energy. In either case, the reference emissions $(\overline{e}_1, \ldots, \overline{e}_n)$ are likely to be higher than the world efficient emissions $(e_1^*, \ldots, e_n^*)$. Thus, reducing the emissions from the reference levels to the world efficient levels requires each country i to reduce its emissions by an amount $\overline{e}_i - e_i^*$. As this entails abatement cost [i.e., $g_i(\overline{e}_i) - g_i(e_i^*)$], as well as benefits [i.e., $d_i \sum_{j=1}^{n}(\overline{e}_j - e_j^*)$], the latter should exceed the former for each i for the emission reductions to be voluntarily acceptable to all countries. However, this is unlikely because unless the countries are identical or quite similar, some countries will have high abatement costs but low benefits, while others will have low abatement costs but high benefits.

To illustrate, consider a country with large emissions $\overline{e}_i$, almost zero damage from climate change (i.e., $d_i \cong 0$), and such that world efficiency requires it to reduce its emissions by a large amount. But why should this country reduce its emissions, as its benefits from doing so would be almost zero but the abatement cost would be high? Because each country is sovereign, there is no way to induce this country to reduce its emissions except by adequate transfers from the countries that will benefit from its

---

14. This is clearly a case of government, and not market, failure.

reductions.[15] Thus, *only* a scheme of transfers that appropriately weighs in the costs and benefits of reducing emissions can induce the countries to reduce their emissions to the world efficient levels. One such scheme of transfers, proposed originally in Chander and Tulkens (1995, 1997) and discussed in detail in chapter 5, is

$$T_i = \left[g_i(\bar{e}_i) - g_i(e_i^*) \right] - \frac{d_i}{\sum_{j=1}^n d_j} \sum_{j=1}^n \left[g_j(\bar{e}_j) - g_j(e_j^*) \right], \; i = 1, \ldots, n, \quad (9.4)$$

where $T_i > 0$ means a receipt by country i, while $T_i < 0$ means a payment by i. The first expression in square brackets on the right-hand side of the equation is equal to country i's total abatement cost, while the summation on the same side of the equation is the total abatement cost of all countries for reducing their emissions from their reference levels to the world efficient levels. Thus, the scheme of transfers does not require each country i to bear its own abatement cost $g_i(\bar{e}_i) - g_i(e_i^*)$ but to bear instead a damage-weighted proportion, $d_i / \sum_{j=1}^n d_j$, of the total abatement cost of all countries.

Clearly, $\sum_{i=1}^n T_i = 0$; that is, the budget is balanced. Thus, the transfer scheme is revenue neutral and can easily be implemented if an international agency were established to administer it. Also, notice that if all countries are identical, then $T_i = 0$ for all i; that is, transfers are not required if all countries are identical.

Notice the role played by the reference emissions $(\bar{e}_1, \ldots, \bar{e}_n)$ in the calculation of the transfers $(T_1, \ldots, T_n)$. In chapter 5, I assumed the reference emissions to be equal to the Nash equilibrium emissions and showed that this transfer scheme enjoys game-theoretic properties. In particular, besides leading to the world efficient emissions, the transfer scheme implies coalitional stability in the sense that not only is each country individually better off but also each coalition of countries

---

15. This should also clarify that transfers are a *necessity* and not a matter of taste or choice. Many studies get around the necessity of transfers by assuming identical countries in which costs and benefits of achieving efficiency can be balanced without transfers. However, if the countries are sufficiently different, then the outcomes of the game forms or mechanisms proposed in these studies must be either inefficient or not "individually rational" for all participants: the sovereign countries that are free to reject any outcome that is not individually rational.

is better off compared to what it can get by adopting any scheme of emissions and transfers among its members, given the emissions of nonmembers.

But what if the reference emissions are not equal to the Nash equilibrium emissions? In particular, what if they are equal to the business-as-usual emissions discussed earlier? It turns out that the game-theoretic properties of the scheme are robust with regard to the reference emissions. If $(\overline{e}_1, \ldots, \overline{e}_n)$ are equal to the business-as-usual emissions, then the corresponding transfers $(T_1, \ldots, T_n)$ have the same game-theoretic property as when they are equal to the Nash equilibrium emissions. This is seen intuitively as follows: (a) the business-as-usual emissions are generally higher than the Nash equilibrium emissions, and (b) given (a), the payoff that a coalition can achieve for itself is lower, as the emissions of members not in the coalition are higher. The core is thus larger and, therefore, includes more imputations, including the one described by (9.4).

The first row of table 9.2 provides an example of a vector of reference emissions. These were estimated by Ellerman and Decaux (1998) on the basis of MIT's Emissions Prediction and Policy Analysis (EPPA) multiregional and multisector computable general equilibrium model of economic activity, energy use, and carbon emissions. I use these estimated reference emissions in my arguments below and for obvious reasons refer to them as the business-as-usual emissions. It may be noted that for the former Soviet Union,[16] the reference emissions $\overline{e}_i$ have been taken to be equal to their Kyoto commitment of 873.32, though the actual emissions were estimated to be only 762.79. This is equivalent to giving credit for emissions reductions that would have happened anyway. For non-Annex B countries, both reference emissions and Kyoto quotas of permitted emissions e_i^0 have been taken to be equal to their estimated business-as-usual (BAU) emissions in 2010, as it was agreed that their emissions need not be restricted in the first commitment period.

---

16. *Annex B countries*: the United States, Japan, the European Union, other OECD countries, Eastern economies in transition, and the former Soviet Union. *Non–Annex B countries*: energy-exporting countries, China, India, dynamic Asian economies, Brazil, and the rest of the world.

TABLE 9.2 Characterization of a Worldwide Competitive Emissions Trading Equilibrium

	USA	JPN	EEC	OOE	EET	FSU	EEX	CHN	IND	DAE	BRA	ROW	WORLD
Reference emissions in 2010 (megatons) $\bar{e}_i$	1,838.25	424.24	1,063.72	472.04	394.76	873.32*	927.39	1,791.96	485.76	308.32	97.27	531.61	9,208.63
Kyoto quotas of permitted emissions (megatons), e_i^0	1,266.67	280.05	756.51	300.66	247.45	873.32	927.39	1,791.96	485.76	308.32	97.27	531.61	7,866.95
Post trading emissions reductions (megatons) $\bar{e}_i - \hat{e}_i$	186.22	12.33	74.96	60.07	52.98	213.36	52.54	447.93	104.87	42.78	2.50	91.07	1,341.61
Emissions permits (megatons) imported (+) / exported (−), $\hat{e}_i - e_i^0$	385.36	131.86	232.25	111.31	94.33	−213.36	−52.54	−447.93	−104.87	−42.78	−2.50	−91.07	0.07
Marginal cost of abatement (US$/ton), $\hat{\gamma} = g_i'(\hat{e}_i)$	24.75**	24.75**	24.75**	24.75**	24.75**	24.75**	24.75**	24.75**	24.75**	24.75**	24.75**	24.75**	24.75**
Total cost of own abatement (billion US$), $g_i(\bar{e}_i) - g_i(\hat{e}_i)$	1.77	0.15	0.76	0.44	0.46	0.86	0.57	4.49	1.01	0.47	0.03	0.86	11.86
Cost (+) / receipt (−) of emission permits exports / imports (billion US$), $\hat{\gamma}(\hat{e}_i - e_i^0)$	9.54	3.26	5.75	2.75	2.33	−5.28	−1.30	−11.09	−2.60	−1.06	−0.06	−2.25	0.00

Note: The following abbreviations for countries and regions are used: *Annex B countries*: the United States (USA), Japan (JPN), the European Union (EEC), other OECD countries (OOE), Eastern economies in transition (EET), and the former Soviet Union (FSU). *Non–Annex B countries*: energy-exporting countries (EEX), China (CHN), India (IND), dynamic Asian economies (DAE), Brazil (BRA), and the rest of the world (ROW).

*Reference, not actual, emissions

**1985 US Dollars

Source: Ellerman and Decaux (1998, table G) (August version).

9.3.1. Competitive Emission Trading

Unlike the scheme of transfers specified in (9.4), the Kyoto Protocol did not propose any direct transfers among the countries. It only proposed ceilings or caps on the emissions of some countries, and these caps were most likely not equal to the world efficient emissions. Yet, as I argue below, the caps on emissions proposed in the Protocol, together with the trade that they induce, can be interpreted as a scheme of transfers and, therefore, its whole architecture as a step toward reaching world efficient emissions. To see this, I now redefine the above scheme of transfers in terms of emission quotas and trades. This requires me to first introduce the concept of a "competitive emission trading equilibrium."

A *competitive emissions trading equilibrium* with respect to emission quotas $(e_1^0, \ldots, e_n^0)$ is a vector of emissions $(\hat{e}_1, \ldots, \hat{e}_n)$ and a price $\hat{p}$ (expressed in units of the consumption good per unit of emissions) such that for each country $i = 1, \ldots, n$,

$$\hat{e}_i = \mathrm{argmax}\left[g_i(e_i) + \hat{p}(e_i^0 - e_i) \right] \tag{9.5}$$

and

$$\sum_{i=1}^{n} \hat{e}_i = \sum_{i=1}^{n} e_i^0. \tag{9.6}$$

Assuming an interior solution, the first-order conditions for maximization imply $g_i'(\hat{e}_i) = \hat{p}$, $i = 1, \ldots, n$. This implies that competitive trade in emissions enables the countries to reallocate production and emissions so as to maximize their total output while keeping their total emissions restricted to the sum total $\sum_{i=1}^{n} e_i^0$, as by definition of the competitive equilibrium $\sum_{i=1}^{n} \hat{e}_i = \sum_{i=1}^{n} e_i^0$ and $g_i'(\hat{e}_i) = g_j'(\hat{e}_j)$ for all $i, j = 1, \ldots, n$.

In a competitive emission trading equilibrium, countries trade their "pollution rights" (which are equal to their emission quotas $(e_1^0, \ldots, e_n^0)$), taking as given the market price $\hat{p}$ such that the demand and supply for the pollution rights are equal. For each i, the amount $\hat{p}(e_i^0 - \hat{e}_i)$ is the value of payment, in units of the consumption good, for the purchase of pollution rights at the competitive international price $\hat{p}$ if $(e_i^0 - \hat{e}_i)$ is negative or receipt from the sale of pollution rights if $(e_i^0 - \hat{e}_i)$ is positive.

Tradable emission quotas $(e_1^0, \ldots, e_n^0)$, which are defined from the world efficient emissions $(e_1^*, \ldots, e_n^*)$ and the reference emissions $(\bar{e}_1, \ldots, \bar{e}_n)$ such that for each country i,

$$(e_i^0 - e_i^*)\sum_{j=1}^{n} d_j = \left[g_i(\bar{e}_i) - g_i(e_i^*)\right] - \frac{d_i}{\sum_{j=1}^{n} d_j}\sum_{j=1}^{n}\left[g_j(\bar{e}_j) - g_j(e_j^*)\right], \quad (9.7)$$

are similarly equivalent to a transfer scheme. The left-hand side of (9.7) is what country i pays (or receives) if it buys (sells) pollution rights in amount $(e_i^0 - e_i^*)$ at price $p^* = \sum_{j=1}^{n} d_j$. In view of (9.2), $p^* = g_i'(e_i^*) = g_j'(e_j^*)$ for all $i, j = 1, \ldots, n$, and $\sum_{i=1}^{n} e_i^0 = \sum_{i=1}^{n} e_i^*$, by definition of $(e_1^0, \ldots, e_n^0)$. This means that $(e_1^*, \ldots, e_n^*)$ and p^* are actually the competitive emission trading equilibrium relative to the pollution quotas $(e_1^0, \ldots, e_n^0)$ that result in after-trade transfers exactly equal to the transfers T_i, $i = 1, \ldots, n$, advocated above for inducing each country i to emit an amount equal to its world efficient emissions e_i^*.

It may be noted that while the world efficient emissions $(e_1^*, \ldots, e_n^*)$, as defined in (9.2), are independent of the reference emissions $(\bar{e}_1, \ldots, \bar{e}_n)$, the pollution quotas $(e_1^0, \ldots, e_n^0)$, as defined in (9.7), are not. In fact, because the world efficient emissions are independent of the reference emissions and thus fixed, there is a one-to-one correspondence between the quotas $(e_1^0, \ldots, e_n^0)$ and reference emissions $(\bar{e}_1, \ldots, \bar{e}_n)$. This, as in chapters 5 and 8, means that if the countries are agreeable to the reference emissions $(\bar{e}_1, \ldots, \bar{e}_n)$, then, after some thought and analysis, they would also be agreeable to the assignment of emissions quotas $(e_1^0, \ldots, e_n^0)$ that are tradable on a competitive international market, as by definition these would not only lead to the world efficient emissions $(e_1^*, \ldots, e_n^*)$ but also result in after-trade transfers that make each country or coalition of countries better off relative to the reference emissions. This shifts the argument for seeking an agreement on emissions quotas $(e_1^0, \ldots, e_n^0)$ to an agreement on reference emissions $(\bar{e}_1, \ldots, \bar{e}_n)$.

Fairness and Emissions Quotas

Homo sapiens are often willing to forgo some amount of economic benefits in exchange for fairness, and a number of studies on ultimatum games (see, e.g., Bearden 2001) establish that a reductionist approach that simplifies decision making to a simple study of incentives is incomplete. But fairness is a tricky issue—while people are indeed concerned about fairness, what is perceived as fair is malleable. For this reason, issues of perceived fairness have often caused logjams in climate change negotiations. For instance, one of the sticking points in climate change negotiations has been that of fairness—developing countries want developed countries to take historical responsibility for the existing GHG stock in the atmosphere, even though it is clear that efficiency requires some developing countries to reduce their emissions.

But if the countries agree on what is "fair," then a "fair" outcome can be achieved without compromising on efficiency. All that is needed to achieve a fair outcome in that case is a suitable redistribution of the tradable emission quotas $(e_1^0, \ldots, e_n^0)$ as determined by the equalities (9.7) or equivalently a redistribution of the reference emissions $(\bar{e}_1, \ldots, \bar{e}_n)$, because, as noted earlier, there is one-to-one correspondence between emissions quotas and reference emissions. For instance, if *all* countries agree that it is fair for the developed countries to take historical responsibility for the existing GHG stock, then an outcome that is considered fair by all countries can be achieved by suitably modifying the tradable emissions quotas without compromising on efficiency. In particular, the assignment of tradable emissions quotas can be modified, say to $(e_1^{00}, \ldots, e_n^{00})$ such that $e_i^{00} < e_i^0$ if i is a developed country, $e_i^{00} > e_i^0$ if i is a developing country, and $\sum_{i \in N} e_i^{00} = \sum_{i \in N} e_i^0$. As can easily be verified, such a modified assignment of tradable emissions quotas will also lead each country i to emit—after trade in emissions quotas—the efficient amount e_i^* such that each developing country would be better off than if the assignment of tradable emissions quotas were $(e_1^0, \ldots, e_n^0)$. The resulting outcome would be efficient and may be acceptable to all countries or groups of countries *only* because they all consider the outcome resulting from the modified tradable emissions quotas $(e_1^{00}, \ldots, e_n^{00})$ as fair. In sum, there is no conflict between efficiency and fairness, and the two can be treated separately. I discuss this further by means of a simple model in section 9.4.3.

9.3.2. Agreement on Reference Emissions

Reaching an agreement on reference emissions, however, may not be easy. This is because of the following two problems: first, the current Nash or business-as-usual reference emissions $(\bar{e}_1, \ldots, \bar{e}_n)$ (that determine the tradable pollution quotas $(e_1^0, \ldots, e_n^0)$ and the after-trade transfers $(T_1, \ldots, T_n)$ may be considered unfair, especially by those countries that are in the early stages of their economic development. They currently have low emissions, while the developed countries have high emissions. Because the developing countries will have higher emissions in the future, they can argue that those should be used as the reference emissions instead of the current ones. Thus, the scheme of transfers, though Pareto improving compared to the *current* Nash or business-as-usual reference emissions, may be considered unfair by the developing countries. For instance, as seen from the first row of table 9.2, India's estimated reference emissions were about one-fourth of those of the United States or China.[17] Obviously, India cannot be expected to accept such low reference emissions compared to those of China and the United States.[18]

Second, if the required reductions $(e_i^0 - e_i^*)$ are very large, they may not be politically feasible, at least in the short run. The Kyoto Protocol seems to address both these concerns. Because the emissions of developing countries in general and of India and China in particular were not subjected to ceilings, their emissions would have risen, but those of Annex B countries would have fallen as a result of the required abatements and remained fixed at the levels agreed upon at Kyoto until at least 2012. Thus, the emissions of developing countries would have become comparable to those of Annex B countries—sooner in the case of China than India—and these could then have been subjected to ceilings. Furthermore, the Kyoto Protocol only required relatively small reductions for the immediate future, leaving further reductions to later periods. In other words, the

17. Table 9.2 uses estimates from MIT's EPPA model 1998, which to my knowledge has not been updated. But China and India have actually grown faster than estimated in the EPPA model, especially China, whose emissions now exceed those of the United States.

18. The then prime minister of India made an announcement at the G-8 summit held in November 2007 that the developing countries would never undertake anything that does not match the per capita emissions of developed countries.

Kyoto Protocol was not really inconsistent with the ultimate goal of reaching an agreement regarding country-wise reference emissions $(\overline{e}_1, \ldots, \overline{e}_n)$ in some future round of negotiations, typically in the second or a later commitment period.

Many equity principles for determining the reference emissions have been proposed and discussed in various international forums and meetings. As discussed in the beginning of this subsection, the currently considered baselines of business-as-usual or historically grandfathered emissions are problematic.[19] Similarly, the uniform per capita emissions, being advocated by India and China, also may not be acceptable: if emissions cannot be grandfathered, then by the same logic population size cannot be grandfathered either. A scheme of differential standards of emissions per unit of GDP, independently of how much is the GDP of a country, seems more likely to be acceptable, but it would not fully resolve the problem. That is because as the world economy grows and the total world emissions rise, the standard will have to be revised from time to time and made more stringent even for countries whose emissions may not have risen at all. A rule that is time consistent and free from normative principles is needed such as the one proposed in chapter 8.[20]

But it seems unlikely from the figures in the first and second rows of table 9.2 that the minimal emission reductions or nonreductions implied by the Kyoto Protocol would have been inconsistent with whatever normative criteria were followed for determining the reference emissions. This is especially true in case of India, which unlike China has rather low emissions. In policy terms, this means the developing countries would have been better off if the Kyoto Protocol was fully implemented and extended beyond its expiry date in 2012.

9.3.3. Coalitional Stability of the Trading Equilibrium

If each country were to meet its Kyoto commitment, to be denoted by e_i^0 on its own, the world output would have been equal to $\sum_{i=1}^n g_i(e_i^0)$, which,

---

19. Historically, grandfathered emissions are business-as-usual emissions of some fixed year, not necessarily of the current year.

20. The interested reader may want to refer back to the discussion of this issue in chapter 8.

by definition, is less than $\sum_{i=1}^{n} g_i(\hat{e}_i)$, where $\hat{e}_i$ represents the corresponding competitive trading equilibrium emissions, as defined in (9.5) and (9.6). In fact, as can easily be seen, competitive emission trading allows the countries to restrict their total emissions to their aggregate Kyoto commitment $e^0 = \sum_{i=1}^{n} e_i^0$ at least cost.

As seen above, each country or coalition of countries gains from competitive trade in emissions. However, this does not imply that each country or coalition of countries would be willing to participate in competitive emission trading. For that to be true, we must show further that no country or coalition of countries can gain more by forming a separate bloc and confining trade only among the members of the bloc. An argument based on the theory of market games indeed shows that no coalition of countries can be better off compared to the competitive emission trading equilibrium by forming a separate trading bloc. This is seen as follows.

Let $S \subset N$ be a coalition of countries whose members decide, given their aggregate emission quota $\sum_{i \in S} e_i^0$, to adopt some joint policy of their own such as trading only among themselves or engaging in some other bilateral/multilateral trading agreements among themselves. The maximum payoff that such a coalition of countries can achieve is then

$$w(S) = \max \sum_{i \in S} g_i(e_i) \text{ subject to } \sum_{i \in S} e_i = \sum_{i \in S} e_i^0. \qquad (9.8)$$

We can ignore the damage as it remains the same, as the aggregate emission quota $\sum_{i=1}^{n} e_i^0$ is fixed. Consider again the competitive trading equilibrium emissions $(\hat{e}_1, \ldots, \hat{e}_n)$ with respect to the quotas $(e_1^0, \ldots, e_n^0)$. I show that the payoff of members of S under the competitive equilibrium is not lower than their payoff when they form a separate trading bloc [as defined in (9.8)]. This would establish that no country or coalition of countries will have incentives to form a separate trading bloc and not participate in worldwide competitive emission trading.

To that effect, I show that $\sum_{i \in S}[g_i(\hat{e}_i) + \hat{p}(e_i^0 - \hat{e}_i)] \geq w(S)$, where the $\hat{e}_i$'s are the international competitive trading equilibrium emissions, as in (9.5). This inequality is equivalent to $\sum_{i \in S}[g_i(\hat{e}_i) + \hat{p}(e_i^0 - \hat{e}_i)] \geq \sum_{i \in S} g_i(\tilde{e}_i)$, where $(\tilde{e}_i)_{i \in S}$ is the solution to (9.8). Because $\sum_{i \in S} \tilde{e}_i = \sum_{i \in S} e_i^0$, I must show that $\sum_{i \in S} g_i(\hat{e}_i) + \hat{p}(\sum_{i \in S} \tilde{e}_i - \sum_{i \in S} \hat{e}_i)] \geq \sum_{i \in S} g_i(\tilde{e}_i)$. This inequality is true, as each g_i is concave and $\hat{p} = g_i'(\hat{e}_i)$ in competitive emission trading equilibrium. Therefore, $g_i(\hat{e}_i) + \hat{p}(\tilde{e}_i - \hat{e}_i) \geq g_i(\tilde{e}_i), i \in S$, irrespective of whether $(\tilde{e}_i - \hat{e}_i)$ is positive or negative.

This leads to the conclusion that no country or coalition of countries will have an incentive to form a separate bloc and not participate in worldwide competitive emission trading.

Thus, the outcome of worldwide competitive trade in emissions among the countries cannot be improved upon by the formation of separate coalitions of countries, such as separate blocs for trading in emissions. We are thereby rediscovering a general property of competitive markets known as their "core" property, which says that competitive equilibria belong to the core of an appropriately defined cooperative game.[21]

9.3.4. Desirability of Free Trade in Emissions

While the Kyoto Protocol allowed trade in emissions among the Annex B countries, it left unspecified the extent and nature of such trading. Economic and game-theoretic considerations can be further called upon to deal with this issue. As to the extent of trading, that is, the number of participants in the trade, the market equilibrium theory generally favors trade among the largest number of economic agents. This is also implied by the argument in the preceding section against formation of separate trading blocs or any other form of "coalitions" that restrict trade. Indeed, it is not to the benefit of any country or group of countries to form a coalition and trade separately and independently of other countries.

Thus, it would have been in the world's overall economic interest had the Protocol somehow allowed the non–Annex B countries, whose emissions were however not subject to quotas, to participate in the emission trading process. I argue below that the CDM actually contained provisions to that effect. A policy implication is that this mechanism should have been designed so as to make it as open as possible to the largest number

---

21. This game is a pure market game where externalities play no role because once the emission quotas are fixed, the public good aspect of the problem disappears. One is left with only the private goods–like problem of allocating the total fixed amount of emissions among the countries. Note, however, that this game represents a production economy and not a standard pure exchange economy.

of countries. The fact that no quotas were assigned to some countries was irrelevant if the full benefits of trade in emissions were to be realized.[22]

As to the nature of trading, the same body of theory advocates that the institutions governing trades in emissions be designed so as to ensure that they are as competitive as possible—competitiveness here meaning that all participants behave as price takers. It is indeed only for markets with this property that efficiency, coalitional stability, and worldwide maximal benefits are established. Regulatory provisions that restrict competitiveness in the emissions trading process thus should be avoided, such as, for instance, provisions allowing for market power to be exerted by some traders so as to influence price formation to their advantage, as well as regulatory controls that would impede sufficient price flexibility; or still, as proposed by some countries, limiting the quantities that can be traded. As is well known, the larger the number of participants, the more competitive the market: my argument favoring a larger market thus also favors competition.[23] Larger numbers are admittedly neither the only way nor a sufficient condition to ensure the competitiveness of a market, but they are a powerful factor.

Table 9.2 gives a numerical illustration of the outcome of worldwide competitive trade in emissions.[24] The competitive equilibrium price of emissions $\hat{p}$ is estimated to be equal to US\$24.75 per ton in 1985 dollars. Country i is an *exporter* of emissions credits if $e_i^0 > \hat{e}_i$ and an *importer* if $e_i^0 < \hat{e}_i$. Country i's gain from emissions trade is equal to $\hat{p}(e_i^0 - \hat{e}_i) - (g_i(e_i^0) - g_i(\hat{e}_i))$ if it is an exporter of emissions credits and $g_i(\hat{e}_i) - g_i(e_i^0) - \hat{p}(\hat{e}_i - e_i^0)$ if it is an importer—both are positive, as the

---

22. One might even argue that it was similarly irrelevant whether or not a country ratified the Protocol or did not meet its commitment under the Protocol. Excluding a country from trade in emissions for any reason hurts *all*.

23. My argument on the role of markets to achieve coalitional stability is also reinforced by a central result in economic theory (Debreu and Scarf 1963; Edgeworth 1881) according to which *only* competitive equilibria are coalitionally stable if the number of traders is large.

24. In this table, the quotas, e_i^0, as well as the reference emissions, $\bar{e}_i$, for the non–Annex B countries are taken to be equal to their estimated BAU emissions in 2010. For the former Soviet Union, the reference emissions, $\bar{e}_i$, are taken to be equal to their Kyoto commitment (873.32), although the actual emissions were estimated to be only (762.79). This is equivalent to giving credit for emission reductions that would happen in any case.

price $\hat{p}$ is equal to the *marginal* cost of abatement at $\hat{e}_i$ and g_i is concave. Exporting country i will not gain from trade if it is paid only its actual cost of abatement, that is, $g_i(\hat{e}_i) - g_i(e_i^0)$; all the gains from trade in this case would go to the importing countries. Competitive emission trading thus distributes the gains from trade among the exporters and importers in exactly the same way as it does in competitive markets for any commodity.

Among the developing countries, as seen from table 9.2, China turns out to be the single largest exporter of emissions credits followed by India. Among the Annex B countries, the United States turns out to be the single largest importer of emissions credits followed by the European Union. But all countries gain from emission trading, and the gains are substantial for both sides. The illustration confirms the need for cooperation among the developed and developing countries by institutionalizing such trade. Yet both developed and developing countries were opposed to trade in emissions, though for different reasons.

9.3.5. The Clean Development Mechanism

For the reasons mentioned in section 9.3.4, confining trade in emissions to Annex B countries alone would have affected not only the non–Annex B countries but also the Annex B countries themselves. This raises the question of how the non–Annex B countries could have been involved in emission trading while respecting their insistence to not commit to any emission quotas.[25] This was difficult, but not impossible.[26] In fact, the CDM could have served that purpose to a large extent. In essence, trade under CDM could occur only through "certified project activities"[27] located in

---

25. One colleague expressed this problem as follows: "Should we allow Mexico to 'sell' permits to the US if it is not guaranteed that Mexico will really reduce emissions accordingly?" Also, see Bento, Ho, and Ramirez-Basora (2015), among others, for a similar view.

26. For example, one could calculate the impact of an increase in the tax on fossil fuels in a developing country and offer to transfer to the developing country an amount that is equal to the market value of the consequent estimated reduction in the use of fossil fuels and emissions of the developing country.

27. This is as per the vocabulary of Article 12 of the Kyoto Protocol.

non–Annex B countries. The certification could determine the amount of reductions in emissions ("the certified units") that the project generated, in comparison to a baseline that specified what the emissions would have been in the absence of the project. The amount of the reduction so achieved could then be sold by the initiator of the project to any economic agent belonging to an Annex B country, with the certified amounts being credited to meet the commitment of the country to which the purchasing party belonged.

The price at which the certified units could be sold and purchased would have been determined by supply and demand, which in turn would have been determined by the supply of project activities and the demand for the same by those Annex B countries for which buying the certified units would have been cheaper than reducing their own emissions. Creating competitive markets for the formation of this price would have been as necessary as in the formation of the price for the quotas in the emissions trading scheme of section 9.3.1.

However, the developing countries perhaps feared that participation in any form of trade in emissions will amount to some kind of acceptance of emission quotas on their part. Developing countries such as India and China were of the view that the problem of climate change has been created by the industrialized countries, and therefore it is these countries that should first reduce their emissions, no matter how, before the developing countries can consider accepting any quotas.

In addition, the CDM is often interpreted by the developing countries as a form of trading that distributes gains from trade entirely to the emissions credits–importing (read Annex B) country and none to the emissions credits–exporting (read non–Annex B) country.[28] More so, as it is often proposed that rather than paying an exporting developing country i the market value at the competitive price, that is, $\hat{p}(e_i^0 - \hat{e}_i)$, the importing countries may pay only the actual cost of abatement, that is, $g_i(e_i^0) - g_i(\hat{e}_i)$, which (given the strict concavity of the function g_i) is strictly less than $\hat{p}(e_i^0 - \hat{e}_i)$. This form of trade in emissions can be given

---

28. It is ironic that the countries that generally extol the virtues of competitive markets should look for other forms of trading when it suits them.

effect by the importing countries by systematically "offering" to cover the cost, and the cost alone, of abatement activities in developing countries on a project-by-project basis. This form of trading can indeed be sustained if emissions credits importers collude so as to behave monopolistically.

To conclude this subsection, regardless of whether or not a competitive trade in emissions was established, the developing countries would have benefited from implementation of the Kyoto Protocol. Had the Annex B countries fully met their Kyoto commitments, the international prices of fossil fuels would have been lower, which could have further accelerated the economic growth in developing countries.[29] The energy-exporting non–Annex B countries, however, would have suffered some economic losses because of (a) less revenue from energy exports and (b) higher prices of energy-intensive imports from Annex B countries.[30] But other non–Annex B countries such as India and China with a different mix of imports and exports would have been better off, as shown by Babiker, Reilly, and Jacoby (2000).

9.3.6. The Post-Kyoto Period

The Kyoto Protocol officially came into force on February 16, 2005, which was the date on which the required minimum number of ratifications by countries was reached.[31] But its demise was predicted right from the start. The United States signed the Protocol but, with stiff opposition in Congress from both the House of Representatives and the Senate, never ratified it. In the absence of the United States, the European Union took the lead and pursued the Protocol with some vigor, but it was not a big

---

29. In fact, the non–Annex B countries would have benefited even more if, as some Annex B countries had suggested, no trade in emissions was established among Annex B countries and each country was to meet its Kyoto commitment on its own. That is so because then Annex B countries would have no access to the Russian "hot air" and therefore their actual total reductions in emissions would have been much larger.

30. The prices of energy-intensive goods may be higher in Annex B countries because of higher taxes on fossil energy to discourage their use.

31. Recall from footnote 6 that it is 55 countries, representing 55 percent of the world total emissions.

success. Russia, Canada, Australia, and Japan did less than what they had promised to do under the Protocol. Though the European Union succeeding in meeting its own target of an 8 percent reduction (mostly because it entered into recession and had lower than expected growth), the overall world emissions continued to grow at a fast pace. To begin with, the targets to reduce emissions were set rather low, and, in the absence of the United States, the countries realized the futility of implementing the Protocol. As a result, only a handful of countries were willing to make commitments after the expiry date of 2012.[32]

Inspired by the results generated by the theory of internal-external stability of coalitions—which question the logical likelihood of a world-wide agreement on climate change—a stream of thought has spread the thesis,[33] from the late 1990s onward, that only an agreement among a small number of countries can ever emerge. In chapter 6, I have shown the weaknesses of the theoretical underpinnings of this internal-external stability thesis. Now it also stands empirically disproved as the Paris Agreement is an agreement among at least 195 countries, not a small number of countries. In fact, the Paris Agreement is best explained in terms of the coalition formation game proposed in chapter 6 that asserts that the grand coalition of all countries will form if the players are farsighted and if the coalition formation game is played repeatedly. Indeed, not many countries were willing to continue with emission reductions beyond the expiry date of the Kyoto Protocol unless the United States—the world's biggest economy and the second biggest emitter—and some other countries were also willing to do so. It may be noted in this regard that if a country is a small emitter, then it does not really matter whether or not it joins the other countries to form the grand coalition because its equilibrium strategy/emissions would more or less be the same. In other words, formation of a coalition of all major emitters is effectively equivalent to formation of the grand

---

32. To be precise, a second commitment period was agreed on in 2012, known as the Doha Amendment to the protocol. But only 55 states accepted the Doha Amendment, while entry into force required the acceptance of 144 states.

33. See, for instance, Carraro (2008), Buchner and Carraro (2009), and Rubio and Ulph (2007). Nordhaus (2015) is a more recent subscriber to this thesis.

coalition. To put it another way, *only* a coalition that does not include all major emitters is not equivalent to the grand coalition. Thus, as the coalition formation game in chapter 6 predicts, there were only two possibilities left after the expiry of the first period of commitment of the Kyoto Protocol: either the United States will ratify an extension of the Kyoto Protocol and agree to reduce its emissions or the world will return to the pre-Kyoto situation and negotiations will start afresh and continue until a new agreement that is acceptable to all countries including the United States and the developing countries is reached at a future date. That future date turned out to be December 12, 2015, at Paris. On that day, France—the host for the Conference of Parties (COP)-21—had prepared, after much deliberation and consultation with various countries and groups, a "take it or leave it" draft agreement. The countries chose to take it! Rejection by any one of the major emitters—China, the United States, the European Union, or India—would have, as the coalition formation theory in chapter 6 predicts, led to further rounds of negotiations in the future.

The United States, under the Obama administration, ratified the Paris Agreement, but has recently decided, under the Trump administration, to withdraw from it. However, there are indications that the United States may reverse its decision and rejoin the Paris Agreement on terms that are "fair to the United States." In view of the uncertainty regarding the U.S. participation in the Paris Agreement, I first study the Paris Agreement as if the United States had not withdrawn, and then discuss implications of the U.S. withdrawal if the United States does not reverse its decision and indeed withdraws.

9.4. THE PARIS AGREEMENT AND THE ROAD AHEAD

It is said that the Paris Agreement, on its own, will not solve global warming. At best, scientists who have analyzed it say that it will cut global GHGs emissions by about half enough required to stave off an increase in atmospheric temperature of 2°C, or 3.6°F. But it is too early to judge how effective the Paris Agreement will be, as it has not yet been translated into actions for its entire duration. Rather, it should be seen as an agreement that, as of now, only includes an indication of actions to be taken by *each* country in the commitment periods 2020–2025 and 2025–2030. The real test of the Paris Agreement will be in how it is implemented in the future commitment periods. As the then United Nations chief Ban Ki-moon put it, "We have an agreement. It is a good agreement. You should be proud of it. Now we must stay united—and bring the same spirit to the crucial test of implementation. That work starts tomorrow."

In this section, I note the main features of the Paris Agreement that are relevant for my purposes and interpret and analyze it in light of the environmental games in the previous chapters. I then discuss what may be done for its successful implementation.

9.4.1. The Long-Term Goals

The Paris Agreement has the following long-term goals: (i) Restrict the global temperature rise to "well below" 2°C above the pre-industrial level by the end of the century with a promise to "pursue" a tougher goal of limiting the increase to 1.5°C. (ii) All countries are to stop the rise in GHGs emissions "as soon as possible," but developing countries may take more time. (iii) Achieve by some point after 2050 a situation whereby all man-made emissions can be absorbed by forests, soil, and oceans.

Because global warming is a function of the GHGs stock z, the long-term goal of restricting climate change to +2°C, as also noted in chapter 8, is effectively equivalent to a goal of stabilizing the GHGs stock at a specific level, say $\hat{z}$, which is determined by the ecological phenomenon, independently of not only the current stock z_0 but also of how, when, and who may contribute to it.[34] The implicit goal of stabilizing the GHGs stock at $\hat{z}$ has been proposed by climate scientists and others because damage from climate change of more than +2°C will be devastating for many regions of the world. But from an economics perspective, we must ask whether limiting global warming to +2°C is socially optimal. I now identify conditions under which it is. Formally, as shown in chapter 8, the socially optimal

---

34. It has been estimated that GHG concentration of no more than 450 parts per million (ppm) in the atmosphere can limit global warming to 2°C.

steady-state GHGs stock z^* and emissions $(e_1^*, \ldots, e_n^*)$ are characterized by the following equations:

$$(r + \delta)\left(\sigma - h_i'(e_i^*)\right) = \sum_{i \in N} v_i'(z^*) \text{ and } \sum_{i \in N} e_i^* = \delta z^*, i = 1, 2, \ldots, n, \quad (9.9)$$

where r is the discount rate; δ is the natural rate of decay of the GHGs stock; $\sigma > 0$ is the "technology parameter," which indicates how much of the consumption good is produced for each unit of emissions in a country;[35] h_i is the local damage function, which specifies damage to country i from its own emissions; and v_i is the damage function, which specifies damage from the GHGs stock.[36] As noted in chapter 8, these equations are actually the dynamic counterpart of the familiar Lindahl-Samuelson condition for optimal public good provision (see chapter 8 for more details).

If each *local* damage function h_i is strictly convex and each damage function v_i is convex, these equations admit a unique solution, which depends on the magnitude of the technology parameter σ. In fact, the equations imply that the higher the value of the technology parameter σ, the higher the socially optimal steady-state GHGs stock z^* and emissions $(e_1^*, \ldots, e_n^*)$. This is because a higher σ implies a higher cost of abatement for each country. In other words, the GHGs stock $\hat{z}$ for stabilizing global warming at +2°C is socially optimal (i.e., $\hat{z} = z^*$) only for a specific value of the technology parameter σ. But since the *world* damage from climate change, as many scientific studies show, rises very steeply once the GHGs stock reaches the threshold $\hat{z}$, it follows that $\hat{z}$ is also "nearly" socially optimal for all higher values of the technology parameter σ. Put another way, once the GHGs stock reaches the threshold $\hat{z}$, it will no longer be possible to achieve a higher world welfare without adopting technologies that can produce additional amounts of the consumption good without increasing the emissions and the GHGs stock. This is because any increase in the GHGs stock then will result in damage that is more than the additional amount of the consumption good produced. Thus, stabilizing climate

---

35. It is the inverse of "emission intensity" of production; a higher value of σ means a cleaner production technology. For simplicity, the discount rate r and the technology parameter σ are taken to be the same for all countries.

36. Besides releasing GHGs, burning of fossil fuels can also cause damage locally, such as smog.

change at +2°C, if socially optimal, implies that any additional output of the consumption good may not lead to higher world welfare unless the additional output is produced with a cleaner technology such that the emissions and the GHG stock are not higher. In other words, the world output of the consumption good and welfare may stagnate unless there is continuous innovation and adoption of cleaner technologies, leading eventually to innovation and adoption of almost "carbon-free" technologies.

The difference between the base year GHGs stock z_0 and the targeted stock $\hat{z}$ determines what has come to be known as the "carbon budget": it is the amount of GHGs that can be emitted such that the GHGs stock will not exceed $\hat{z}$ by the end of the century. As per the Fifth Assessment Report (AR5) of the IPCC (2013), for an even chance of limiting temperature increase to 2°C, the total amount of GHGs that can be emitted between the base year 2011 until the target year 2100 is the equivalent of 900 gigatons of CO_2 equivalent. The UNFCCC's *Synthesis Report on the Aggregate Effect of Intended Nationally Determined Contributions (INDCs)* estimates that the CO_2 equivalent emissions for 2011–2025 will aggregate to 542 gigatons and for 2011–2030 to 748 gigatons, which leaves a meager carbon budget of 122 gigatons of CO_2 equivalent for the world for the remaining 70 years from 2030 to 2100. Therefore, unless there are great scientific discoveries and adoption of significantly cleaner technologies, the world faces the prospect of an economic collapse and hardship either because of devastation due to climate change or because of lower production of the consumption goods to avoid the devastation due to climate change.

Besides the size of the carbon budget, the time profile of emissions also matters. To see why, figure 8.1 in chapter 8 illustrates two alternative trajectories for the GHGs stock, both of which lead to the threshold $\hat{z}$ by the end of the century. The upper trajectory implies implicitly that immediately before the stock rises to $\hat{z}$, the world emissions are at a rate that is just a little more than the rate of natural decay of the GHGs stock, that is, $\sum_{i \in N} e_i(t) \cong \delta \hat{z}$, and then become equal to it. But this is not so in the case of the lower trajectory. In both cases the stock rises to $\hat{z}$ in the same year, but in the case of the lower trajectory, unless there is a sudden reduction in the rate of world emissions, the GHGs stock may continue to rise beyond $\hat{z}$. The point is that if the stock is to be stabilized at $\hat{z}$, then the countries will need not only to agree on how the carbon budget is to be shared but also to choose the time profiles of their emissions such that they will not face a sudden economic collapse. At the minimum, they need to fix a year in which their emissions will peak, then decline, and eventually stabilize at the socially optimal levels.

9.4.2. The 5-Year National Action Plans

The Paris Agreement has something for every country and grouping and is flexible on when a country has to make the necessary reductions in its emissions and by how much. That is both its strength and its weakness. To be specific, it proposes the following architecture for the actions to be taken by the countries to achieve its stated long-term goals: (i) The developed countries are to undertake absolute reduction of emissions. (ii) The developing countries are to increase efforts to reduce emissions and to be "encouraged" to move to absolute reductions. (iii) Beginning in 2020, when the agreement will come into force, every country is required to submit a national action plan every 5 years. (iv) Every successive national plan must be a progression beyond the previous one. (v) A common time frame for national plans will be discussed at the next round of climate talks in 2018.

Already 147 countries have submitted their national action plans, known in United Nations climate lingo as intended nationally determined contributions (INDCs). The remaining countries have been requested to submit their plans. These plans will become part of a public registry maintained by the climate arm of the United Nations. Their implementation will be monitored and reviewed regularly to encourage countries to raise their targets over time. For the first round, a majority of countries have provided 10-year plans for the period 2020–2030. The accompanying decisions suggest that those countries are free to resubmit their plans or revise them after 5 years. The INDCs submitted by most countries, however, are insufficient and will have to be adapted to the new level of ambition in the next few years. But this is probably of less importance than the very fact that these commitments are being made by every country, made publicly known, and can eventually be converted into effective reductions. Currently, they may not be sufficient to tackle climate change, but as will be seen below, they have the potential to tackle it.

The Paris Agreement is formulated in terms of a precise sequence of back-to-back commitment periods of 5 years each—which will be 16 in total by the end of the century. Each one of these back-to-back commitment periods may be considered as being covered by an agreement of its own as part of a well-structured sequence that constitutes an overall agreement, which allows for redefining strategies at the beginning of each commitment period depending on the level of the GHGs stock at that point of time. On the face of it, the architecture of these back-to-back commitment periods seems to be of the same type as that of the first commitment period (2008–2012) of the Kyoto Protocol. While this is true, there are some important differences that give the Paris Agreement a better chance of success.

First, China and India have agreed to control the rates at which their emissions may increase by adopting cleaner technologies. In contrast, the Kyoto Protocol required the developing countries to undertake such actions only as CDM projects, which could at best shift the responsibility for reducing emissions—from the Annex B countries of the Protocol to the developing countries—but not lead to additional reductions in the aggregate. Second, the back-to-back 5-year national action plans going far into the future can set in place a stable demand and a global market for cleaner energy and technology. Therefore, the task of developing cleaner energy and technology may be carried out largely by businesses and investors operating under emission-reduction policies that will be in place with more certainty for a longer period of time.

The back-to-back 5-year national action plans fit well with the framework of the dynamic environmental game in discrete time studied in chapter 7, and the emissions proposed via the INDCs submitted by the countries are similar to the Nash equilibrium emissions of the dynamic environmental game. This means there may be further scope for cooperation and emission reductions by instituting another form of emission trading that was suggested in chapter 8. To be specific, let $(\bar{e}_{1t}, \ldots, \bar{e}_{nt})$ denote the emissions proposed by the countries via the INDCs at the beginning of a 5-year commitment period starting at time t. Then, each country may benefit by offering to reduce its proposed emissions for the 5-year commitment period in exchange for similar reductions by other countries, leading them in this way to agree to limit their emissions to, say $(e_{1t}^0, \ldots, e_{nt}^0)$, such that $\sum_{i \in N} e_{it}^0 < \sum_{i \in N} \bar{e}_{it}$. This form of trading is known as barter or non-tatonnement exchange in economics. In contrast to trade in emissions rights on a competitive international market that can only lead to a reallocation of the rights but not to reduction in the aggregate emissions, barter trade in emissions can induce the countries to reduce the emissions proposed by them via the INDCs such that the sum total is reduced and each country is better off.

Role of Cleaner Technologies

Unlike in the case of the Kyoto Protocol, development and adoption of cleaner technologies is crucial for meeting the goals of the Paris Agreement. Because the Agreement proposes deeper cuts in total world emissions, which are possible only with adoption of cleaner technologies, it conveys a critical message to the marketplace. Hours after the Paris Agreement, then U.S. President Barack Obama said that the international agreement on climate change reached in Paris has set in place a demand for clean energy. To quote: "We now have a global marketplace for clean energy that is stable and accelerating over the course of next decade."

However, relying on market forces alone to develop and adopt cleaner technologies may not be enough. In particular, the developing countries may not be able to adopt cleaner technologies because of intellectual property rights barriers to transfer of technology. But intellectual property rights cannot just be wished away. They need to be protected to strengthen private incentives to innovate. For this reason, it seems the Paris Agreement proposes cooperation in development, rather than transfer, of cleaner technologies. In fact, a number of cooperative initiatives for development of cleaner technologies have been already taken. These include the International Solar Alliance of 121 countries initiated by Indian Prime Minister Narendra Modi and the Breakthrough Energy Coalition initiated by Bill Gates and Mark Zuckerberg. Both were launched at an event during COP-21.

But given the importance of cleaner technologies in limiting global warming to 2°C and that the development of cleaner technologies cannot be left merely to markets, a global agreement in parallel with the

Paris Agreement, and similar to the International Solar Alliance of 121 countries, to fund research for development of cleaner technologies on a global scale may be needed. An agreement to create such a fund to which governments of all countries as well as private individuals and entities may contribute can boost the efforts to develop cleaner technologies and overcome the problem of intellectual property rights barriers, as the cleaner technologies so developed can be freely transferred.

Trade in Emissions

The Paris Agreement has similar provisions for trade in emissions as the Kyoto Protocol. Thus, the same analysis applies in that trade in emissions should not be restricted in any way, and policies and institutions should be designed to make it as competitive as possible so that GHGs emissions can be reduced at least cost. But no CDM is needed, as the INDCs of the developing countries can be used as the baseline for the purpose of international trade in emissions rights. Also, the long-term nature of the Paris Agreement opens newer possibilities, as trade in emissions across 5-year commitment periods can also be considered to further reduce the cost of meeting the commitments. More specifically, country i may agree to reduce its emissions beyond the commitment made via its INDCs in some commitment period t by an amount Δ so that country j can emit more than its commitment made via its INDCs in period t by the same amount Δ in exchange for country j agreeing to reduce its emissions beyond its commitment via its INDCs in some future period $t + k$ by an amount Δ' so that country i can emit more than its commitment made via its INDCs in period $t + k$ by the amount Δ'. Such bilateral trades across periods can also reduce costs of fulfilling the commitments made by countries i and j via their INDCs and benefit them both.

9.4.3. How Much Is Too Much?

Though the INDCs submitted by the countries for the first commitment periods 2020–2025 and 2025–2030 are not sufficient to achieve even the relatively less ambitious goal of restricting global warming to 2°C, the more ambitious goal of limiting temperature rise to 1.5°C has also been included in the Paris Agreement after a push by the small island states such as Kiribati, who say that limiting temperature rise to 1.5°C could allow them to survive. Thus, the inclusion of the 1.5°C target is a huge win for the small island states and the other nations that are more vulnerable to climate change effects, particularly to sea-level rise. It has also helped in keeping the grand coalition of all countries intact.[37] In pursuance of the more ambitious goal, the IPCC has been tasked to provide a special report in 2018 on the implications of limiting global warming to 1.5°C, and a final decision regarding the target for temperature rise may be taken after that.

But is it really socially optimal to restrict climate change to 1.5°C rather than to 2°C? Social optimality requires weighing the damage that the small island states will suffer against the economic gains net of additional damage for the rest of the world if the temperature is allowed to rise by 2°C rather than 1.5°C. If the gains net of additional damage for the rest of the world are more than the damage suffered by the small island states, then allowing a temperature rise of 2°C rather than 1.5°C is welfare improving, though not a Pareto improvement.

The carbon budget corresponding to a 1.5°C target is so limited that developed countries will have to reach net zero emissions in the next 5–10 years. The developing countries will have some more time, but they will not be able to meet their basic development and poverty-alleviation goals with the reduced carbon budget unless they will be provided massive support in terms of finance, technologies, and capacity building by the developed countries at a time when the developed countries will themselves be engaged in making huge investments to drastically reduce their own emissions.

If the economic cost of limiting climate change to 1.5°C is indeed too high for the rest of the world but fairness is a real concern, then a better policy may be to not compromise on the temperature rise but to offer assistance to the small island states to meet the cost of adaptation such as relocation of their populations or building of dikes for protection against

---

37. When a decision to include the more ambitious target of 1.5°C seemed unlikely at COP-21, the Marshall Islands' Foreign Minister Tony De Brum is reported to have commented: "I think we're done here."

sea rise. This may also help keep the grand coalition intact and achieve efficiency, but at a much lower cost to the rest of the world.[38]

Fairness in a Simple Model of Climate Change

A simple model of climate change may further help to justify this policy. In this model there is one pollutee, the low-lying small island states and the least developed small countries, and one polluter, the rest of the world. The polluter can produce a composite consumption good according to an increasing and concave production function $g_1(e)$, where e is the amount of GHGs emissions over a period of time. For simplicity, I take both the initial stock of GHGs and the natural rate of decay of GHGs to be zero. Thus, e is also the GHGs stock at the end of the period. The utility of the polluter is given by $u_1 = w_1 + g_1(e) - v_1(e)$, where $w_1 > 0$ is its initial endowment of the composite consumption good (or wealth), and $v_1(e)$ is the damage from climate change for the polluter, assumed to be increasing and convex in e. Notice that unlike the classic Coasian model of one polluter and one pollutee discussed in chapter 3, the polluter in the present model is also affected by its own pollution. The utility of the pollutee is given by $u_2 = w_2 - v_2(e)$. That is, the pollutee neither produces nor pollutes. The optimal emissions are given by e^*, where e^* is the unique solution to the optimality condition $g_1'(e) - v_1'(e) = v_2'(e)$ or $g_1'(e) = v_1'(e) + v_2'(e)$. That is, the marginal cast of abatement is equal to the sum of marginal damage from climate change for the polluter and the pollutee.

This simple world economy is represented in figure 9.1. The initial endowment w_1 is increasing to the right of point 0_1 and w_2 to the left of point 0_2. Then, point H on the horizontal line in the box of length $w_1 + w_2$ represents the initial endowments of the consumption good of both countries, and thus $0_1 H = w_1$ and $0_2 H = w_2$.

---

38. However, this policy may be considered "unfair" and discriminatory by some because there are also other low-lying small islands that are similarly vulnerable to climate change but are not independent sovereign states as they fall under the jurisdictions of some larger countries. In addition, there are some disputed islands that are not inhabited by humans but have significant existence and option values.

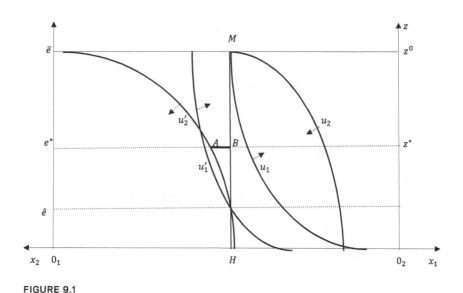

FIGURE 9.1

Fairness in a one polluter–one pollutee economy

I assume that the polluter has (by default) the right to pollute and in the absence of any agreement will emit $\bar{e}$ to maximize its utility; that is, $\bar{e}$ is the solution of $g_1'(e) = v_1'(e)$. The set of points (e, x_1) that satisfy the equation $x_1 + g_1(e) - v_1(e) = w_1 + g_1(\bar{e}) - v_1(\bar{e})$ constitutes an indifference curve for the polluter yielding the utility level $w_1 + g_1(\bar{e}) - v_1(\bar{e})$; it is represented by the curve labeled u_1 passing through point M. The convexity of this curve results from the assumed strict concavity of the production function $g_1(e)$ and convexity of the damage function $v_1(e)$. Letting w_1 vary, a family of congruent indifference curves is generated, to the left of the indifference curve u_1 for lower values of w_1 and to the right of u_1 for higher values. Similarly, the set of points that satisfy the equation $x_2 - v_2(e) = w_2 - v_2(\bar{e})$ constitutes an indifference curve passing through point M and yields utility level $w_2 - v_2(\bar{e})$; it is represented by the curve labeled u_2, and the convexity of this curve results from the convexity of the damage function v_2. Letting w_2 vary, a family of congruent indifference curves is generated, to the left of the indifference curve u_2 for higher values of w_2 and to the right of u_2 for lower values. The points on the line segment e^*z^* are the points of tangency of the indifference curves of the polluter and the pollutee and, therefore, represent the socially optimal outcomes.

Suppose that out of concern for fairness, the polluter agrees to emit no more than $\hat{e} < e^*$. With this self-imposed constraint, the maximum utilities that the polluter and the pollutee can achieve are u_1' and u_2', as shown in figure 9.1. However, both the polluter and the pollutee would be better off if the polluter were to emit instead the socially optimal amount e^* and make a transfer equal to the length of the line segment AB to the pollutee.

Because temperature rise is a function of the GHG stock, this can be interpreted to mean that rather than compromising—for the sake of fairness—on the optimal level of temperature rise, a Pareto superior policy is for the polluter (i.e., the rest of the world) to not compromise on the socially optimal level of temperature rise, but to offer instead transfers to adequately compensate the pollutee (i.e., the low-lying island states and the least developed small countries) that can be used by these countries to meet the costs of adaptation.

9.5. THE DEVELOPING COUNTRIES AND THE PARIS AGREEMENT

The Kyoto Protocol was based on the principle of "common but differentiated responsibilities" (CBDR), which put the obligation to reduce current emissions on the developed countries on the basis that they are historically responsible for the accumulated GHGs stock in the atmosphere. It did not require the developing countries to emit below their BAU levels. However, they could undertake CDM projects such as wind farms or solar panels and earn credits for every ton of carbon avoided due to the projects undertaken. These could be sold to Annex B countries to help them meet their Kyoto targets. In contrast, the Paris Agreement requires all parties—including the developing nations—to take actions to cut emissions. Though the CBDR principle is also mentioned in the text of the Paris Agreement, some experts say it has been diluted after tremendous pressure from the United States—which is facing issues with domestic politics—and an umbrella group of developed nations. It may be recalled that the United States did not ratify the Kyoto Protocol for the same reason, as it faced opposition from the House of Representatives and the Senate to the idea of participating in a global agreement that required the United States, but not the developing countries, to curb its high greenhouse gas emissions. Hours after the 196 countries arrived at the agreement in Paris, a senior Obama administration official is reported to have said that differentiation in climate responsibility will now be a forward-looking concept, as opposed to the earlier backward-looking notion. It revises the architecture of the climate change system with the means of differentiation that look forward, not back. In contrast, the Indian Minister for Environment and Forests, Prakash Javadekar, said that the agreement could have been more ambitious as the proposed actions of the developed nations are "far below" their historical responsibilities and fair shares.

9.5.1. The Actual INDCs of the Developing Countries

A look at the actual INDCs submitted by China and India for the period 2020–2030 can further clarify how the Paris Agreement differs from the Kyoto Protocol.

China

By 2030, China proposes to peak its emissions to lower the carbon intensity of GDP by 60 to 65 percent below 2005 levels, to increase the share of renewables in the total energy supply to around 20 percent, and to increase its forest stock volume by 4.5 billion cubic meters compared to 2005 levels.

India

By 2030, India proposes to lower the emissions intensity of GDP by 33 to 35 percent below 2005 levels, to increase the share of non-fossil-based power generation capacity to 40 percent of installed capacity (equivalent to 26 to 30 percent of power generation in 2030), and to create an additional (cumulative) carbon sink of 2.5 to 3.0 gigatons through additional forest and tree cover.

Reducing the emissions intensity of GDP and increasing the forest cover to absorb more CO_2 by India and China would have qualified as certified CDM projects under the Kyoto Protocol and would have led to significant financial transfers to them. Thus, voluntarily making these commitments without compensation to facilitate the Paris Agreement constitutes a generous gesture by both these developing countries. To get an idea of the costs of meeting the obligations of the Paris Agreement by the developing countries, it may be noted that the Indian government in 2015 said that the country would need U.S.$2.5 trillion to reduce GHGs emissions per unit of GDP by 33 to 35 percent by 2030 from 2005 levels, and for other mitigation and adaptation efforts.

Besides, it is worth noting that India's emissions in 2013 were just 5.7 percent of the global total. In comparison, the United States, the European Union, Russia, Japan, and Canada contributed 40.6 percent of global emissions in 2012. According to the consortium Climate Action Tracker, India's emissions will increase 90 percent by 2030, if current policies are any indication, and will be about 9 percent of the global total by 2030, still well below those of the United States or the European Union. In addition, India has a relatively clean past. But because India is industrializing later than the developed countries, there has been maximum global interest in how it plans to manage its economic growth. In fact, the timing of India's economic growth is such that it is sometimes seen, even blamed, as an obstacle to putting the world on a path away from dependence on fossil fuels.

Though India has been able to successfully deflect this misplaced criticism, many experts believe that it will be under constant pressure to take on a greater burden for mitigating climate change by 2020 and beyond, especially when the next review of the INDCs of all countries takes place. But accepting any emissions reductions by India without adequate compensation to enable it to invest in cleaner technologies and energy would amount to saying goodbye to the efforts to lift its people out of extreme poverty. In fact, India's per capita consumption of fossil energy is the lowest in the world—less than one-eighth of that of the United States and one-fourth of that of China. No democratically elected government in India can survive if it has to make policies—such as imposing higher taxes on fossil fuels—that further discourage fossil energy consumption by an already deprived population without providing access to alternative and economical sources of energy. Thus, the best that India can contribute toward controlling climate change is to invest in cleaner technologies and energy. Indeed, India has initiated an ambitious program to generate solar, wind, nuclear, and hydro energy.[39] It has no choice but to refuse for the next few decades any reduction in its emissions that is more than what is possible with cleaner technologies and use of cleaner energy. Thus, deeper cuts in emissions by the developed countries are the only reasonable way forward by which the Paris Agreement can be fully implemented, especially as that would unleash stronger market forces for innovation and adoption of cleaner technologies for which the developed countries are far better equipped.

Implications of the U.S. Withdrawal

After ratifying the Paris Agreement, the United States has recently decided to withdraw from it, but has offered to negotiate to re-enter either the Paris Agreement or enter an entirely new agreement on terms that are "fair to the United States." The remaining signatories to the Agreement have taken the high moral ground and publicly announced to do even more to implement the Agreement. Announcements to this effect were made even before the decision to withdraw was taken; ostensibly to put moral pressure on the United States to not withdraw. However, such announcements, as the coalition formation theory in chapter 6 predicts, may have encouraged, rather than deter, the United States to withdraw, as the announcements amounted to assuring the United States of a free ride. The United States would have had second thoughts had the other countries announced instead that they would also withdraw from the Agreement if the United States does, as that would have made the withdrawal costly for the United States and could have discouraged it from withdrawing.

The United States, under the Bush administration, had mentioned China and India as the reason for not ratifying the Kyoto Protocol. It has again cited China and India as the reason for withdrawing from the Paris Agreement. Thus, history is likely to repeat itself in that the Paris Agreement may fizzle out for the same reason as the Kyoto Protocol, unless the United States soon reverses its decision to withdraw.

---

39. India's renewable energy capacity has already crossed the 50 gigawatts mark, doubling in just 5 years. And solar power capacity, which was hardly anything 5 years back, reached 12 gigawatts. Going by capacities awarded and live tenders, this number could well double in 2 years.

It was noted earlier that there is little room to push India to do more; and as far as China is concerned, the United States and China had entered into a bilateral agreement on climate change ahead of the negotiations in Paris. If the United States is really serious about re-entering the Paris Agreement on terms that are fair to the United States, then perhaps it needs to first negotiate a new bilateral agreement with China. Clubbing India with China seems to be a strategic mistake, since, as noted earlier, the case against India is rather weak. All that the clubbing does is push India to join hands with China against the U.S. policy on climate change.

To summarize, the thinking in the developing countries is that the industrialized or developed countries have already used up much of the limited carbon space and, thus, they are the ones who should take the responsibility for controlling climate change, and leave most of the remaining carbon space for the developing countries. This thinking was implicitly accepted in the Kyoto Protocol, but was somewhat diluted in the Paris Agreement. In contrast, the current U.S. administration believes that it is not fair that the U.S. emissions of greenhouse gases should begin to decline soon, while those of the developing countries may continue to rise, even if at a slower pace, for some more time. But this thinking has not found any takers and, as of now, the United States seems to be isolated in this regard.

As noted earlier, the Paris Agreement is little more than a declaration of intentions. In fact, the Paris Agreement is flexible on when a country has to make reductions in its emissions and by how much. Thus, there was no need for the United States to withdraw from the Agreement. It could have achieved its goal of an agreement that is "fair to the United States" by negotiating within the procedures and mechanisms of the Paris Agreement. For this reason many negotiators and policymakers believe that the United States withdrawal is just an attempt to fulfill the campaign rhetoric of the recent presidential election in the United States. Thus, most countries seem to think that it is just a matter of waiting out the Trump administration before the United States rejoins the Paris Agreement.

10

INTERNATIONAL TRADE AND CLIMATE CHANGE

B esides the negotiations on climate change, most countries are, at the same time, also engaged in negotiations on trade liberalization in goods through the World Trade Organization (WTO).[1] Though these negotiations have been taking place independently of each other, it has been claimed, especially by the environmentalists, that trade liberalization will contribute to climate change as it will, on the one hand, increase the scale of economic activities and therefore the accompanying emissions (see, e.g., Audley 1993) and, on the other hand, shift production of the pollution-intensive goods from countries with strict environmental regulations toward those with lax ones.[2] The response from the proponents of free international trade has been the argument that environment is a normal good and hence trade-induced income gains would lead to stricter environmental regulation and neutralize the effect of trade liberalization on climate change (see, e.g., Bhagwati 1993). Thus, the current debate centers on the strength of the income effect. Empirical work (e.g., Grossman and Krueger 1993) and theoretical models (e.g., Copeland and Taylor 1995) suggest that income gains from trade can indeed have a substantial impact on emissions. This argument, though important, does not

---

1. The Paris Agreement was concluded just a few days before the WTO Ministerial Conference began in Nairobi.
2. See the United Nations Sustainable Programme publication *Environment and Trade* for a detailed discussion of how the environment and international trade in goods may be linked. Other related papers include Low and Yeats (1992), Barrett (1994b), and Jaffe et al. (1995). A classic reference is Markusen (1975).

challenge but only qualifies the environmentalists' claim in that it seems to concede that but for the income effects, trade liberalization would aggravate climate change. Furthermore, if that is so, then why was the Paris Agreement on climate change concluded without waiting for the outcome of the WTO conference?

In this chapter, I study the relationship between international trade in goods and climate change. In a trade model with two consumption goods and specific utility and production functions, I show that world emissions may not be higher under free international trade in goods even in the complete absence of income effects. The environmentalists' argument against free international trade, as well as that of the proponents of free international trade, overlooks the fact that the much emphasized differences in environmental concerns across the countries lead to differences in the pollution intensity of production and hold the possibility of increasing world output and income without raising world emissions by shifting production of the polluting good from countries with higher pollution-intensity of production to those with lower pollution-intensity of production. I show that free international trade in goods and emissions can induce precisely such a shift in production. This is just the opposite of the environmentalists' claim that free international trade in goods will lead to higher world emissions and shift the production of pollution-intensive goods toward countries with lax environmental regulation.

The emphasis in this chapter is not so much on cooperation and efficient control of climate change as it is on the effect of international trade in goods on climate change. However, I shall briefly return to discuss possibility of cooperation for achieving efficiency later in the chapter, as I show that the model with two consumption goods can be reduced to a model that is identical to the basic model with only one consumption good in chapter 5, and therefore it is amenable to similar game-theoretic analysis. Furthermore, I show that if the damage functions are linear, then the equilibrium prices of the consumption goods do not depend on the total amount of emissions. This means if the damage functions are linear, then the agreements on climate change do not depend on trade policy and can be entered into independent of the trade policy.

To introduce international trade in goods and study its impact on climate change in a simple manner, I consider a model that consists of two primary factors of production, capital and labor, and two composite consumption goods of which good 1 is polluting and good 2 is nonpolluting. Capital is mobile across the two sectors, but labor is specific to the production of the nonpolluting good. The utility function of each country is assumed to be linear in good 1, so that income effects are absent, but log-linear in good 2 so that substitution effects are not ruled out. This makes it possible to consider international trade in goods and analyze the relationship between patterns of trade and the total amount of emissions.

The model in this chapter is in one respect more general than those in the previous chapters, as there are two consumption goods, but in another respect it is less general, as I assume specific utility and production functions. Furthermore, as is often assumed in conventional trade models, I assume the countries to differ only in terms of their endowments of primary factors of production. Also, to highlight comparative welfare gains, I will sometimes assume that there are only two types of countries: North and South.

I first characterize the autarky equilibrium when there is no international trade in goods so that it can be used as a benchmark against the equilibrium with free international trade in goods. I show that if the South is "sufficiently" abundant in the factor specific to production of the nonpolluting good, then North exports the polluting good and imports the nonpolluting good. The total amount of emissions falls below the autarky level, but the world output of the polluting good and income rise. The welfare of the North may rise and that of the South may fall.

I then introduce trade in pollution permits in addition to free international trade in goods[3] and show that the total amount of emissions under free international trade in both goods and pollution permits falls below the autarky level whatever the factor endowments of the countries. Furthermore, North exports the polluting good if it has a smaller endowment of the factor specific to the production of the nonpolluting good.

---

3. The Paris Agreement proposes to establish trade in emissions independently of free trade in goods.

Though free trade in goods and emissions reduces the total amount of emissions below the autarky level, the total amount of emissions is still above the efficient level. The last section of the chapter shows how the analysis in chapter 5 can be extended to the trade model. It is shown that if the damage functions are linear, an agreement on climate change may not depend on trade policy.

This chapter is organized as follows. Section 10.1 introduces the model. Section 10.2 analyzes the relationship between patterns of trade in goods and the total amount of emissions. Section 10.3 analyzes the impact of trade in emissions or pollution permits in addition to free international trade in goods on the pattern of trade and the total amount of emissions. Section 10.4 returns to the issue of efficiency in controlling climate change. It extends the analysis in chapter 5 to the trade model with linear damage functions.

10.1. A MODEL WITH TWO CONSUMPTION GOODS

There are n countries each endowed with two primary factors of production: labor and capital. There are two consumption goods. Production of good 1 requires use of capital and generates GHGs as a by-product. Accordingly, I refer to good 1 as the *polluting good*. Good 2 is produced by combining both capital and labor, and its production does not lead to emissions. Accordingly, I refer to it as the *nonpolluting good*. More specifically,

$$y_{i1} = \begin{cases} e_i^{\alpha} k_{i1}^{1-\alpha} & \text{if } e_i < ak_{i1} \\ a^{\alpha} k_{i1} & \text{if } e_i \geq ak_{i1}; \end{cases}$$

$$y_{i2} = l_i^{\alpha} k_{i2}^{1-\alpha};$$

$$k_i = k_{i1} + k_{i2}, \ i = 1, \ldots, n, \tag{10.1}$$

where y_{ij} is the output of good $j(=1, 2)$ by country i, and k_i and l_i are country i's total capital stock and labor endowment, respectively. As in the previous chapters, e_i denotes the amounts of both fossil energy input and emissions of country i. The production functions assume implicitly that capital as a production factor is mobile across the two sectors while fossil energy is specific to the production of good 1 and labor is specific to the production of good 2. The two sectors correspond roughly to the capital-intensive industrial sector and the labor-intensive agriculture sector. The production function for good 1 implies that for any *given* capital stock, the marginal product of fossil energy falls to zero if the input of fossil energy is higher than a certain proportion. I assume that $0 < \alpha < 1$. The utility of country i is

$$u_i(x_{i1}, x_{i2}, z) = x_{i1} + \log x_{i2} - v_i(z),$$

where $z = \sum_{i \in N} e_i$ is the total amount of emissions, and x_{i1} and x_{i2} are the consumptions of goods 1 and 2. This form of utility function implies that pollution affects only the level of utility and plays no role in the determination of country i's preferences between the two consumption goods. As before, I assume that $v_i'(z) > 0$ and $v_i''(z) \geq 0$.

Each country chooses its emissions to maximize its utility, taking as given the emissions of other countries. In the absence of any international trade in goods, the utility of country i is maximized for *given* emissions e_i if

$$\frac{d}{dk_{i1}}\left(e_i^\alpha k_{i1}^{1-\alpha} + \log l_i^\alpha (k_i - k_{i1})^{1-\alpha} - v_i(z)\right) = 0.$$

This leads to the first-order condition

$$\left(\frac{e_i}{k_{i1}}\right)^\alpha (k_i - k_{i1}) = 1. \tag{10.2}$$

The condition implies that higher amounts of emissions are related to higher allocations of the capital stock to the production of good 1.

10.2. INTERNATIONAL TRADE IN GOODS

To understand the impact of free international trade in goods on climate change, I first characterize the total amount of emissions in autarky, that is, in the absence of international trade in goods, which is then used as the benchmark.

10.2.1. The Autarky Equilibrium

Choosing good 1 as the numeraire in each country, let p_i denote the market-clearing domestic price of good 2 in country i, and let $I_i = y_{i1} + p_i y_{i2}$ denote its aggregate income. I assume perfect competition in each country. Thus, for any given emissions e_i, the value of the marginal product of capital must be equal in equilibrium of the market for inputs, that is,

$$e_i^\alpha k_{i1}^{-\alpha} = p_i l_i^\alpha k_{i2}^{-\alpha}. \qquad (10.3)$$

Substituting for p_i from (10.3) and using (10.1), we can rewrite

$$I_i = y_{i1}\left(1 + \frac{k_{i2}}{k_{i1}}\right) = \frac{y_{i1}}{k_{i1}} k_i^{1-\alpha} k_i^\alpha$$

$$= k_i^{1-\alpha} e_i^\alpha \left(1 + \frac{k_{i2}}{k_{i1}}\right)^\alpha$$

$$= k_i^{1-\alpha}\left(e_i + p_i^{1/\alpha} \ell_i\right)^\alpha, \qquad (10.4)$$

as $k_i = k_{i1} + k_{i2}$. Because each country maximizes utility, $p_i x_{i2} = 1$ and thus $x_{i1} = I_i - 1$. Because the production function is Cobb-Douglas and firms in each country maximize profits, $\alpha y_{i1} = e_i \, \partial I_i / \partial e_i$ and in an autarky equilibrium $x_{ij} = y_{ij}$, $i, j = 1, 2$, equality (10.4) implies $k_i^{1-\alpha}(e_i + p_i^{1/\alpha}\ell_i)^\alpha - 1 = e_i k_i^{1-\alpha}/(e_i + p_i^{1/\alpha}\ell_i)^{1-\alpha}$. This equality can be simplified and rewritten as

$$k_i^{1-\alpha} p_i^{1/\alpha}\ell_i = (e_i + p_i^{1/\alpha}\ell_i)^{1-\alpha}. \qquad (10.5)$$

It is useful to display (10.5) diagrammatically. Let $s = p_i^{1/\alpha} l_i$.

For each fixed e_i, the function $(e_i + s)^{1-\alpha}$ in figure 10.1 is shown to be concave in s, as $0 < \alpha < 1$. Similarly, the function $k_i^{1-\alpha} s$ is shown to be linear in s. While the function $(e_i + s)^{1-\alpha}$ is higher if e_i is higher, the function $k_i^{1-\alpha} s$ is independent of e_i. The point where the two functions intersect represents the solution s to equality (10.5). As seen from the diagram representing equality (10.5), the higher the emissions e_i, the higher the solution s and, thus, the domestic price p_i.

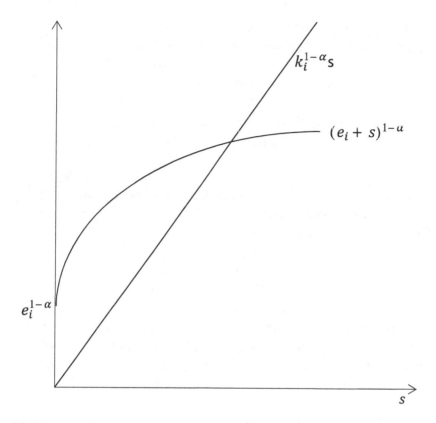

FIGURE 10.1

Determination of the domestic equilibrium price of the nonpolluting good in country i

Because $p_i x_{i2} = 1$ and $x_{i1} = I_i - 1$, the indirect utility function of country i is

$$\tilde{u}_i = I_i - 1 - \log p_i - v_i(z)$$

$$= k_i^{1-\alpha}(e_i + p_i^{1/\alpha} l_i)^\alpha - 1 - \log p_i - v_i(z), \qquad (10.6)$$

by using (10.4). I assume that each country i chooses its emissions e_i to maximize the utility, treating its domestic price p_i and the emission levels of all other countries as fixed. Taking the domestic price p_i as fixed amounts to assuming that country i does not take into account the impact of its environmental policy on domestic prices. Thus, in an autarky equilibrium,

$$v_i'(z) = \frac{\partial I_i}{\partial e_i} = \alpha k_i^{1-\alpha} / (e_i + p_i^{1/\alpha} \ell_i)^{1-\alpha}. \tag{10.7}$$

Equations (10.5) and (10.7) together imply

$$v_i'(z) \ell_i p_i^{1/\alpha} = \alpha. \tag{10.8}$$

This means that in an autarky equilibrium, the price of good 2 is smaller in the country with larger $v_i'(z)\ell_i$. Furthermore, using the fact that $\alpha y_{i1} = e_i \partial I_i / \partial e_i$, it follows from (10.7) that

$$\frac{\alpha y_{i1}}{e_i} = v_i'(z), \ i \in N. \tag{10.9}$$

The equilibrium condition (10.9) implies that the pollution intensity of production e_i / y_i is lower in countries with higher willingness to pay for controlling climate change (i.e., in the countries with stricter environmental regulation). As mentioned in the introduction to the chapter, I shall sometimes assume that there are only two countries (i.e., $n = 2$). Country 1 represents the industrialized North and country 2 the agricultural South. I assume that the marginal willingness to pay for controlling climate change is higher in the North [i.e., $v_1'(z) > v_2'(z)$ for all $z \geq 0$], but the endowment of the nonpolluting production factor labor is sufficiently smaller such that $v_1'(z)l_1 < v_2'(z)l_2$.

Proposition 10.1

There exists an autarky equilibrium that is unique. If $v_1'(z) > v_2'(z)$ and $v_1'(z)l_1 < v_2'(z)l_2$ for all $z \geq 0$, then $p_1 > p_2$; that is, the domestic price of the nonpolluting good is lower in country 2.

Proof of Proposition 10.1

Existence of an autarky equilibrium follows from standard arguments. I prove that it is unique. Suppose contrary to the assertion that there are two autarky equilibriums. Let $(y_1, y_2; e_1, e_2; z)$ and $(y_1', y_2'; e_1', e_2'; z')$, where $y_i = (y_{i1}, y_{i2})$, $y_i' = (y_{i1}', y_{i2}')$, $i=1, 2$, denote the two equilibria and (p_1, p_2) and (p_1', p_2') the corresponding domestic prices of the two goods. Without loss of generality, let $z \leq z'$. If $z < z'$, then $v_i'(z) \leq v_i'(z')$, $i=1, 2$. Therefore, from (10.8), $p_i \geq p_i'$, $i=1, 2$. It follows from (10.5) and figure 10.1 that $e_i \geq e_i'$, $i=1, 2$. This contradicts the supposition that $z < z'$. Similarly, if $z = z'$, $p_i = p_i'$ and $e_i = e_i'$, $i=1, 2$. Hence, the autarky equilibrium is unique. It is seen from (10.8) that if $v_1'(z)l_1 < v_2'(z)l_2$, then $p_1 > p_2$; that is, the domestic price of good 2 is lower in country 2. ■

10.2.2. International Trade in Goods

Proposition 10.1 implies that if free international trade in goods is allowed, then country 2 will export the nonpolluting good. I now allow free international trade in goods and analyze its impact on the total amount of emissions. Let p denote the international price of good 2 under free international trade. Then, equality (10.3) must continue to hold with p_i replaced by p. Equalities (10.6) and (10.7) must hold similarly. However, equality (10.5) may no longer hold, as it is now world and not domestic demands and supplies that must be equal. Accordingly, it must be replaced by

$$\sum_{i \in N} \frac{k_i^{1-\alpha} p^{1/\alpha} \ell_i}{(e_i + p^{1/\alpha} \ell_i)^{1-\alpha}} = n.$$

Using (10.7) with p_i replaced by p, this equality is equivalent to

$$p^{1/\alpha} \sum_{i \in N} v_i'(z) l_i = n\alpha. \tag{10.10}$$

Proposition 10.2

Consider a move from autarky to free international trade in goods. If $v_1'(z) > v_2'(z)$ and $v_1'(z)l_1 < v_2'(z)l_2$ for all $z \geq 0$, (i) the total amount of emissions will fall, but the world output of the polluting good and income will rise; (ii) output of the polluting good and emissions will rise in country 1, but fall in country 2; and (iii) country 1 might be better off and country 2 might be worse off.

It may look surprising, as the proposition shows, that the country with stronger concern for climate change, as measured by $v_i'(z)$, will export the polluting good. However, it comes from the fact that differences in concerns for climate change across the countries lead to differences in the pollution intensity of production and hold the possibility of increasing the world output without increasing the world pollution by shifting the production of the polluting good to the country with the lower pollution-intensity of production. Free trade in goods induces precisely such a shift.

Proof of Proposition 10.2

Let $(\bar{x}; \bar{e}; \bar{z})$, where $\bar{x} = (\bar{x}_1, \bar{x}_2)$ and $\bar{e} = (\bar{e}_1, \bar{e}_2)$, be the autarky equilibrium, and let $(\bar{p}_1, \bar{p}_2)$ be the corresponding domestic prices of good 2 in the two countries. Let $(\tilde{x}, \tilde{e}, \tilde{z})$ be the equilibrium under free international trade in goods, and let $\tilde{p}$ be the international price of good 2. Then from (10.7), which holds both under autarky and free international trade,

$$\frac{(\bar{e}_i + \bar{p}_i^{1/\alpha}\ell_i)^{1-\alpha}}{(\tilde{e}_i + \tilde{p}^{1/\alpha}\ell_i)^{1-\alpha}} = \frac{v_i'(\tilde{z})}{v_i'(\bar{z})}, i = 1, 2. \tag{10.11}$$

From (10.10), $\tilde{p}^{1/\alpha} = \frac{2\alpha}{v_1'(\tilde{z})\ell_1 + v_2'(\tilde{z})\ell_2}$. From (10.8), $\bar{p}_i^{1/\alpha} = \frac{\alpha}{v_i'(\bar{z})\ell_i}, i = 1, 2$. Thus,

$$\left(\frac{\bar{e}_i + \dfrac{\alpha}{v_i'(\bar{z})}}{\tilde{e}_i + \dfrac{2\alpha\ell_i}{v_1'(\tilde{z})\ell_1 + v_2'(\tilde{z})\ell_2}}\right)^{1-\alpha} = \left(\frac{v_i'(\tilde{z})}{v_i'(\bar{z})}\right)^{1-\alpha}\left(\frac{v_i'(\tilde{z})}{v_i'(\bar{z})}\right)^{\alpha}, i = 1, 2, \tag{10.12}$$

after substituting for $\tilde{p}^{1/\alpha}$ and $\bar{p}_i^{1/\alpha}$. Suppose contrary to the assertion that $\tilde{z} \geq \bar{z}$ and, therefore, $v_i'(\tilde{z}) \geq v_i'(\bar{z})$. Then, from (10.12),

$$\frac{\bar{e}_i v_i'(\bar{z})}{v_i'(\tilde{z})} + \frac{\alpha}{v_i'(\tilde{z})} \geq \tilde{e}_i + \frac{2\alpha\ell_i}{v_1'(\tilde{z})\ell_1 + v_2'(\tilde{z})\ell_2}, i = 1, 2.$$

Because $v_i'(\tilde{z}) \geq v_i'(\bar{z})$,

$$\bar{e}_1 + \bar{e}_2 + \frac{\alpha}{v_1'(\tilde{z})} + \frac{\alpha}{v_2'(\tilde{z})} \geq \tilde{e}_1 + \tilde{e}_2 + \frac{2\alpha(\ell_1 + \ell_2)}{v_1'(\tilde{z})\ell_1 + v_2'(\tilde{z})\ell_2}.$$

Because $v_1'(\tilde{z}) > v_2'(\tilde{z})$ and $v_1'(\tilde{z})\ell_1 < v_2'(\tilde{z})\ell_2$, the inequality implies $\bar{e}_1 + \bar{e}_2 > \tilde{e}_1 + \tilde{e}_2$. But this contradicts my supposition that $\tilde{z} \geq \bar{z}$. Hence, $\tilde{z} < \bar{z}$ and $v_i'(\tilde{z}) \leq v_i'(\bar{z})$.

This proves the first part of (i). For the remaining part of (i), because $v_i'(\tilde{z}) \leq v_i'(\bar{z})$ as shown, it is seen from (10.7) and (10.4), which must hold both under autarky and under free international trade, that the world income $(= I_1 + I_2)$ and therefore the world output of good 1 $(= I_1 + I_2 - 2)$ must be higher under free international trade in goods. I next prove (ii). Because $v_i'(\tilde{z}) \leq v_i'(\bar{z})$, $i = 1, 2$, as shown, it is seen from (10.12) that

$$\bar{e}_i v_i'(\bar{z}) + \alpha \leq \tilde{e}_i v_i'(\tilde{z}) + \frac{2\alpha v_i'(\tilde{z})\ell_i}{v_1'(\tilde{z})\ell_1 + v_2'(\tilde{z})\ell_2}, i = 1, 2.$$

Because $v_1'(\tilde{z})\ell_1 < v_2'(\tilde{z})\ell_2$ and $v_1'(\tilde{z}) \leq v_1'(\bar{z})$, the above inequality implies $\bar{e}_1 < \tilde{e}_1$. Because $\bar{e}_1 + \bar{e}_2 > \tilde{e}_1 + \tilde{e}_2$, as shown above, $\tilde{e}_2 < \bar{e}_2$. The above inequality also implies $\bar{e}_1 v_1'(\bar{z}) < \tilde{e}_1 v_1'(\tilde{z})$, as $v_1'(\tilde{z})l_1 < v_2'(\tilde{z})l_2$. Therefore, $\tilde{y}_{11} > \bar{y}_{11}$, as equality (10.9) implies, $\alpha \bar{y}_{11} = \bar{e}_1 v_1'(\bar{z})$ and $\alpha \tilde{y}_{11} = \tilde{e}_1 v_1'(\tilde{z})$, where $\bar{y}_{11}$ and $\tilde{y}_{11}$ denote the outputs of good 1 in country 1 under autarky and free international trade equilibrium, respectively. Because $\tilde{e}_2 < \bar{e}_2$ and $v_2'(\tilde{z}) \leq v_2'(\bar{z})$, as shown above, $\tilde{e}_2 v_2'(\tilde{z}) < \bar{e}_2 v_2'(\bar{z})$, which implies that $\tilde{y}_{21} < \bar{y}_{21}$, where $\bar{y}_{21}$ and $\tilde{y}_{21}$ denote the outputs of good 1 in country 2 under autarky and free international trade. Moreover, $\bar{e}_1 v_1'(\bar{z}) + \bar{e}_2 v_2'(\bar{z}) < \tilde{e}_1 v_1'(\tilde{z}) + \tilde{e}_2 v_2'(\tilde{z})$; that is, $\bar{y}_{11} + \bar{y}_{21} < \tilde{y}_{11} + \tilde{y}_{21}$.

I now prove (iii). By definition, $\bar{p}_2^{1/\alpha} = \frac{\alpha}{v_2'(\bar{z})\ell_2}$ and $\tilde{p}^{1/\alpha} = \frac{2\alpha}{v_1'(\tilde{z})\ell_1 + v_2'(\tilde{z})\ell_2}$. Because $v_2'(\tilde{z})\ell_2 > v_1'(\tilde{z})\ell_1$ and $v_2'(\tilde{z}) \leq v_2'(\bar{z})$, it follows that $\tilde{p} > \bar{p}_2$. From utility maximization, $\tilde{x}_{22} = 1/\tilde{p}$ and $\bar{x}_{22} = 1/\bar{p}_2$. Hence, $\tilde{x}_{22} < \bar{x}_{22}$. Because $v_i'(\tilde{z}) \leq v_i'(\bar{z})$, inequality (10.4), which must hold also under free international trade, implies

$$\tilde{I}_2 = k_2^{1-\alpha}(\tilde{e}_2 + \tilde{p}^{1/\alpha}l_2)^\alpha > k_2^{1-\alpha}(\bar{e}_2 + \bar{p}_2^{1/\alpha}l_2)^\alpha = \bar{I}_2.$$

Therefore, from utility maximization, $\tilde{x}_{21} = \tilde{I}_2 - 1 > \bar{x}_{21} = \bar{I}_2 - 1$. However, the incidence of free international trade on the welfare of country 2 is ambiguous, as $\tilde{x}_{22} < \bar{x}_{22}$, but $\tilde{x}_{21} > \bar{x}_{21}$. ■

Linear Damage Functions

I can drive a sharper result if the damage functions are linear. The utility function of country i is then $u_i(x_i, z) = x_{i1} + \log x_{i2} - \bar{\pi}_i z$, $i = 1, 2$, where $\bar{\pi}_i > 0$ is the constant marginal willingness to pay of country i. Keeping all other assumptions the same and following the same steps as in Proposition 10.2, $\tilde{x}_{11} = \bar{x}_{11}$, $\tilde{x}_{21} = \bar{x}_{21}$, but

$$\tilde{x}_{12} = \frac{1}{\tilde{p}} > \frac{1}{\bar{p}_1} = \bar{x}_{12} \quad \text{and} \quad \tilde{x}_{22} = \frac{1}{\tilde{p}} < \frac{1}{\bar{p}_2} = \bar{x}_{22}.$$

Hence country 2 is worse off, but country 1 is better off.

10.3. INTERNATIONAL TRADE IN BOTH GOODS AND EMISSIONS

Notice that, as seen from (10.9), free international trade in goods does not eliminate the gap in pollution intensity of production, as $v_1'(z) > v_2'(z)$ for all z, and may not even narrow it. Thus, free trade in goods does not fully exhaust the possibility of increasing the world output without increasing the total amount of emissions. Moreover, the results above are reversed if $v_1'(z)\ell_1 > v_2'(z)\ell_2$. This means that free international trade in goods can indeed lead to the feared "pollution havens" if the difference in the willingness to pay is sufficiently large. Of course, this is an empirical question, and equation (10.9) can be used to estimate differences in the willingness to pay for controlling climate change across the countries.

Alternatively, South may not be sufficiently abundant in labor (i.e., in the factor specific to the production of the nonpolluting good). I now show that even in these cases, the total amount of emissions falls if trade in emissions along with free trade in consumption goods is allowed, as the gap in the pollution intensities of production is then fully eliminated.

Proposition 10.3

(i) If $v_1'(z) > v_2'(z)$ and $v_1'(z)\ell_1 > v_2'(z)\ell_2$ for all $z \geq 0$, then under free international trade in goods, the total amount of emissions is higher, and country 1 exports the nonpolluting good and country 2 the polluting good.
(ii) Suppose trade in emissions is introduced along with free international

trade in goods. Then the total amount of emissions falls, but the world output of the polluting good and income rise. If $l_1 < l_2$, the pattern of trade will be reversed; that is, country 1 will export the polluting good and country 2 the nonpolluting good, and the output and pollution of country 1 will rise.

Proof of Proposition 10.3

The proof of part (i) is along the same lines as in Proposition 10.2 and hence it is not included. I prove part (ii).

Let $(\hat{x}, \hat{e}, \hat{z})$ be the equilibrium under free international trade in emissions and goods, and let $\hat{p}$ be the corresponding price of good 2. Then, given free international trade in emissions, the following inequality must hold:

$$v_1'(\hat{z}) = \frac{\alpha k_1^{1-\alpha}}{(\hat{e}_1 + \hat{p}^{1/\alpha} \ell_1)^{1-\alpha}} = \frac{\alpha k_2^{1-\alpha}}{(\hat{e}_2 + \hat{p}^{1/\alpha} \ell_2)^{1-\alpha}} > v_2'(\hat{z}).$$

This is because in equilibrium, the price of the pollution permits must be equal to the highest marginal willingness to pay of the countries. Equality (10.10) is then replaced by

$$\hat{p}^{1/\alpha} = \frac{2\alpha}{v_1'(\hat{z})(\ell_1 + \ell_2)}.$$

It follows from (10.7) and these inequalities that

$$\left(\frac{\overline{e}_1 + \dfrac{\alpha}{v_1'(\overline{z})}}{\hat{e}_1 + \dfrac{2\alpha \ell_1}{v_1'(\hat{z})(\ell_1 + \ell_2)}} \right)^{1-\alpha} = \frac{v_1'(\hat{z})}{v_1'(\overline{z})}$$

$$\left(\frac{\overline{e}_2 + \dfrac{\alpha}{v_2'(\overline{z})}}{\hat{e}_2 + \dfrac{2\alpha \ell_2}{v_1'(\hat{z})(\ell_1 + \ell_2)}} \right)^{1-\alpha} = \frac{v_1'(\hat{z})}{v_2'(\overline{z})}. \tag{10.13}$$

Suppose contrary to the assertion that $\hat{z} = \hat{e}_1 + \hat{e}_2 \geq \overline{e}_1 + \overline{e}_2 = \overline{z}$. Then, $v_1'(\hat{z}) \geq v_1'(\overline{z}) > v_2'(\overline{z})$. The two equalities in (10.13) imply that

$$\frac{\overline{e}_2 v_2'(\overline{z}) + \alpha}{\overline{e}_1 v_1'(\overline{z}) + \alpha} > \frac{\hat{e}_2 v_1'(\hat{z}) + 2\alpha \dfrac{l_2}{l_1 + l_2}}{\hat{e}_1 v_1'(\hat{z}) + 2\alpha \dfrac{l_1}{l_1 + l_2}}.$$

Thus,

$$\overline{e}_1 v_1'(\overline{z}) + \overline{e}_2 v_2'(\overline{z}) + 2\alpha > \left(\hat{e}_1 v_1'(\hat{z}) + \hat{e}_2 v_1'(\hat{z}) + 2\alpha\right) \left(\frac{\overline{e}_1 v_1'(\overline{z}) + \alpha}{\hat{e}_1 v_1'(\hat{z}) + 2\alpha \dfrac{l_1}{l_1 + l_2}} \right).$$

The first equality in (10.13) and $v_1'(\hat{z}) \geq v_1'(\overline{z})$ then imply $\overline{e}_1 v_1'(\overline{z}) + \overline{e}_2 v_2'(\overline{z}) > \hat{e}_1 v_1'(\hat{z}) + \hat{e}_2 v_1'(\hat{z})$. But, because $v_1'(\hat{z}) \geq v_1'(\overline{z}) > v_2'(\overline{z})$, this contradicts my supposition that $\overline{e}_1 + \overline{e}_2 \leq \hat{e}_1 + \hat{e}_2$. Hence, $\overline{e}_1 + \overline{e}_2 > \hat{e}_1 + \hat{e}_2$ and $\overline{e}_1 v_1'(\overline{z}) + \overline{e}_2 v_2'(\overline{z}) < \hat{e}_1 v_1'(\hat{z}) + \hat{e}_2 v_1'(\hat{z})$; that is, $\overline{y}_{11} + \overline{y}_{21} < \hat{y}_{11} + \hat{y}_{21}$.

This proves that world pollution is lower but the world output of the polluting good and, therefore, income is higher. It is also seen from (10.13) that if $l_1 < l_2$, then $\overline{e}_1 v_1'(\overline{z}) < \hat{e}_1 v_1'(\hat{z})$; that is, the output of the polluting good of country 1 is higher. Proof of the proposition is now completed by showing that if $l_1 < l_2$, then country 2 will export the nonpolluting good. In view of (10.4), let $\hat{I}_i = k_i^{1-\alpha} (\hat{e}_i + \hat{p}^{1/\alpha} l_i)^\alpha$. Then,

$$\frac{\partial \hat{I}_i}{\partial \ell_i} = \frac{\alpha k_i^{1-\alpha} \hat{p}^{1/\alpha}}{(\hat{e}_i + \hat{p}^{1/\alpha} \ell_i)^{1-\alpha}},$$

and

$$\frac{\partial \hat{I}_i}{\partial k_i} = \frac{(1-\alpha)(\hat{e}_i + \hat{p}^{1/\alpha} \ell_i)^\alpha}{k_i^\alpha}.$$

Using (10.13), it follows that the factor prices and therefore factor intensities of production are equalized across the countries under free international trade in emissions and consumption goods. Because $l_1 < l_2$ and the factor intensities of production are equal, $\hat{y}_{22} > \hat{y}_{12}$. However, from utility maximization, $\hat{x}_{22} = \hat{x}_{12} = \frac{1}{\hat{p}}$ and $\hat{x}_{22} + \hat{x}_{12} = \hat{y}_{22} + \hat{y}_{12}$. This proves that country 2 must export good 2.∎

Proposition 10.3 shows that the "pollution-havens" effect may obtain, if at all, not because of free international trade in goods, but because of lack of free international trade in emissions or pollution permits.

10.4. ENVIRONMENTAL AGREEMENTS AND INTERNATIONAL TRADE IN GOODS

I now consider whether trade policy can have an impact on agreements on climate change. In other words, do the conditions for efficient control of climate change depend on trade policy?

10.4.1 Environmental Efficiency under Autarky

We can write the utility of country i as $u_i(e_1,\ldots,e_n) = g_i(e_i) - v_i(z)$ where, in view of (10.1), $g_i(e_i) = e_i^\alpha k_{i1}^{1-\alpha} + \log l_i^\alpha (k_i - k_{i1})^{1-\alpha}$ and, as (10.2) implies, k_{i1} is a function of e_i. Let $E = E_1 \times \cdots \times E_n$, $E_i = \{e_i : 0 \le e_i \le ak_i\}$, $u = (u_1,\ldots,u_n)$, and $u_i(e_1,\ldots,e_n) = g_i(e_i) - v_i(z)$. Then, (N, E, u) is a strategic game that is similar to the environmental game in chapter 5. Furthermore, $g_i'(e_i) = \alpha e_i^{\alpha-1} k_{i1}^{1-\alpha} + (1-\alpha)e_i^\alpha k_{i1}^{-\alpha} k_{i1}' + \frac{(1-\alpha)(-k_{i1}')}{k_i - k_{i1}} = \alpha e_i^{\alpha-1} k_{i1}^{1-\alpha}$, as differentiation of equation (10.2) implies

$$k_{i1}' = \frac{\dfrac{\alpha}{e_i}(k_i - k_{i1})}{1 + \dfrac{\alpha}{k_{i1}}(k_i - k_{i1})}.$$

Thus, $g_i'(e_i) > 0$. Next, I show that $g_i''(e_i) < 0$. Differentiating g_i' and using equation (10.2) and substituting for the derivative k_{i1}', we have

$$g_i''(e_i) = \alpha(\alpha - 1)e_i^{\alpha-2} k_{i1}^{1-\alpha} + \alpha(1-\alpha)e_i^{\alpha-1} k_{i1}^{-\alpha} \left[\frac{\dfrac{\alpha}{e_i}(k_i - k_{i1})}{1 + \dfrac{\alpha}{k_{i1}}(k_i - k_{i1})} \right]$$

$$= \alpha(1-\alpha)e_i^{\alpha-2} k_{i1}^{1-\alpha} \left[\frac{\alpha(k_i - k_{i1})}{k_{i1} + \alpha(k_i - k_{i1})} - 1 \right] < 0.$$

Thus, the environmental game (N, E, u) satisfies the conditions of Proposition 4.1 and admits a Nash equilibrium. As in (10.9), the first-order conditions for a Nash equilibrium are

$$\alpha e_i^{\alpha-1} k_{i1}^{1-\alpha} = v_i'(z), \, i=1,\ldots,n.$$

Because the left-hand side is decreasing in e_i, as just shown, similar arguments as in Proposition 5.1 imply that the game has a unique Nash equilibrium. Similarly, first-order conditions for efficiency are $\alpha e_i^{\alpha-1} k_{i1}^{1-\alpha} = \sum_{i\in N} v_i'(z), \, i=1,\ldots,n$. By the same arguments as in Proposition 5.1, these equalities also have a unique solution. Thus, all conditions for the existence of a nonempty γ-core of the environmental game (N, E, u) are satisfied.

10.4.2. Environmental Efficiency and International Trade in Goods

In autarky, the income of each country i depends entirely on its emissions e_i. Accordingly, we could write its income simply as a function of its emissions, that is, as $g_i(e_i)$. This is no longer true if there is free international trade in goods. As seen from definition (10.4), the income of country i is $I_i = k_i^{1-\alpha}(e_i + p^{1/\alpha} l_i)^\alpha$, which depends on the international price p of good 2. As seen from (10.10), the price p depends not only on the emissions of country i but also on the emissions of all other countries. However, as will be shown below, it does not if the damage functions are linear.

Linear Damage Functions

If the damage functions are linear, equation (10.10) implies $p^{1/\alpha} = \frac{n\alpha}{\sum_{i\in N} \bar{\pi}_i l_i}$. That is, the international price of good 2 is independent of emissions of any country. In other words, the trade and environmental policies can be treated separately. Thus, define

$$g_i(e_i) = k_i^{1-\alpha}\left(e_i + \frac{n\alpha}{\sum_{i\in N} \bar{\pi}_i l_i}\right)^\alpha.$$

Clearly, $g_i'(e_i) > 0$ and $g_i''(e_i) < 0$, and the environmental game is (N, E, u), where $u = (u_1, \ldots, u_n)$ and $u_i(e_1, \ldots, e_n) = g_i(e_i) - \bar{\pi}_i z$. Because (N, E, u) has exactly the same structure as the environmental game in chapter 5, it has a nonempty γ-core. Thus, an agreement for efficient control of climate change is possible independent of trade policies. This is because if the damage functions are linear, the marginal willingness to pay for the environment is constant for each country, and the international prices of goods do not depend on the level of pollution.

Whether this result can be extended to the more general case in which marginal willingness to pay (and therefore the prices) depends on the total amount of emissions remains an open problem.

To conclude, this chapter has extended the analysis in chapter 5 to a model with two consumption goods. It shows that the claim of environmentalists that trade liberalization can aggravate climate change is questionable and demonstrates that an agreement to control climate change is possible independent of trade policies. The chapter shows that it is not free international trade in goods but lack of trade in emissions or pollution permits that can aggravate climate change, if at all. As seen from Propositions 10.1 and 10.3, this result is independent of the relative factor endowments of the countries. We could also analyze the incidence of trade in emissions on the welfare of the countries.

I showed that if the damage functions are linear, then trade and climate policies can be treated separately. Thus, it is not surprising that the Paris Agreement on climate change was concluded without waiting for the outcome of the trade negotiations. In particular, my analysis implies that bilateral trade deals or formation of regional trading blocs (e.g., TPP, NAFTA) may not hinder the countries from entering a worldwide agreement on climate change.

Several assumptions limit my analysis. As in earlier chapters, I assumed that countries or governments choose their environmental policy rationally (see Grossman and Helpman 1995 for an analysis of the consequences of relaxing this assumption on free trade agreements). I also assumed that the countries or governments do not use environmental regulation as a strategic trade policy. My results will obviously be diluted if either of these behavioral assumptions does not hold. I assumed labor and capital to be immobile across the countries. Beladi, Chau, and Khan (1997) and Rauscher (1991) study the effect of capital mobility on climate change. As seen from the proof of Proposition 10.3, pollution permit prices as well as prices of capital and labor will all be equalized across the countries if either capital or labor is mobile. Thus, the message of Proposition 10.3 is unlikely to change if factor mobility is assumed.

CONCLUSION

The book has developed a game-theoretic framework for analyzing and proposing policies to control climate change. It can be divided into four distinct parts. The first part consists of chapters 2–6, which focus on a single-period model of climate change. The second part consists of chapters 7 and 8, which focus on dynamic and long-run models of climate change, respectively. The third part consists of chapter 9, which focuses on interpretation and analysis of the two actual agreements—the failed Kyoto Protocol and the more recent Paris Agreement—by combining the cooperative and noncooperative approaches to game theory. The fourth part consists of chapter 10, which critically examines the claim of environmentalists that more free trade in goods and services can aggravate climate change. In this concluding chapter, I briefly summarize the theory and policy implications of this book. I also discuss some open problems that are to be addressed in future work.

I assumed throughout the book that the countries behave rationally and choose their emissions to maximize their payoffs whether in a static or dynamic formulation of climate change. Though this is a standard assumption in both economics and game theory, there is little empirical evidence that shows that the countries indeed behave in this manner. But this seems to be changing with the involvement of experts and informed policymakers and as both climate change and international negotiations become more intense and competitive.

I began my analysis by arguing that financial transfers to balance costs and benefits of controlling climate change among sufficiently heterogeneous sovereign countries are a necessity and not a matter of approach. Some studies get around the necessity of transfers by assuming identical countries, but the solutions proposed in these studies when applied to models with sufficiently heterogeneous countries cannot be both efficient and individually rational. However, because the countries are sovereign and free to reject any proposed agreement, an agreement must be at least individually rational. Similarly, sacrificing efficiency though possible is not desirable, as that may leave the countries exposed to extreme effects of climate change and lead to attempts for further negotiations. Thus, any agreement on climate change must aim at achieving both efficiency and at the minimum individual rationality.

It was argued that the Coase theorem as such does not apply to the climate change problem because there is no supranational authority to assign and enforce pollution rights, whereas such an authority is a prerequisite for any bargaining or markets for pollution permits to work. However, I showed that in the absence of a supranational authority, the countries themselves will assign rights to each other such that competitive international trade in those rights and exercise of the after-trade rights would result in an efficient control of climate change, and every country or group of countries would be better off.

The analysis of the infinitely repeated game motivated and introduced in chapters 4 and 6 showed that a partial climate agreement (i.e., an agreement only among a subset of countries), though an improvement over no agreement, can be a hurdle to a full agreement (i.e., an agreement among all countries). That is, if some countries were not farsighted and were to go ahead with a partial agreement, it will be an improvement over no agreement, but would change the incentives such that the countries left out of the agreement would have even weaker incentives to join it at a future date. Thus, a partial agreement may continue to be a partial agreement in perpetuity unless it is abandoned. In other words, rather than quickly settling for a partial agreement, it may be better for the countries to continue negotiations until a full agreement is reached. In fact, the recently concluded Paris Agreement among 196 countries and discontinuation of the Kyoto Protocol can be explained precisely in these terms. Indeed, the Kyoto Protocol was doomed right from the day the United States—the biggest economy and the second biggest emitter of GHGs—decided not to ratify it. There were only two possibilities left after the expiry of the half-heartedly implemented first commitment period of the Kyoto Protocol in

2012: either there will be a new agreement similar to the Kyoto Protocol, but with active participation of the United States, or the Protocol will be abandoned and negotiations will start afresh and continue until a full agreement is reached. As the theory in this book predicts, it turned out to be the latter possibility. A plain extension of the Kyoto Protocol without effective participation of the United States was obviously not acceptable to the other countries and resulted in non-ratification of the Doha Amendment to the Protocol.

I showed that climate agreements can be of two types: an agreement may be defined either in terms of both emissions and financial transfers or in terms of emissions alone that can be traded as pollution rights on a competitive international market. Though the latter type requires creation of institutions and competitive markets for trading pollution rights, it may be preferred as it does not require direct financial transfers, which may be vulnerable to domestic politics. Thus, the analysis in this book supports international trade in pollution rights and that policies should be made and institutions should be created to make it as competitive as possible.

The Paris Agreement has set a target of limiting global warming to "well below 2 degrees Celsius" and to "pursue efforts to limit the temperature increase to 1.5 degrees Celsius above pre-industrial levels." The more ambitious goal to limit global warming to 1.5°C has been included in the agreement after a push by the low-lying small island states, and the IPCC has been tasked to provide a special report in 2018 on its implications. But is such a goal really socially optimal? I showed that social optimality requires weighing the additional damage the world will suffer against the additional economic gains for the world if the temperature rise is to be 2°C rather than 1.5°C. If the additional economic gains are more than the additional damage, then allowing a temperature rise of 2°C is socially optimal. Because the differential game model proposed in chapter 8 is computable, simulations of the model can be used to find an answer to this question. However, this would require more accurate information regarding the damage functions and other relevant parameters. Though several studies have estimated damage from climate change for individual countries, the estimates differ widely[1] as the science of measuring

---

1. See, for example, Fankhauser (1995) and Stern et al. (2006).

the physical and economic effects of climate change, despite the excellent work by the IPCC, is still in the early stages. The task is complicated by the fact that both climate science and economics must work hand in hand: climate science is needed to identify the physical damage caused by climate change, and economics is needed to estimate the economic value of the physical damage.

In any case, if the economic cost of limiting climate change to 1.5°C is too high, but altruism and fairness are real concerns, then a better policy might be to offer assistance to the low-lying small island states to meet the cost of adaptation such as relocation of their populations or building dikes for protection against sea rise.

The approach in this book shows that the socially optimal climate policy does not depend on who polluted how much in the past. What matters is the current stock of GHGs and not how that stock was created or who contributed how much to it. This means the issue of "historical responsibilities" is purely an equity issue and can be treated separately.[2] For example, after an agreement that ignores the historical responsibilities has been reached, it may be modified at the implementation stage to take account of the historical responsibilities in terms of appropriately reduced emission rights for past heavy emitters and higher emission rights for those with a relatively clean past—the total emission rights remaining the same.

The theory in this book supports free international trade in goods and pollution permits. It was shown that more free international trade in goods may not worsen climate change. The intuitive argument behind this result is quite general; namely, total world production of consumption goods can be increased without increasing total world emissions by shifting production of the polluting goods from countries with higher pollution intensities to those with lower ones. Free international trade in consumption goods and pollution permits induces precisely such a shift, as it tends to equalize the pollution intensities in production across all countries.

---

2. Though the issue is purely distributional, it acquires strategic overtones when linked with negotiations on climate change.

I have deliberately not considered the proposal to impose a uniform "carbon tax" on the GHGs emissions of every country which is often proposed in discussions for tackling climate change. Though a uniform tax seems fair and equitable, it ignores the fact that countries are sovereign and the economic costs and benefits of a uniform carbon tax are different for different countries. For instance, the small island states that are most vulnerable to climate change and hardly emit any GHGs themselves would be in favor of imposition of a high uniform carbon tax, whereas the countries that are least affected by climate change would favor a low carbon tax, if at all. In fact, an agreement on a uniform carbon tax is equivalent to an agreement on limiting emissions to specific levels by each country. But this book has already considered such agreements more generally.

I have also not considered axiomatic solutions to the climate change problem, not because I think that axiomatic approaches to game theory are less important, but because I do not think that negotiations on climate change are guided by normative criteria.[3] It is the self-interest of the countries that dominates climate negotiations—normative criteria have at best a secondary role.

It was argued, on the basis of the analysis in this book, that the Paris Agreement has a better chance of success. First, China and India have committed to control the rates at which their emissions may increase by adopting cleaner technologies. In contrast, the Kyoto Protocol required the developing countries to undertake such actions only as CDM projects, which could at best shift the responsibility for reducing emissions—from developed countries to developing countries—but not lead to additional reductions overall. Second, the back-to-back, 5-year national action plans extending far into the future that are proposed in the Paris Agreement can set in place a stable demand and a global market for clean energy. Therefore, the task of developing and adopting cleaner technologies may be carried out largely by businesses and investors operating under

---

3. Normative criteria, however, can play a role in determining liability and compensation for past emissions, which, as noted above, is purely an issue of equity and can be treated separately from socially optimal climate policy.

emission-reduction policies that will be in place with more certainty for a much longer period of time.

The analysis in this book showed that development and adoption of cleaner technologies is the key to meeting the goals of the Paris Agreement and the future growth of the world economy. Though the Paris Agreement indeed conveys a critical message to the marketplace for development and adoption of cleaner technologies, reliance on market forces alone may not be sufficient. In particular, the developing countries may not be able to adopt cleaner technologies because of intellectual property rights barriers. However, because intellectual property rights cannot be wished away and are, in fact, needed to protect and strengthen private incentives to innovate and adopt, a global agreement in parallel with the Paris Agreement—and similar to the International Solar Alliance of 121 countries—to fund research for development of cleaner technologies on a global scale seems necessary. Such a fund to which governments of all countries as well as private individuals and organizations may contribute can boost the efforts to develop and adopt cleaner technologies and overcome intellectual property rights barriers to transfer of technology.

I briefly discussed implications of the U.S. decision to withdraw from the Paris Agreement. It is feared that the Paris Agreement, like the Kyoto Protocol, may fizzle out unless the United States soon reverses its decision and rejoins the Agreement.

The γ-core of a *general* strategic game introduced and applied in this book is an important game theory concept, as, unlike in the previous literature, it is now defined for a general, not just a specific, strategic game. This means its applications are not restricted to the environmental game alone, and it can be applied to any other model that can be represented by a strategic game. One such application that was discussed in chapter 5 is to the problem of seasonal haze in Southeast Asia. Another, which for obvious reasons was not discussed in the book, is the Cournot oligopoly (see, e.g., Chander 2017c; Lardon 2012; Stamatopoulos 2016). Furthermore, I showed that the γ-core of a general strategic game can be interpreted both as a cooperative and as a noncooperative solution concept if the partition function representation of the strategic game is partially superadditive. In fact, the book creates bridges to reunify cooperative and noncooperative approaches to game theory. In that sense, the book might be of interest not just to students but also to researchers who focus on noncooperative game theory. Though we are still quite far from a grand unified model of cooperative and noncooperative game theory, this book does bring us one step closer.

I showed that the γ-core of the environmental game is nonempty, and each γ-core imputation is equivalent to an allocation of pollution rights that can be traded on a competitive international market. This equivalence result was interpreted as a generalization of the Coase theorem. I also formulated and analyzed climate change as a dynamic game. I showed that the dynamic environmental game admits both open-loop and subgame-perfect equilibria and noted that transfers, as in the short-run model, are necessary to balance the costs and benefits of controlling climate change. But a new issue that arises is that of subgame perfection of the transfers. This motivated me to introduce the concept of a subgame-perfect cooperative agreement such that no coalition will have incentive to withdraw from the agreement in any subgame.

As to future research, in a working paper Chander and Wooders (2016) introduce a core concept for a *general* extensive game that incorporates both cooperation and subgame perfection in that it takes into account that the payoff a coalition can achieve may change along the history generated by a play of the game. This new concept, labeled the subgame-perfect core of an *extensive* game, is a refinement of the γ-core of the strategic form representation of the extensive game in the same sense as the set of subgame-perfect Nash equilibria is of the set of Nash equilibria.[4]

The book has intentionally ignored some additional features of international negotiations, as one book can only do so much. These include legitimacy, trust, procedural fairness, transaction costs, bargaining,

---

4. A number of previous studies have explored core concepts in dynamic games without externalities. Gale (1978) explores the issue of time consistency in the Arrow-Debreu model with dated commodities when agents distrust the forward contracts signed at the first date. Gale introduces the sequential core, which consists of allocations that cannot be improved upon by anyone at any date. Similarly, Forges, Mertens, and Vohra (2002) propose the ex-ante incentive compatible core. Becker and Chakrabarti (1995), as noted in chapter 1, propose the recursive core as an allocation such that no coalition can improve upon its consumption stream at any time.

within-country differences of opinion, and so forth. These are all questions that are being explored in the current game theory literature, which includes the work by Ostrom on game theory of international cooperation, work in behavioral economics on trust and norms and fairness and altruism, and work on noncooperative game theory of norms (Kandori 1992) or institutional economics (North 1990).

REFERENCES

Adda, J., and R. Cooper. (2003). *Dynamic Economics: Quantitative Methods and Applications*. Cambridge, MA: MIT Press.

Aivazian, V. A., and J. L. Callen. (1981). "The Coase theorem and the empty core." *Journal of Law and Economics* 24: 175–181.

Alcamo, J., R. Shaw, and L. Hordijk. (1990). *The RAINS Model of Acidification: Sciences and Strategies for Europe*. Boston: Kluwer Academic.

Aldy, J. E., S. Barrett, and R. N. Stavins. (2003). "Thirteen plus one: A comparison of global climate policy architectures." *Climate Policy* 3: 373–397.

Arrow, K. J. (1970). "The organization of economic activity: Issues pertinent to the choice of market versus non-market allocation." In *Public Expenditure and Policy Analysis*, edited by R. H. Haveman and J. Margolis, 1–16. Chicago: Markham Publishing.

Arrow, K. J., and G. Debreu. (1954). "The existence of an equilibrium for a competitive economy." *Econometrica* 22: 265–290.

Arrow, K. J., and F. Hahn. (1971). *General Competitive Analysis*. Edinburgh: Oliver and Boyd.

Asheim, G. B., and B. Holtsmark. (2009). "Renegotiation-proof climate agreements with full participation: Conditions for Pareto efficiency." *Environmental and Resource Economics* 43: 519–533.

Asheim, G. B., C. B. Froyn, J. Hovi, and F. C. Menz. (2006). "Regional versus global cooperation for climate control." *Journal of Environmental Economics and Management* 51: 93–109.

Audley, J. J. (1993). "Why environmentalists are angry about the North American Free Trade Agreement." In *Trade and the Environment: Law, Economics and Policy*, edited by D. Zaelke, P. Orbuch, and R. F. Housman, 191–202. Washington, DC: Island Press.

Aumann, R. J. (1959). "Acceptable points in general cooperative n–person games." In *Contributions to the Theory of Games*, Vol. IV, 287–324. Princeton, NJ: Princeton University Press.

Aumann, R. J. (1987). "Game theory." In *The New Palgrave: A Dictionary of Economics*, Vol. 2, edited by J. Eatwell, M. Milgate, and P. Newman, 1–53. London: Macmillan.

Aumann, R .J., and J. Dreze. (1974). "Cooperative games with coalition structures." *International Journal of Game Theory* 3: 217–237.

Babiker, M., J. M. Reilly, and H. D. Jacoby. (2000). "The Kyoto Protocol and developing countries." *Energy Policy* 28: 525–536.

Bagnoli, M., and B. L. Lipman. (1988). "Successful takeovers without exclusion." *Review of Financial Studies* 1: 89–110.

Barrett, S. (1994a). "Self-enforcing international environmental agreements." *Oxford Economic Papers* 46: 804–878.

Barrett, S. (1994b). "Strategic environmental policy and international trade." *Journal of Public Economics* 54: 325–338.

Barrett, S. (2003). *Environment and Statecraft: The Strategy of Environmental Treaty-Making.* Oxford: Oxford University Press.

Barrett, S. (2013). "Climate treaties and approaching catastrophes." *Journal of Environmental Economics and Management* 66: 235–250.

Barrett, S., and R. Stavins. (2003). "Increasing participation and compliance in international climate change agreements." 3: 349–376.

Basar, T., and G.-J. Olsder. (1995). *Dynamic Non-cooperative Game Theory.* London: Academic Press.

Baumol, W., and W. Oates. (1975). *The Theory of Environmental Policy.* Englewood Cliffs, NJ: Prentice Hall.

Bearden, J. N. (2001). "Ultimatum bargaining experiments: The state of art." SSRN. http://ssrn.com/abstract=626183.

Becker, R. A., and S. K. Chakrabarti. (1995). "The recursive core." *Econometrica* 63: 401–423.

Beladi, H., N. H. Chau, and M. A. Khan. (1997). "North-South investment flows and strategic environmental policies." [Mimeo] Baltimore: Johns Hopkins University.

Belleflamme, P. (2000). "Stable coalition structures with open membership and asymmetric firms." *Games and Economic Behaviour* 30: 1–21.

Benchekroun, H., and H. M. Yildiz. (2011). "Free trade, autarky and sustainability of an international environmental agreement." *BE Journal of Economic Analysis and Policy* 11.

Bento, A., B. Ho, and M. Ramirez-Basora. (2015). "Optimal monitoring and offset prices in voluntary emissions markets." *Resource and Energy Economics* 41: 202–223.

Bergstrom, T., L. Blume, and H. Varian. (1986). "On the private provision of public goods." *Journal of Public Economics* 25–49.

Bernheim, B. D., B. Peleg, and D. Whinston. (1987). "Coalition-proof Nash equilibria: I. Concepts." *Journal of Economic Theory* 42: 1–12.

Bhagwati, J. (1993). "Trade and the environment: The false conflict." In *Trade and the Environment: Law, Economics and Policy,* edited by D. Zaelke, P. Orbuch, and R. F. Housman, 159–190. Washington, DC: Island Press.

Biancardi, M., and G. Villani (2014). "International environmental agreements with developed and developing countries in a dynamic approach." *Natural Resource Modelling* 27: 338–359.

Bloch, F. (1996). "Sequential formation of coalitions in games with externalities and fixed payoff division." *Games and Economic Behavior* 14: 90–123.

Boadway, R., and N. Bruce. (1984). *Welfare Economics*. Oxford: Basil Blackwell.

Bondareva, O. N. (1963). "Some applications of linear programming methods to the theory of cooperative games." [in Russian] *Problemy Kibernetiki* 10: 119-139.

Breton, M., L. Sbragia, and G. Zaccour. (2010). "A dynamic model for international environmental agreements." *Environmental and Resource Economics* 45: 25–48.

Brewer, T. L. (2004). "The WTO and the Kyoto Protocol: Interaction issues." *Climate Policy* 4: 3–12.

Buchanan, J. (1965). "An economic theory of clubs." *Economica* 32: 1–14.

Buchner, B., and C. Carraro. (2009). "Parallel climate blocs: Incentives to cooperation in international climate negotiations." In *The Design of Climate Policy*, edited by R. Guesnerie and H. Tulkens, 137–164. Cambridge, MA: MIT Press.

Carraro, C. (2008). "Incentives and institutions: A bottom up approach to climate policy." [Comment on the paper "Fragmented carbon markets and reluctant nations: Implications for the design of effective architectures" by David Victor] In *Architectures for Agreement: Addressing Global Climate Change in the Post-Kyoto World*, edited by J. Aldy and R. Stavins, 161–172. Cambridge, UK: Cambridge University Press.

Carraro, C., and D. Siniscalco. (1993). "Strategies for the international protection of the environment." *Journal of Public Economics* 52: 309–328.

Chander, P. (1993). "Dynamic procedures and incentives in public good economies." *Econometrica* 61: 1341–1354.

Chander, P. (1999). "International treaties on global pollution: A dynamic time-path analysis." In *Trade, Growth and Development: Essays in honor of T. N. Srinivasan*, edited by G. Ranis and L. K. Raut, 353–362. Amsterdam: Elsevier Science.

Chander, P. (2003). "The Kyoto Protocol and developing countries: Strategy and equity issues." In *India and Global Climate Change: Perspectives on Economics and Policy from a Developing Country*, edited by M. A. Toman, U. Chakravorty, and S. Gupta, 271–282. Washington, DC: Resources for the Future.

Chander, P. (2007). "The gamma core and coalition formation." *International Journal of Game Theory* 35: 539–556.

Chander, P. (2010). *Cores of Games with Positive Externalities*. CORE Discussion Paper No. 2010/4. Louvain-la-Neuve, Belgium: Center for Operations Research and Econometrics, LLN.

Chander, P. (2011). "The Coase theorem and climate change." Paper presented at the Singapore Economic Review Conference 2011. August 2–4.

Chander, P. (2014, December 14). "How to talk climate change in Paris." [Op-Ed] *The Hindu*.

Chander, P. (2016). "Climate change and international trade." *Reflets et Perspectives de la Vie economique* 55: 89–97.

Chander, P. (2017a). "Subgame-perfect cooperative agreements in a dynamic game of climate change." *Journal of Environmental Economics and Management* 84: 173–188.

Chander, P. (2017b, January 25). "The complex political economy of the Southeast Asian haze and some solutions." Distinguished Public Lecture, S. Rajaratnam School of International Studies.

Chander, P. (2017c). "Will an oligopoly become a monopoly?" [Mimeo] Singapore: Nanyang Technological University.

Chander, P., and M. A. Khan. (2001). "International treaties on trade and global pollution." *International Review of Economics and Finance* 10: 303–324.

Chander, P., and S. Muthukrishnan. (2015). "Green consumerism and pollution control." *Journal of Economic Behavior and Organization* 114: 27–35.

Chander, P., and E. Quah. (2014, June 13). "Tackling haze with cost sharing." *The Straits Times.*

Chander, P., and H. Tulkens. (1992). "Theoretical foundations of negotiations and cost sharing in transfrontier pollution problems." *European Economic Review* 36 (2/3): 288–299.

Chander, P., and H. Tulkens. (1995). "A core-theoretic solution for the design of cooperative agreements on transfrontier pollution." *International Tax and Public Finance* 2: 279–293.

Chander, P., and H. Tulkens. (1997). "The core of an economy with multilateral environmental externalities." *International Journal of Game Theory* 26: 379–401.

Chander, P., and H. Tulkens. (2008). "Cooperation, stability, and self-enforcement in international environmental agreements: A conceptual discussion." In *The Design of Climate Policy*, edited by R. Guesnerie and H. Tulkens, 165–186. Cambridge, MA: MIT Press.

Chander, P., and M. Wooders. (2016). *The Subgame Perfect Core.* Working Paper 16–00006. Department of Economics, Vanderbilt University.

Chander, P., H. Tulkens, J.-P. Van Ypersele, and S. Willems. (1998). "The Kyoto Protocol: An economic and game theoretic interpretation." In *Economic Theory for the Environment: Essays in honour of Karl-Göran Mäler*, edited by B. Kriström, P. Dasgupta, and K.-G. Löfgren, 98–117. Cheltenham, UK: Edward Elgar.

Chiang, A. (1992). *Elements of Dynamic Optimization.* New York: McGraw-Hill.

Clarke, E. (1971). "Multipart pricing of public goods." *Public Choice* (Fall): 17–31.

Cline, W. R. (1992). *The Economics of Global Warming.* Washington, DC: Institute for International Economics.

Common, M., and S. Stagl. (2005). *Ecological Economics: An Introduction.* Cambridge, UK: Cambridge University Press.

Compte, O., and P. Jehiel. (2010). "The coalitional Nash bargaining solution." *Econometrica* 78: 1593–1623.

Cornes, R., and T. Sandler. (1996). *The Theory of Externalities, Public Goods and Club Goods*, 2nd ed. Cambridge, UK: Cambridge University Press.

Coase, R. H. (1960). "The problem of social cost." *Journal of Law and Economics* 3: 1–44.

Coase, R. H. (1981). "The Coase theorem and the empty core: A comment." *Journal of Law and Economics* 24: 183–187.

Cooter, R. D. (1987). "Coase theorem." In *The New Palgrave: A Dictionary of Economics*, edited by J. Eatwell, M. Milgate, and P. Newman. London: Macmillan.

Copeland, B. R., and M. S. Taylor. (1994). "North-South trade and the environment." *Quarterly Journal of Economics* 109: 755–787.

Copeland, B. R., and M. S. Taylor. (1995). "Trade and transboundary pollution." *American Economic Review* 85: 716–737.

Copeland, B. R., and M. S. Taylor. (1997). *A Simple Model of Trade, Capital Mobility, and the Environment*. NBER Working Paper 5898. Cambridge, MA: National Bureau of Economic Research.

Cornes, R., and T. Sandler. (1995). *The Theory of Externalities, Public Goods and Club Goods*, 2nd ed. Cambridge, UK: Cambridge University Press.

Crawford, V. P., and H. Haller. (1990). "Learning how to cooperate: Optimal repeated coordination games." *Econometrica* 58: 571–595.

D'Aspremont, C., and J. J. Gabszewicz. (1986). "On the stability of collusion." In *New Developments in the Analysis of Market Structure*, edited by J. E. Stiglitz and G. F. Mathewson, 243–264. Cambridge, MA: MIT Press.

D'Aspremont, C., J. Jacquemin, J. J. Gabszewicz, and J. A. Weymark. (1983). "On the stability of collusive price leadership." *Canadian Journal of Economics* 16: 17–25.

Debreu, G. (1959). *Theory of Value*. New York: Wiley.

Debreu, G., and H. E. Scarf. (1963). "A limit theorem on the core of an economy." *International Economic Review* 4: 235–246.

De Canio, S. J. (2003). *Economic Models of Climate Change*. New York: Palgrave Macmillan.

de Clippel, G., and R. Serrano. (2008). "Marginal contributions and externalities in the value." *Econometrica* 76: 1413–1436.

de Zeeuw, A., and F. van der Ploeg. (1992). "International aspects of pollution control." *Environmental and Resource Economics* 2: 117–139.

Diamantoudi, E., and E. S. Sartzetakis. (2006). "Stable international environmental agreements: An analytical approach." *Journal of Public Economic Theory* 8: 247–263.

Dixit, A. (1990). *Optimization in Economic Theory*, 2nd ed. Oxford: Oxford University Press.

Dixit, A., and M. Olson. (2000). "Does voluntary participation undermine the Coase theorem?" *Journal of Public Economics* 76: 309–335.

Dockner, E. J., and N. V. Long. (1993). "International pollution control: Cooperative versus non-cooperative strategies." *Journal of Environmental Economics and Management* 25: 13–29.

Dockner, E. J., N. V. Long, and G. Sorger. (1996). "Analysis of Nash equilibrium in a class of capital accumulation games." *Journal of Economic Dynamics and Control* 20: 1209–1235.

Dutta, P. K., and R. Radner. (2004). "Self-enforcing climate change treaties." *Proceedings of the National Academy of Sciences of the United States of America* 101: 5174–5179.

Dutta, P. K., and R. Radner. (2009). "A strategic analysis of global warming: Theory and some numbers." *Journal of Economic Behavior and Organization* 71: 187–209.

Ecchia, G., and M. Mariotti. (1998). "Coalition formation in international environmental agreements and the role of institutions." *European Economic Review* 42: 573–582.

Edgeworth, F. Y. (1881). *Mathematical Psychics*. London: Kegan Paul.

Eichner, T., and R. Pethig. (2013). "Self-enforcing environmental agreements and international trade." *Journal of Public Economics* 102: 37–50.

Ellerman, A. D. (2005). "A note on tradeable permits." *Environmental and Resource Economics* 31: 123–131.

Ellerman, A. D., and A. Decaux. (1998). *Analysis of Post-Kyoto CO_2 Emissions Trading Using Marginal Abatement Curves*. MIT Joint Program on the Science and Policy of Global Change, Report No. 40. Cambridge, MA: Massachusetts Institute of Technology.

Eyckmans, J. (1997). *On the Incentives of Nations to Join International Environmental Agreements*. PhD thesis, Katholieke Universteit Leuven.

Eyckmans, J., and H. Tulkens. (2003). "Simulating coalitionally stable burden sharing agreements for the climate change problem." *Resource and Energy Economics* 25: 299–327.

Fankhauser, S. (1995). *Valuing Climate Change: The Economics of Greenhouse*. London: Earthscan Publications.

Filar, J. A., and L. A. Petrosjan. (2000). "Dynamic cooperative games." *International Game Theory Review* 2: 47–65.

Finus, M. (2001). *Game Theory and International Environmental Cooperation*. Cheltenham, UK: Eward Elgar.

Foley, D. (1970). "Lindahl solution and the core of an economy with public goods." *Econometrica* 38: 66–72.

Forges, F., J.-F. Mertens, and R. Vohra. (2002). "The ex-ante incentive compatible core in the absence of wealth effects." *Econometrica* 70: 1865–1892.

Frankel, J. A. (2005). "Climate and trade: Links between the Kyoto Protocol and WTO." *Environment: Science and Policy for Sustainable Development* 47: 8–19.

Frenkel, J. A. (2009). "An elaborated global climate policy architecture: Specific formulas and emission targets for all countries and all decades." Working Paper 14876. Cambridge, MA: National Bureau of Economic Research.

Friedman, J. (1990). *Game Theory with Applications to Economics*, 2nd ed. Oxford: Oxford University Press.

Froyn, C. B., and J. Hovi. (2008). "A climate agreement with full participation." *Economics Letters* 99: 317–319.

Funaki, Y., and T. Yamato. (1999). "The core of an economy with a common pool resource: A partition function form approach." *International Journal of Game Theory* 28: 157–171.

Gale, D. (1978). "The core of a monetary economy without trust." *Journal of Economic Theory* 19: 456–491.

Geden, O. (2016, April). "An actionable climate target." *Nature Geoscience* 9: 340–342.

Germain, M., Ph. Toint, H. Tulkens, and A. de Zeeuw. (2003). "Transfers to sustain dynamic core-theoretic cooperation in international stock pollutant control." *Journal of Economic Dynamics and Control* 28: 79–99.

Gillies, D. B. (1953). "Discriminatory and bargaining solutions to a class of symmetric n-person games." In *Contributions to Theory of Games*, Vol. 2, edited by H.W. Kuhn and A.W. Tucker, 325–342. Princeton, NJ: Princeton University Press.

Grossmann, G., and E. Helpman. (1995). "The politics of free international trade agreements." *American Economic Review* 85: 667–690.

Grossman, G., and A. Krueger. (1993). "Environmental impacts of a North American free international trade agreement." In *The Mexico-U.S. Free Trade Agreement*, edited by P. Garber, 361–389. Cambridge, MA: MIT Press.

Grossman, S. J., and M. Perry. (1986). "Perfect sequential equilibrium." *Journal of Economic Theory* 39: 97–119.

Groves, T., and J. Ledyard. (1977). "Optimal allocation of public goods: A solution to the 'free rider' problem." *Econometrica* 45: 783–810.

Harstad, B. (2012). "Climate contracts: A game of emissions, investments, negotiations, and renegotiations." *Review of Economic Studies* 79: 1527–1557.

Haeckel, E. (1866). *Generelle Morphologie der Organismen*. Berlin: G. Reimer.

Hafalir, I. E. (2007). "Efficiency in coalitional games with externalities." *Games and Economic Behavior* 61: 242–258.

Harsanyi, J. (1974). "An equilibrium-point interpretation of stable sets and a proposed alternative definition." *Management Science* 20: 1472–1495.

Harsanyi, J. C. (1977). *Rational Behavior and Bargaining Equilibrium in Games and Social Situations*, chap. 11. Cambridge, UK: Cambridge University Press.

Hart, S., and M. Kurz. (1983). "Endogenous formation of coalitions." *Econometrica* 51: 1047–1064.

Helm, C. (2000). *Theories of International Environmental Cooperation*. Cheltenham, UK: Edward Elgar.

Helm, C. (2001). "On the existence of a cooperative solution for a coalitional game with externalities." *International Journal of Game Theory* 30: 141–147.

Helm, C. (2003). "International emissions trading with endogenous allowance choices." *Journal of Public Economics* 87: 2737–2747.

Henda, B. (2016, November 26). "Despite tough talk, Indonesia's government is struggling to stem deforestation." *Economist*.

Hoel, M. (1992). "Emission taxes in a dynamic international game of CO_2 emissions." In *Conflicts and Cooperation in Managing Environmental Resources, Microeconomic Studies*, edited by R. Pethig, 39–70. Berlin: Springer-Verlag.

Hoel, M. (1993). "Intertemporal properties of an international carbon tax." *Resource and Energy Economics* 15: 51–70.

Hoel, M. (1994). "International environmental conventions: The case of uniform reductions of emissions." *Environmental and Resource Economics* 2: 141–159.

Hoel, M., and K. Schneider. (1997). "Incentives to participate in an international environmental agreement." *Environmental and Resource Economics* 9: 153–170.

Holmstrom, B., and B. Nalebuff. (1992). "To the raider goes the surplus? A reexamination of the free-rider problem." *Journal of Economics and Management Strategy* 1: 37–62.

Hong, F., and L. Karp. (2012). "International environmental agreements with mixed strategies and investment." *Journal of Public Economics* 96: 685–697.

Houba, H., and A. de Zeeuw. (1995). "Strategic bargaining for the control of a dynamic system in state-space form." *Group Decision and Negotiation* 4: 71–97.

Hourcade, J. C., and P. R. Shukla. (2009). "Untying the climate-development Gordian Knot: Economic options in a politically constrained world." In *The Design of Climate Policy*, edited by R. Guesnerie and H. Tulkens, 75–102. Cambridge, MA: MIT Press.

Huang, C. Y., and T. Sjöström. (2006). "Implementation of the recursive core for partition function form games." *Journal of Mathematical Economics* 42: 771–793.

Ichiishi, T. (1981). "A social equilibrium existence lemma." *Econometrica* 49: 369–377.

International Energy Agency. (1996). *Energy Prices and Taxes*. Paris: Author.

IPCC. (1990, 1995, 2001, 2007, 2013). *Climate Change*, First, Second, Third, Fourth, and Fifth Assessment Reports [four volumes each]. Cambridge, UK: Cambridge University Press.

IPCC. (2007). *Climate Change 2007: Synthesis Report. Summary for Policymakers.* Fourth Assessment Report. Cambridge, UK: Cambridge University Press.

IPCC. (2013). *Climate Change*, Fifth Assessment Report (AR 5). Cambridge, UK: Cambridge University Press.

Jaffe, A., S. Peterson, P. Portney, and R. Stavins. (1995). "Environmental regulation and the competitiveness of U.S. manufacturing: What does the evidence tell us?" *Journal of Economic Literature* 33: 132–163.

Jorgensen, S., G. Martin-Herran, and G. Zaccour. (2003). "Agreeability and time consistency in linear-state differential games." *Journal of Optimization Theory and Applications* 119: 49–63.

Jorgensen, S., G. Martin-Herran, and G. Zaccour. (2005). "Sustainability of cooperation in linear-quadratic differential games." *International Game Theory Review* 7: 395–406.

Jotzo, F. A. (2013). "Decentralization and avoiding deforestation: The case of Indonesia." In *Federal Reform Strategies*, edited by S. Howes and M. G. Rao, 273–301. Oxford: Oxford University Press.

Kandori, M. (1992). "Social norms and community enforcement." *Review of Economic Studies* 59: 63–80.

Kaneko, M. (1977). "The ratio equilibria and the core of the voting game $G(N,W)$ in a public good economy." *Econometrica* 45: 1589–1594.

Karp, L. (2005). "Global warming and hyperbolic discounting." *Journal of Public Economics* 89: 261–282.

Karp, L. (2017). *Natural Resources as Capital: Theory and Policy*. Cambridge, MA: MIT Press.

Karp, L., and J. Zhang. (2005). "Regulation with anticipated learning about environmental damages." *Journal of Environmental Economics and Management* 32: 273–300.

Kaul, I., I. Grunberg, and M. Stern, eds. (1999). *Global Public Goods: International Cooperation in the 21st Century*. Oxford: Oxford University Press.

Kemfert, C., W. Lise, and R. S. Tol. (2004). "Games of climate change with international trade." *Environmental and Resource Economics* 28: 209–232.

Kolstad, C. D. (2000). *Environmental Economics*. New York: Oxford University Press.

Kolstad, C. D. (2010). "Equity, heterogeneity, and international environmental agreements." *B.E. Journal of Economic Analysis & Policy* 10.

Kranich, L., A. Perea, and H. Peters. (2005). "Core concepts for dynamic TU games." *International Game Theory Review* 7: 43–61.

Laffont, J.-J. (1977a). *Effets externes et théorie économique*. Monographies du Seminaire d'économétrie. Paris: Editions du Centre national de la recherché scientifique (CNRS).

Laffont, J.-J. (1977b). *Fundamentals of Public Economics*. Cambridge MA: MIT Press.

Lardon, A. (2012). "The γ-core in Cournot oligopoly TU-games with capacity constraints." *Theory and Decision* 72: 387–411.

Lehrer, E., and M. Scarsini. (2013). "On the core of dynamic cooperative games." *Dynamic Games and Applications* 3: 359–373.

Long, N. V. (1992). "Pollution control: A differential game approach." *Annals of Operations Research* 37: 283–296.

Low, P., and A. Yeats. (1992). "Do 'dirty' industries migrate?" In *International Trade and the Environment*, edited by P. Low, 89–103. World Bank Discussion Paper No. 159. Washington, DC: World Bank.

Mäler, K.-G. (1989a). "International environmental problems." *Oxford Review of Economic Policy* 6: 80–108.

Mäler, K.-G. (1989b). "The acid rain game." In *Valuation Methods and Policy Making in Environmental Economics*, edited by H. Folmer and E. van Ierland, 231–252. Amsterdam: Elsevier.

Mäler, K.-G., and A. de Zeeuw. (1998). "The acid rain differential game." *Environment and Resource Economics* 12: 167–184.

Malinvaud, E. (1971). "Procedures for the determination of a program of collective consumption." *European Economic Review* 2: 187–217.

Markusen, J. R. (1975). "Cooperative control of international pollution and common property resources." *Quarterly Journal of Economics* 89: 618–632.

Marx, L. M., and S. A. Matthews. (2000). "Dynamic voluntary contribution to a public project." *Review of Economic Studies* 67: 327–358.

Mas-Colell, A., M. Whinston, and J. Green. (1995). *Microeconomic Theory*. New York: Oxford University Press.

Maskin, E. (2003). "Bargaining, coalitions and externalities." Presidential Address to the Econometric Society. Princeton, NJ: Institute for Advanced Study, Princeton University.

Mauleon, A., and V. Vannetelbosch. (2004). "Farsightedness and cautiousness in coalition formation games with positive spillovers." *Theory and Decision Sciences* 56: 291–324.

McQuillin, B. (2009). "The extended and generalized Shapley value: Simultaneous considerations of coalitional externalities and coalitional structures." *Journal of Economic Theory* 144: 696–721.

Moreno, D., and J. Wooders. (1996). "Coalition-proof equilibrium." *Games and Economic Behavior* 17: 80–112.

Moulin, H. (1987). "Egalitarian-equivalent cost sharing of a public good." *Econometrica* 55: 963–976.

Musgrave, R. (1969). "Provision for social goods." In *Public Economics* [proceedings of the IEA-CNRS conference held in 1966 at Biarritz, France], edited by J. Margolis and H. Guitton. London: Macmillan; New York: St. Martin's Press.

Myerson, R. B. (1978). "Threat equilibria and fair settlements in cooperative games." *Mathemaics of Operations Research* 3: 265–274.

Myerson, R. B. (1991). *Game Theory: Analysis of Conflict.* Cambridge, MA: Harvard University Press.

Nash, J. (1953). "Two-person cooperative games." *Econometrica* 21: 128–140.

Nordhaus, W. (2015). "Climate clubs: Overcoming free-riding in international climate policy." *American Economic Review* 105: 1339–1370.

Nordhaus, W., and J. Boyer. (1999). "Requiem for Kyoto: An economic analysis of the Kyoto Protocol." *Energy Journal* 20: 93–130.

Nordhaus, W. D., and J. Boyer. (2000). *Warming the World: Economic Models of Global Warming.* Cambridge, MA: MIT Press.

Nordhaus, W., and Z. Yang. (1996). "A regional dynamic general equilibrium model of alternative climate change strategies." *American Economic Review* 86: 741–765.

North, D. C. (1990). *Institutions, Institutional Change and Economic Performance.* Cambridge, UK: Cambridge University Press.

Olmstead, S. M., and R. N. Stavins. (2006). "An international policy architecture for the post-Kyoto era." *American Economic Review* 96: 35–38.

Olmstead, S. M., and R. Stavins. (2012). "Three key elements of a post-2012 international climate policy architecture." *Review of Environmental Economics and Policy* 6: 65–85.

Osborne, M. J., and A. Rubinstein. (1994). *A Course in Game Theory.* Cambridge, MA: MIT Press.

Ostrom, E. (1998). "A behavioral approach to the rational choice theory of collective action." [Presidential address to the American Political Science Association] *American Political Science Review* 92: 1–22.

Palfrey, T. R., and J. E. Prisbrey. (1997). "Anomalous behavior in public goods experiments: How much and why?" *American Economic Review* 87: 829–846.

Palfrey, T. R., and H. Rosenthal. (1984). "Participation and the provision of discrete public goods: A strategic analysis." *Journal of Public Economics* 24: 171–193.

Pérez-Castrillo, D. (1994). "Cooperative outcomes through non-cooperative games." *Games and Economic Behavior* 7: 428–440.

Perry, M., and P. J. Reny. (1994). "A non-cooperative view of coalition formation and the core." *Econometrica* 62: 795–817.

Petrakis, E., and A. Xepapadeas. (1996). "Environmental consciousness and moral hazard in international agreements to protect the environment." *Journal of Public Economics* 60: 95–110.

Pigou, A. C. (1920). *The Economics of Welfare*. London: Macmillan Publishers.

Purnomo, H., and B. Shantiko. (2015). "The political economy of fire and haze: Root causes." Discussion Forum: Long-term Solutions to Fires in Indonesia: Multi-Stakeholder Efforts and the Role of the Private Sector. Paris: Global Landscapes Forum, December 5–6.

Quah, E. (2002). "Transboundary pollution in Southeast Asia: The Indonesian fires." *World Development* 30: 429–441.

Quah, E. (2015). "Pursuing economic growth in Asia: The environmental challenge." *World Economy* 38: 1487–1504.

Rajan, R. (1989). "Endogenous coalition formation in cooperative oligopolies." *International Economic Review* 30: 863–876.

Rauscher, M. (1991). "National environmental policies and the effects of economic integration." *European Journal of Political Economy* 7: 313–329.

Ray, D., and R. Vohra. (1997). "Equilibrium binding agreements." *Journal of Economic Theory* 73: 30–78.

Ray, D., and R. Vohra. (1999). "A theory of endogenous coalition structures." *Games and Economic Behavior* 26: 286–336.

Ray, D., and R. Vohra. (2001). "Coalitional Power and Public Goods." *Journal of Political Economy* 109: 1355–1384.

Ray, D., and R. Vohra. (2015). "The farsighted stable set." *Econometrica* 83: 977–1011.

Reinganum, J. F. (1982). "A class of differential games for which the open-loop and closed-loop equilibria coincide." *Journal of Optimization Theory and Applications* 36: 253–462.

Reinganum, J., and N. Stokey. (1985). "Oligopoly extraction of a common property natural resource; the importance of the period of commitment in dynamic games." *International Economic Review* 26: 161–173.

"Report of the AWG-LCA plenary meeting in Copenhagen 2009." *Earth Negotiations Bulletin* 12 (459): 18–19. International Institute of Sustainable Development, New York.

Reuter, P. (1995). *Introduction to the Law of Treaties*. London: Kegan Paul.

Roberts, D. J. (1974). "The Lindahl solution for economies with public goods." *Journal of Public Economics* 3: 23–42.

Rosenthal, R. W. (1971). "External economies and cores." *Journal of Economic Theory* 3: 133–188.

Rubinstein, A. (1982). "Perfect equilibrium in a bargaining model." *Econometrica* 50: 97–109.

Rubio, S. J., and B. Casino. (2005) "Self-enforcing international environmental agreements with a stock pollutant." *Spanish Economic Review* 7: 89–109.

Rubio, S. J., and A. Ulph. (2007). "An infinite-horizon model of dynamic membership of international environmental agreements." *Journal of Environmental Economics and Management* 54: 296–310.

Saijo, T., and T. Yamato. (1999). "A voluntary participation game with a non-excludable public good." *Journal of Economic Theory* 84: 227–242.

Scarf, H. E. (1971). "On the existence of a cooperative solution for a general class of N-person games. *Journal of Economic Theory* 3: 169–181.

Schelling, T. C. (1960). *The Strategy of Conflict*. Cambridge, MA: Harvard University Press.

Scotchmer, S. (2002). "Local public goods and clubs." In *Handbook of Public Economics*, Vol. IV, edited by A. Auerbach and M. Feldstein, 1997–2042. Amsterdam: North-Holland Press.

Serrano, R. (2008). "Nash program." In *The New Palgrave Dictionary of Economics*, 2nd ed., edited by S. N. Durlauf and L. E. Blume. London: Palgrave Macmillan.

Shapley, L. S. (1953). "A value for *n*-person games." In *Contributions to the Theory of Games*, Vol. II, edited by H. W. Kuhn and A. W. Tucker, 307–317. Princeton, NJ: Princeton University Press.

Shapley, L. S. (1967). "On balanced sets and cores." *Naval Research Logistics Quarterly* 14: 453–460.

Shapley, L. (1971). "Cores of convex games." *International Journal of Game Theory* 1: 11–26.

Shapley, L. S., and M. Shubik. (1969). "On the core of an economic system with externalities." *American Economic Review* 59: 678–684.

Shubik, M. (1984). *A Game Theoretic Approach to Political Economy*. Cambridge, MA: MIT Press.

Siddique, A. S., and E. Quah. (2004). "Modelling transboundrary air pollution in Southeast Asia: Policy regime and the role of stakeholders." *Environment and Planning A* 36: 1411–1425.

Sidgwick, H. (1883). *Principles of Political Economy*. New York: MacMillan.

Stamatopoulos, G. (2016). "The core of aggregative cooperative games with externalities." *B. E. Journal of Economics* 16: 389–410.

Starrett, D. (1973). "A note on externalities and the core." *Econometrica* 41: 179–183.

Stern, N., S. Peters, V. Bakhshi, A. Bowen, C. Cameron, S. Catovsky, D. Crane, et al. (2006). *Stern Review on the Economics of Climate Change*. London: HM Treasury.

Stokey, N., and R. Lucas. (1989). *Recursive Methods in Economic Dynamics*. Cambridge, MA: Harvard University Press.

Tahvonen, O. (1994). "Carbon dioxide abatement as a differential game." *European Journal of Political Economy* 19: 685–705.

Thoron, S. (1998). "Formation of a coalition-proof stable cartel." *Canadian Journal of Economics* 31: 63–76.

Thrall, R., and W. Lucas. (1963). "N-person games in partition function form." *Naval Research Logistics Quarterly* 10: 281–298.

Tol, R. (2009). "The economic effects of climate change." *Journal of Economic Perspectives* 23: 29–51.

Tulkens, H., and F. Schoumaker. (1975), "Stability analysis of an effluent charge and the 'polluterspay' principle." *Journal of Public Economics* 4: 245–269.

Tulkens, H., and V. van Steenberghe. (2009). *Mitigation, Adaptation, Suffering: In Search of the Right Mix in the Face of Climate Change.* CORE Discussion Paper No. 2009/54. Louvain-la-Neuve, Belgium: Center for Operations Research and Econometrics, LLN.

Tuovinen, J.-P., L. Kangas, and G. Nordlund. (1990). "Model calculations of sulphur and nitrogen depositions in Finland." In *Acidification in Finland*, edited by P. Kauppi, P. Anttila, and K. Kenttämies, 167–197. Berlin: Springer-Verlag.

Ulph, A. (1996). "Environmental policy and international trade when government and producers act strategically." *Journal of Environmental Economics and Management* 30: 256–281.

United Nations (UN). (1987). *Our Common Future.* Report of the World Commission on Environment and Development. Oxford: Oxford University Press.

United Nations Environment Programme. (2005). *Environment and Trade—A Handbook.* New York: International Institute for Sustainable Development.

Varian, H. (1990). *Microeconomic Analysis,* 3rd ed. New York: Norton.

Varian, H. (1994). "A solution to the problem of externalities when agents are well-informed." *American Economic Review* 84: 1278–1293.

von Neumann, J., and O. Morgenstern. (1953). *Theory of Games and Economic Behavior,* 3rd ed., chap. 11. Princeton, NJ: Princeton University Press.

Wang, Y. (2013). "A time-consistent model for cooperation in international pollution control." *Economic Modeling* 33: 500–506.

Weber, S., and H. Wiesmeth. (2003). "From autarky to free trade: The impact on environment and welfare." *Jahrbuch für Regionalwissenschaft* 23: 91–115.

Wiesmeth, H. (2012). *Environmental Economics: Theory and Policy in Equilibrium.* New York: Springer.

Wooders, M., and F. H. Page Jr. (2008). "Coalitions." In *The New Palgrave Dictionary of Economics,* 2nd ed., edited by S. N. Durlauf and L. E. Blume. London: Palgrave Macmillan.

Yang, Z. (2003). "Reevaluation and renegotiation of climate change coalitions—a sequential closed-loop game approach." *Journal of Economic Dynamics and Control* 27: 1563–1594.

Yi, S.-S. (1997). "Stable coalition structures with externalities." *Games and Economic Behavior* 20: 201–237.

Yi, S.-S. (2003). "Endogenous formation of economic coalitions: A survey on the partition function approach." In *The Endogenous Formation of Economic Coalitions*, edited by C. Carraro, 80–127. London: Edward Elgar.

Yi, S.-S., and H. Shin. (2000). "Endogenous formation of research coalitions with spillovers." *International Journal of Industrial Organization* 18: 229–256.

Yong, J. (2004). *Horizontal Monopolization via Alliances, or Why a Conspiracy to Monopolize Is Harder Than It Appears*. Melbourne, Australia: MIAESR, University of Melbourne.

Zhao, J. (1996). "The hybrid solutions of an *n*-person game." *Games and Economic Behavior* 4: 145–160.

Zhao, J. (1999). "A β-core existence result and its application to oligopoly markets." *Games and Economic Behavior* 27: 153–168.

Zhao, J. (2000). "Trade and environmental distortions: coordinated intervention." *Environment and Development Economics* 5:361–375.

AUTHOR INDEX

Aivazian, V. A., 125n16

Alcamo, J., 23n14

Arrow, K. J., 18–19, 18n5, 19n9, 225n16

Audley, J. J., 273

Aumann, R. J., 17n1, 72

Babiker, M., 255

Bagnoli, M., 142

Barrett, S., 1n2, 56n18, 139, 211n6, 273n2

Basar, T., 210n4

Baumol, W., 18

Bearden, J. N., 246–247

Becker, R. A., 14n14, 194n18, 205, 297n4

Beladi, H., 290

Bento, A., 253n25

Bergstrom, T., 218

Bernheim, B. D., 73, 75n11

Bhagwati, J., 273

Bloch, F., 156

Blume, E., 218

Boadway, R., 54n15

Bondareva, O. N., 112

Boyer, J., 30, 235n4

Bruce, N., 54n15

Buchanan, J., 18n4, 20

Buchner, B., 256n33

Callen, J. L., 125n16

Carraro, C., 139, 159–160, 256n33

Chakrabarti, K., 14n14, 194n18, 205, 297n4

Chander, P., 7n8, 14n14, 16n15, 26, 30n17, 32n18, 34, 36, 48n6, 52n12, 63, 68n4, 77n15, 80, 91, 113, 113n8, 117n11, 125n14, 130n21, 134, 136, 183n13, 200, 201n21, 225n16, 234n3, 240, 242, 296, 297

Chau, H., 290

Clarke, E., 9

Coase, R. H., 6, 16n15, 18n4, 51n10, 52n12, 56n18, 58–60, 63, 124, 124n12, 125n16

Common, M., 4n5

Compte, O., 95n37

Cooter, R. D., 124n12

Copeland, B. R., 273

Cornes, R., 18n4, 18n5

Crawford, V. P., 142n10

d'Aspremont, C., 139, 159

Debreu, G., 18, 252n23

De Brum, T., 265n37

Decaux, A., 16n15, 243, 244

de Clippel, G., 89n26, 90–93, 90n28

de Zeeuw, A., 185n14, 202, 210n4

Diamantoudi, E., 139n5, 160

Dockner, E. J., 166, 186, 201–202, 210n4

Dutta, P. K., 166, 173–174, 187, 192, 201n19, 202, 212n7

Edgeworth, F. Y., 252n23
Ellerman, A. D., 16n15, 62n23, 243, 244
Eyckmans, J., 111n6, 116n10

Fankhauser, S., 236n7, 293n1
Filar, J. A., 205
Foley, D., 80
Forges, F., 297n4
Friedman, J., 87, 112
Funaki, Y., 91

Gabszewicz, J., 139, 159
Gale, D., 297n4
Gates, B., 263
Germain, M., 183n13, 203–204
Green, J., 8n10, 17n2, 18n4
Grossman, G., 273, 289
Grossman, S. J., 77n14
Groves, T., 7n8, 9
Grunberg, I., 20n10

Haeckel, E., 22n12
Hafalir, I. E., 90–91, 90n28, 90n29, 91n31, 92, 158n20
Hahn, F., 225n16
Haller, H., 142n10
Harsanyi, J. C., 82, 141n6
Harstad, B., 201–202, 210n19
Hart, S., 94n34
Helm, C., 55n17, 112n7, 113, 113n8
Helpman, E., 289
Henda, B., 133
Ho, B., 253n25
Hoel, M., 180n10, 210n4
Holmstrom, B., 142, 143n11
Hong, F., 201n19
Hordijk, L., 23n14
Huang, C. Y., 68n3

Ichiishi, T., 88

Jacoby, D., 255
Jaffe, A., 273n2
Javadekar, P., 219n14, 269
Jehiel, P., 95n37
Jorgensen, S., 180n9
Jotzo, F. A., 133

Kandori, M., 298
Kaneko, M., 80
Karp, L., 4n4, 201n19, 212n7, 213n10
Kaul, I., 20n10
Khan, A., 30n17, 36, 290
Ki-moon, B., 258
Kolstad, C. D., 4n5, 19n6, 23n13, 55n17, 63n24
Kranich, L., 205
Krueger, A., 273
Kurz, M., 94n34

Laffont, J. J., 18n4, 80
Lardon, A., 83n20, 296
Ledyard, J., 7n8, 9
Lehrer, E., 12n12
Lipman, L., 142
Long, N. V., 166, 186, 202, 210n4, 222
Low, P., 273n2
Lucas, W., 88

Mäler, K. G., 23n14, 37, 68, 80, 185n14, 210n4
Markusen, J. R., 273n2
Marshall, A., 18n4
Martin-Herran, G., 180n9
Marx, L. M., 193n16, 201–202, 201n22
Mas-Colell, A., 8n10, 17n2, 18n4
Maskin, E., 68n3, 89, 89n26, 90, 92, 135, 135n1, 143, 156–158
Matthews, A., 193n16, 201–202
Mauleon, A., 141n6
Meade, 18n4
Mertens, F., 297n4
Modi, N., 263

Moreno, D., 75n11
Morgenstern, O., 68, 79
Moulin, H., 80
Musgrave, R., 19–20
Muthukrishnan, S., 48n6
Myerson, R. B., 8n10

Nalebuff, B., 142, 143n11
Nordhaus, W. D., 30, 181n11, 202, 212n7, 235n4, 236n7, 239, 256n33
North, D. C., 298

Oates, W., 18
Olsder, J., 210n4
Osborne, M. J., 8n10, 174n7

Page, H., Jr., 89n27
Palfrey, T. R., 218
Peleg, B., 73, 75n11
Perea, A., 205
Pérez-Castrillo, D., 12n12
Perry, M., 12n12, 77n14
Peters, H., 205
Petrosjan, A., 205
Pigou, A. C., 6, 18n4, 56n18
van der Ploeg, F., 202
Prisbrey, E., 218
Purnomo, H., 133n23, 134

Quah, E., 131, 132n22, 133

Radner, R., 166, 173–174, 187, 192, 201–202, 212n7
Ramirez-Basora, M., 253n25
Rauscher, M., 290
Ray, D., 13, 13n13, 68n5, 91, 100, 101n45, 136n2, 137, 137n4, 139, 139n5, 141n6, 141n7, 155, 157n18, 159n23
Reilly, M., 255
Reinganum, J. F., 178n8, 180n10
Reny, J., 12n12

Reuter, P., 64n25
Roberts, D. J., 80
Rosenthal, R. W., 80n18
Rubinstein, A., 8n10, 12n12, 95n37, 174n7
Rubio, S. J., 201, 256n33

Sandler, T., 18n4, 18n5
Sartzetakis, S., 160
Scarf, H. E., 80, 252n23
Scarsini, M., 12n12
Schelling, T. C., 85n22, 147–148
Schoumaker, F., 59n21
Serrano, R., 89n26, 90–93, 90n28
Shantiko, B., 133n23
Shapley, L. S., 20, 68n5, 87, 92, 112
Shaw, R., 23n14
Shubik, M., 20, 68n5
Siddiqui, 132n22, 133
Sidgwick, H, 18n4
Siniscalco, D., 139, 159–160
Sjöström, T., 68n3
Sorger, G., 166
Stamatopoulos, G., 296
Starrett, D., 80
Stern, N., 20n10, 236n7, 293n1

Tahvonen, O., 210n4
Taylor, S., 273
Thoron, S., 159n24
Thrall, R., 88
Tol, R., 236n7
Tulkens, H., 26, 32n18, 41n23, 59n21, 68n4, 113, 113n8, 116n10, 130n21, 234n3, 242

Ulph, A., 201, 256n33

Vannetelbosch, V., 141n6
van Steenberghe, V., 41n23
van Ypersele, P., 234n3
Varian, H., 59n21, 130n20, 218

Vohra, R., 13, 13*n*13, 38*n*5, 91, 100, 101*n*45,
    136*n*2, 137, 137*n*4, 139, 139*n*5, 141*n*6,
    141*n*7, 155, 157m18, 159*n*23, 297*n*4
von Neumann, J., 68, 79

Warming, E., 22*n*12
Whinston, D., 73, 75*n*11
Whinston, M., 8*n*10, 17*n*2, 18*n*4
Willems, S., 234*n*3
Wooders, J., 75*n*11, 297
Wooders, Myrna, 89

Yamato, T., 91
Yang, Z., 181*n*11, 202, 212*n*7, 236*n*7,
    239
Yeats, A., 273*n*2
Yi, S. S., 13*n*13, 91, 100, 136*n*2, 137,
    137*n*4, 149*n*15, 154–155, 160

Zaccour, G., 180*n*9
Zhang, J., 212*n*7
Zhao, J., 88
Zuckerburg, M., 263

SUBJECT INDEX

acid rain, 1–3

action plans: 5-year national, 261–264; back-to-back, 262

adaptation: challenges of, 42; of climate change, 39–40; cost of, 40–42; optimal, 41

ad hoc methods, 182n12

ad infinitum, 95–96, 95n37

aggregate emission quota, 250

aggregate emission target, 7

agriculture, 32

all-or-none expectation, 108, 108n2

alternative energy sources, 237n8, 270–271

alternative symmetric equilibria, 146–147

ambient pollution, 21–22, 25, 54; change in, 65; commodity of, 26; by country, 45; infinitesimal changes of, 66; lowering of, 55; public good and, 50

anarchy, 49, 55

Asian Development Bank, 132

assimilation capacity, 25

assumptions: of basic flow model, 27–33; of γ-coalitional function, 138; of coalitions, 138–139; of production functions, 165–166, 277; of utility functions, 165–166. *See also* behavioral assumptions

attrition war, 156–158

autarky equilibrium, 275, 278–279, 282, 287–288; Proposition 10.1, 280–281

axiomatic approach, 93, 295

backward induction, 183–184, 191, 203

basic flow model, 34n20; Assumption 2.1, 27; Assumption 2.2, 27–29; Assumption 2.3, 30; Assumption 2.4, 32–33; assumptions of, 27–33; compact form of, 37–39; components of, 26–34; variables for, 32

BAU. *See* business as usual

Bayesian equilibria, 202

behavioral assumptions, 43; alternative, 44–47; of α-coalitional function, 81; of β-coalitional function, 81

Bondareva-Shapley theorem, 112; Theorem 5.4, 113–115

boundary condition, 214

Brundtland Report, 4n4

business as usual (BAU), 44–45, 187, 239; of equilibrium, 47–48; estimated, 243; negligence of, 55; as reference emissions, 248; utilities of, 56–57

"cap-and-trade" mechanism, 7, 14, 238

capital, 275; marginal product of, 278; mobility of, 290

capital accumulation, 201

carbon budget, 216; division of, 231–232; size of, 260–261; target of, 265

carbon dioxide (CO_2), 21, 35, 212n8

carbon-intensive goods, 8

carbon tax, 296

CBDR. *See* common but differentiated responsibilities

CDM. *See* clean development mechanism

certified project activities, 253–254

China, 45, 239–240, 253, 262, 269–272

clean development mechanism (CDM), 52n12, 236, 251–252, 268–269; interpretation of, 254–255; trade under, 253–254

climate: agreements on, 200–201; policy on, 232; stability of, 223–232

climate change. *See specific topics*

CO^2. *See* carbon dioxide

coalitional function, 77, 82–87; constrained, 229; in cooperative game theory, 80; definition of, 79; on environmental game, 107–108; as form, 78; strategic game of, 78

γ-coalitional function, 11, 82; assumption of, 138; definition of, 108; of environmental game, 107–112; interpretation of, 108–110; on nonempty core, 203–204; partition function and, 88; Proposition 4.4, 83–86; Proposition 5.3, 109–110; as superadditive, 90, 111–112

α-coalitional function, 79; behavioral assumption of, 81; conception of, 81n19; limitation of, 81–82; Proposition 4.3, 80

β-coalitional function, 79; behavioral assumption of, 81; conception of, 81n19; limitation of, 81–82; Proposition 4.3, 80

coalitional game: convex, 112; core of, 79; definition of, 87; dynamics in, 205; imputation in, 229; with nonempty core, 112–113, 229; of strategic games, 78–79; as superadditive, 87, 205; with transferable utility, 112–113

coalition formation game: approaches to, 155–161; comprehensive survey on, 154–155; endogenous, 156; exogenously, 156; as noncooperative game, 141–143; Paris

Agreement on, 256–257; repeated game of, 135, 137; sequential games of, 156–158; structure of, 136; with transferable utility, 137

coalitions: assumptions of, 138–139; behavior of, 128n19, 190; combining of, 87; commitments of, 158; common property in, 91; γ-core and, 110; of countries, 199–200; credible cooperation by, 105; deviating, 69, 75, 108–110, 193; dissolution of, 99–100, 99n43, 144, 151–154; on emissions, 106; as equilibrium outcome, 98; expectation of, 135–136; formation of, 110, 110n5; freedom of, 73, 75; future of, 110; incentive of, 151–154; internal-external stability of, 201, 256–257; limitations of, 139–140; mergers of, 139–140, 139n5; non-singleton, 99–100, 100n44, 110; nontrivial, 97–98; payoff for, 75–80, 89, 89n25, 136, 227–228; as singletons, 138–139; size of, 159–160; stability of, 160–161, 161n25, 249–251, 252n23; strategies of, 73; unanimity among, 94. *See also* grand coalition; players

coalition S, 190–191

coalition structure, 87–88; efficient, 91; finest, 154; formation of, 108–109; with induced game, 89; inefficiency of, 137n3, 156–157; internal-external stability of, 158–161. *See also* equilibrium coalition structure

Coase theorem, 59; arguments for, 63–64, 292; concerns of, 63n24; γ-core and, 124–131; diagrammatic exposition of, 128–131; extensions of, 14; generalization of, 124; introduction of, 6; limitations of, 6n7; parts of, 124; on pollution, 59–60; on pollution rights, 125; Proposition 5.10, 127–128; reasoning of, 60–61; support for, 61

collective action, 43–44

commodity: ambient pollutant, 26; composite consumption good, 26; emitted pollutant, 26; space of, 18. *See also* two-dimensional commodities

common but differentiated responsibilities (CBDR), 268–269

common property, 91

compensation mechanism, 130n20

competitive emissions trading equilibrium, 245–246; gains from, 249–250; international, 250–251; regulatory provisions on, 252; worldwide, *244*, 251–253

concave function, 185

Conference of the Parties 21 (COP 21), 218, 257

constrained core, 226–228; Lemma 8.1, 229; Theorem 8.2, 229–231

consumption goods, 26n16, 35; alternative distributions for, 54; compensation as, 62; composite, 26, 211; of countries, 51–52; model of, 36–37, 276–277; production of, 215; transfer of, 43–44; value of, 48; world output of, 259–260

consumption profiles: charactization of, 49–51; efficient, 49–51, 108; emission profiles and, 50–52; equality of, 58; equilibria, 43; feasible, 33–34; multiplicity of, 57–58; optima, 43; as spontaneous, 55–56

consumption time profiles, 166–169; connection to, 194n17; efficient, 195; of GPO, 174, 187; Lemma 7.1, 170–171

continuation game, 97

continuous time models, 210n4

convex, 87

cooperative games, 10

cooperative game theory, 10–13, 93

COP21. *See* Conference of the Parties 21

γ-core, 11; alternative to, 12; application of, 103; base of, 69; benefits of, 135; coalition formation and, 110; Coase theorem and, 124–131; cooperative, 296–297; countries on, 125; diagrammatic exposition of, 128–131; of economy, *129*, *131*; in environmental game, 107–109; forms of, 119; imputations of, 115–124, 129–130, 137–138, 297; introduction of, 68n4; motivation for, 68; noncooperative, 296–297; noncooperative foundations of, 93–102; nonempty, 77, 91–92, 97, 103, 111–116, 124–125, 155n17, 204; normative approaches and, 92–93; payoff vectors of, 12, 102; Proposition 4.5, 92; Proposition 5.10, 127–128; rationale for, 137–148; as strategic games, 68, 103; strategy of, 192; on weak linear inequalities, 128

α-core function, 11, 81n19; nonempty, 115n9

β-core function, 11, 81n19; nonempty, 115n9

corner solutions, 183

cost: adaptation of, 40–42; balance of, 51–52; of cleaner technologies, 215n12; of damage, 41, 132; of efficiency, 51–52

cost-benefit literature, 65–66

cottage industry, 77n14

countries: abatement for, 50, 52; abbreviations for, 243n16; agreements of, 124, 133, 195–196, 289; ambient pollution of, 45; Annex B, 253–255, 255n29, 255n30; authority of, 55–56; behaviors by, 46–47, 239–240; on climate change, 281–282; coalitions of, 199–200; common sense of, 58; compensation to, 126; consumption goods of, 51–52; control over, 45; cooperation of, 297–298; on γ-core, 125; damages of, 212; domestic optimum of, 238; emissions in, 46, 46n4, 239; environment of, 48; evaluations of, 182–183; on fair outcomes, 247; heterogeneous sovereign, 52–53, 291–292; incentives for, 196, 218–219; industrialization of, 208–9, 221;

countries (*continued*)
  intervention in, 46; limiting factors for, 207–208; maximization problem of, 220; negotiations among, 58; options for, 292–293; outside support of, 227–228; payoffs of, 189; on policy, 289–290; pollution in, 45, 280; poverty-alleviation goals of, 265; production capacity of, 214–215; production in, 46; rationality of, 289–291; rights of, 49; roles of, 126–127; sovereign, 52–53; as symmetric, 190; utilities for, 54, 57–58; world and, 50. *See also* developing countries

damage: from climate change, 213, 241–242; compensation for, 62; cost of, 41, 132; of countries, 212; domestic, 55; estimates of, 236*n*7; impact of, 231–232; marginal, 170; natural phenomenon as, 63; output compared to, 50; production function on, 35–36. *See also* environmental damage
damage functions, *29*, 35, 45, 111–112; affine, 118–119; convex, 198, 221–222, 224–225, 230, 259–260; identical, 122–124; information on, 293–294; local, 259; longevity of, 207–208; marginal, 177. *See also* linear damage functions; nonlinear damage functions
decay rate, 172
developing countries, 208–209, 254; carbon space of, 272; INDCs of, 269–271; Kyoto Protocol for, 249; Paris Agreement and, 268–272
development mechanism, 234–235
deviations: breaking up of, 138–139; chain of, 140–141; of coalitions, 69, 75, 108–110, 193; from grand coalitions, 151–152; responses to, 193; self-enforcing, 73; stability of, 159, 204–205; types of, 137–138
differential game, 219, 232*n*18

discount factor, 98, 172
discounting, 167–168
disutility function, 40–41
Doha Amendment, 256*n*32
domestic damage, 45–46
domestic equilibrium, *240*
dominant strategy, 70–71, 81, 98
donations, 68*n*6
double market failure, 55
dynamic environmental game, 181–182; Definition 7.1, 168–170; Definition 7.2, 174–175; Definition 7.3, 175; Proposition 7.2, 176–180
dynamic game: on climate change, 210–211; cooperative agreements in, 201–205; core concepts in, 297*n*4; Nash equilibrium and, 183; noncooperative solutions in, 202–203; for public goods, 201; SPNE in, 183*n*13; subgame-perfect cooperative agreement in, 193
dynamic model, 165, 167–174; of climate change, 166; flow model compared to, 166

ecological economics, 210
ecological science, 2–5, 22
ecological surplus, 56–58, 130; equilibrium and, *57*; support for, 61; utility frontier and, *57*
ecology, 4–5, 22
economic-ecological system, 56
economics, 2; analysis of, 3–4; climate science and, 294; ecological, 4–5; on externality, 3; general equilibrium, 5; institutional, 298; market, 4; non-tatonnement exchange, 263; normative, 4; public, 5; reasoning of, 3–4; as resource allocation, 5. *See also* environmental economics; welfare economics
economic theory, 233; basic framework of, 17–18; equity in, 54

economy, 133; climate change on, 265–266; γ-core of, *129, 131*; one polluter-one pollutee, *59*, 59–60, 266–268, *267*; simple world, 266–267
ecosystem, 22
efficiency: balance of, 51–52; in coalition structures, 91; complements to, 54; as condition, 122; of consumption profiles, 49–51, 108; of consumption time profiles, 195; cost of, 51–52; definition of, 53–54; in emissions, 166–172; of emission time profiles, 173, 189; equity and, 53–54; implications of, 51–52; interpretation of, 53; linear damage functions and, 187; mechanisms for, 52–53; objective of, 56; Proposition 3.1, 51–53; purpose of, 58; rights and, 53; time horizon and, 172–174; world, 49–53; world gain of, 55. *See also* energy efficiency; environmental efficiency; Pareto efficiency; world efficiency
efficiency conditions, 50
El Niño weather pattern, 132
emissions, 29–30; coalitions on, 106; constant, 182; control of, 44, 49; in countries, 46, 46n4, 239; domestic, 55; efficient, 166–172; free trade in, 251–253; global split of, 270; grandfathered, 249n19; impact of, 289; international trade in, 284–287; limiting of, 58; of Nash equilibrium, 106, 145–146; nonreduction, 249; options for, 145–146; payoffs and, 209–210; of players, 112; policy on, 39, 295–296; reduction of, 241–242, 248, 269–270; rights of, 171n4; of SO^2, 23–24; socially optimal, 223–224; source, 23; spillover effects of, 55; steady-state, 216n13; of steady state Nash equilibrium, 223–224; time profile of, 216, 260–261; trade in, 7, 200–201, 252–255, 264, 284. *See also*

reference emissions; world efficient emissions
Emissions Prediction and Policy Analysis (EPPA), 243, 248n17
emissions profile: consumption profiles and, 50–52; equilibria, 42; noncooperative, 105; optima, 43; steady-state, 167; unique efficient, 108
emissions quotas: assignment of, 247; fixed, 251n21; tradable, 246
emissions time profile, 164, 167; definition of, 168; determinants of, 171–172; efficient, 173, 189; equalities of, 173–174
emission trading, 245–247
energy efficiency, 45, 241
energy prices, 45n3; policy on, 46–47; in U.S., 240
environment: of countries, 48; degradation of, 39; value of, 4–5, 64–66
environmental agreements, 287–290; international, 44, 58
environmental damage, 163–164
environmental economics, 64–65, 210
environmental efficiency, 287–290
environmental game, 74, 296–297; analysis of, 148–149; γ-coalitional function of, 107–112; coalitional function on, 107–108; γ-core in, 107–109; dynamic version of, 104; framework of, 236–237; induced game of, 106; Nash equilibrium of, 104–105; Proposition 5.1, 104–105. *See also* dynamic environmental game
environmental games, 11
environmental policy, 46–47
environmental pollution, 18
environmental resources, 65
environmental values: existence value, 64–65; option value, 64–65; usage value, 64–65
Environment and Forestry Ministry of Indonesia, 133

EPPA. *See* Emissions Prediction and Policy
Analysis
equalities, 170, 173–174
equilibrium: asymmetric, 142, 142n9;
BAU of, 47–48; characterizing of,
46; concepts of, 44–49, 78; domestic
price, *279*; ecological surplus and,
57; international, 44; long-run,
210–211; mixed-strategy, 148, 151,
154–155; outcome of, 101, 143n12;
payoffs of, 146, 179; payoff vector of,
136–137; per-period payoff of, 97,
98n42; of repeated games, 144–147;
spontaneous, 43; in stationary
strategies, 97; strategically relevant,
101–102, 101n45; strategies of, 98–99;
sustainability of, 142–143; symmetric,
143n11, 145; types of, 210n5; utility
frontier and, *57*. *See also* autarky
equilibrium; focal-point equilibrium;
Markov-perfect equilibrium;
Nash equilibrium; noncooperative
equilibrium; spontaneous equilibria;
trading equilibrium
equilibrium coalition structure, 142–143;
characterization of, 155–156; grand
coalition as, 136–137
equilibrium emission: rise of, 182; weighted
average of, 114
equilibrium strategy: Nash equilibrium as,
178–179; subgame-perfect, 136
equilibrium theory: analysis of, 18;
environmental pollution in, 18
equity, 53; in economic theory, 54;
intergenerational, 168
European Union, 45, 156, 255–256
existence value, 64–65
exogenous upper bound, 181–182
expositional convenience, 32
externality: agents of, 20; dimensions of, 19;
economics on, 3; emitters of, 3;
flow, 3; global, 20; level of, 60n22;
markets for, 67; nature as, 63; real-life,
158n22; recipients of, 3; reciprocal, 20;
social interactions and, 61; transfer
functions and, 30–33; value of, 5.
See also environmental pollution; flow
externality; stock externality

feasible consumption profile. *See*
consumption profiles
financial transfers, 16, 16n15, 290–291
finite T, 197
finite time horizon, 14
fires, 132, 134
first-order conditions (FOCs), 44–45,
109–110, 111, 184
flow externality, 3, 6
flow models, 25–26, 166
focal-point equilibrium, 147–148
FOCs. *See* first-order conditions
Former Soviet Union (FSU), 243
fossil energy, 237
fossil fuels: burning of, 237; combustion
of, 5n6, 35; domestic price of, 239–240;
price of, 45, *240*; in U.S., 45. *See also*
polluting input
framework convention, 234
free riding, 68–69
FSU. *See* Former Soviet Union

game-theoretic analysis, 274
game-theoretic approach, 131–132
game-theoretic properties, 242–243
GDP, 249, 269–270
general equilibrium, 5
general extensive game, 297
GHGs. *See* greenhouse gases
global Pareto optimal (GPO), 173–174,
187
global pollution, 68–69
global pollution model, 80

global warming, 1, 1n1, 213; action for, 233; on agriculture, 32; GHG on, 213n9, 258n34; Paris Agreement on, 257–258

government: corrupt, 134; failure of, 55n16, 241n14; sovereign, 200; subsidies from, 239n1

GPO. *See* global Pareto optimal

grand coalition, 13–14, 79, 100–102; deviation from, 151–152; as equilibrium coalition structure, 136–137; members of, 108; as outcome, 144; stability of, 153–154; in strategic games, 87; Theorem 6.3, 153–154

greenhouse gases (GHGs), 1; accumulation of, 163; base layer, 215; concentration target of, 182; current stock of, 294; on global warming, 213n9, 258n34; rights to, 5; temperature rise on, 268

greenhouse gases (GHGs) stock, 34–35; natural decay of, 216–217; stability of, 217; transition of, 212n7

Hamiltonian, 214

heterogeneous players, 154–155, 160–161

House of Representatives, U.S., 268–269

human behavior, 4

hyperbolic discounting, 213n10

imputations: alternative, 121–122; in coalitional game, 229; core, 79; of γ-core, 115–124, 129–130, 137–138, 297; payoffs of, 129–130

INDCs. *See* intended nationally determined contributions

India, 219n14, 248n18, 253, 262, 269–272; renewable energy of, 271n39

Indonesia, 131–134

induced game: coalition structure with, 89; of environmental game, 106; Nash equilibrium with, 75–77, 88–89, 105–109; "one-player," 193–194; subgame, 192

inequalities, 97

infinite time horizon, 14, 186, 198–201

institutions, 34

intended nationally determined contributions (INDCs), 201, 220, 261; commitment for, 264–266; of developing countries, 269–271; reductions of, 219; voluntarily submitted, 218–219

interactive decision theory, 17n1

Intergovernmental Panel on Climate Change (IPCC), 182, 265

internal-external stability, 201

international agreement, 211n6

international authority, 49, 58

international equilibrium, 47–48

international market, 223–224

international optima, 49–54

International Solar Alliance of 121 countries, 263–264

international trade, 36–37; climate change and, 274–275; in emissions, 284–287; environmental agreements and, 287–290; environmental efficiency and, 288–290; free, 277, 294; in goods, 48n7, 274–275, 277–284, 281–284, 284–287; impact of, 277–278, 281; Proposition 10.3, 284–287

IPCC. *See* Intergovernmental Panel on Climate Change

Jacobian matrix, 222–223

Japan, 45; coal in, 240n13; marginal abatement costs in, 240n12

joint payoff maximization, 79

Kakutani's Fixed-Point Theorem. *See* Lemma

Kyoto Protocol, 1, 7; Annex B of, 235n5; benefits from, 255; challenges of, 234–235; commitment to, 252n22, 252n24, 292–293; for developing countries, 249; extension of, 293; failings of, 15; goals of,

Kyoto Protocol (*continued*)
249; impact of, 234–235, 255–257;
implementation of, 148n14;
interpretation of, 233–241; limitations of,
236, 295–296; main features of, 235–236;
principle of, 268–269; ratifications
to, 235n6; on reductions, 248–249;
requirements of, 262; strategy behind,
235n4; on trade, 251; on transfers, 52n12;
U.S. on, 95n35, 255–256, 271; withdraws
from, 235n6

labor, 284
labor theory of value, 210n3
Lagrange multipliers, 169, 176–177
Lagrangian, 169
Lemma (Kakutani's Fixed-Point Theorem),
71–72
Lindahl-Samuelson condition, 214
linear damage functions, 173–174, 180–181,
183; Assumption 5.1, 116–117; efficiency
and, 187; impact of, 288–289; results
with, 283–284; Theorem 5.5, 117–119;
variables of, 196–199. *See also* nonlinear
damage functions
linear functions, 23
linear productions, 202
long-run analysis, 208

macroeconomics micro-foundations,
12n12
Malaysia, 131–134
marginal abatement cost, 237–239, *240*,
252–253; equalization of, 50–51;
functions of, 45, 55, 166n2; in Japan,
240n12; in U.S., 240–241
market: competitive, 126–127, 254;
economics of, 4; for externality, 67;
failure of, 55n16
market game, 251n21
market solution, 239

Markov-perfect equilibrium (MPE), 166n2,
185–186; alternatives to, 201; reversal to,
192–193
mathematical models: abstract, 15; common
language of, 15; simulation, 15
matrix A, 23
maximin coalitional function, 79–80
maximization problem, 45
microeconomic theory, 44
Middle Ages, 68n6
models, 38–39, 211–217; additional
interpretations of, 34–36; for climate
change, 266–268; compact, *38*;
consumption goods, 36–37, 276–277;
dynamic, 207; flow, 163; institutions
in, 34; long-run, 209–210; of oligopoly,
149n15; for public good, 34, 69; single-
period, 291; static, 207; transfers and,
52n13. *See also* basic flow model;
continuous time models; dynamic
model; flow models; global pollution
model; mathematical models;
simulation models; stock models
Montreal Protocol, 52n12
MPE. *See* Markov-perfect equilibrium
MPE/SPNE payoff, 192–193

Nash equilibrium, 10–11, 297; alternative
definition of, 75–77; as central concept,
70; challenges of, 73; characterizations
of, 12–13; closed loop, 220; coalition-
proof, 68, 73–74, 105n1; conditions
for, 71, 76, 177, 239; Definition 7.3,
175; definition of, 70; of dynamic
environmental game, 175–182; dynamic
game and, 183; emissions of, 106,
145–146; of environmental games,
104–105; equilibrium concepts on,
78; as equilibrium strategy, 178–179;
existence of, 76; generality in, 177;
of induced game, 75–77, 88–89,

105–109; mixed-strategy, 98; multiple, 77; noncooperative, 179–180, 238; noncooperative equilibrium in, 179–180, 238; open-loop, 175–182, 220; partition function with, 88–89; payoffs of, 142–143; production function and, 183; Proposition 4.2, 76–77; Proposition 5.2, 106–7, 122; reference emissions and, 243; strategies of, 111, 138; strong, 68, 73–77; stronger concepts of, 72; subgame-perfect, 164, 175, 182–191; time horizon and, 181–182; unique, 105. *See also* dominant strategy; steady state Nash equilibrium

Nash program, 93

natural carbon sinks, 35

natural disaster, 131–134

natural phenomenon, 63

natural primitive framework, 69–70

natural science, 43. *See also* ecology

nature, 63

negotiating groups, 226–229

negotiations, 67–68; for aggregate emission target, 7; on climate change, 233–234, 237–238, 273; among countries, 58; international, 297–298; among players, 78; unconstrained, 75

noncooperative equilibrium, 47–48; equalization of, 50–51; as foundation, 58; as game-theoretic, 105; inefficiency of, 55, 67; international, 67–68; as Nash equilibrium, 179–180, 238; as spontaneous outcome, 55

noncooperative games, 10

noncooperative game theory, 10–13; coalition formation as, 141–143; intuitive, 108n3; stages of, 95–96; two-stage, 142

nonlinear damage functions, 181n11, 202; Assumption 5.2, 119; Lemma 5.6, 120–122; Theorem 5.7, 120–122;

Theorem 5.8, 123; Theorem 5.9, 123–124

nonpolluting good, 276–277

normative criteria, 295n3

n-player games, 148, 153–155; Lemma 6.1, 149–151; Proposition 6.2, 152

oligopoly game, 77n15, 136n2; models of, 149n15; symmetric, 137n4, 158–159

optimal control literature, 167

optimal control problem, 223n15

optimality condition, 41

option value, 64–65

ozone layer depletion, 1–2

Pareto efficiency, 49n8, 101

Paris Agreement, 1, 7; on coalition formation game, 256–257; conclusion of, 273n1; developing countries and, 268–272; flexibility of, 272; future of, 257–268, 295; on global warming, 257–258; implementation of, 216–217; interpretation of, 15, 233; long-term goals of, 258–261; provision of, 128n17; ratifying of, 257; target of, 231–232, 264–265, 265n37, 293–294; terms of, 262; on transfers, 52n12; U.S. on, 296; withdrawal from, 271–272, 296

partition function game: γ-coalitional function and, 88; finest, 96, 96n38, 101, 101n45; framework of, 135n1; induction to, 150–151; Nash equilibrium with, 88–89; nontrivial, 96–97, 96n40; on payoff, 89; primitive framework of, 38n3; representation of, 111; stages of, 93, 95–96; strategically relevant, 100; strategic game into, 88, 90–91; as superadditive, 90–91, 97, 99–100; symmetric, 92; 3-player, 156–157

payoff functions, 73–74, 77n16, 130, 203–204

payoff matrix, *99*

payoffs: after-trade, 126, 224–225; of
    coalitions, 75–80, 89, 89n25, 136,
    227–228; emissions and, 209–210; of
    equilibrium, 146, 179; of imputations,
    129–130; of players, 109; of SPNE, 190; in
    strategic games, 69–70; trade, 126
payoff sharing game: definition of, 94;
    equal, 94; monotonic, 94; outcome of,
    101–102; proof of, 97–102; rules of, 93
payoff vector, 96
per-period payoffs, 96–98
Pigouvian argument, 6–7
players, 95n36, 99–100; agreements of,
    182–183; commitments of, 158n21; in
    complement, 110n5; contribution
    from, 193n16; cooperation of, 179–180;
    emissions of, 112; ex-ante symmetric,
    154–155; identical, 123, 137, 160–161;
    negotiating among, 78; options of,
    136, 142n8; payoffs of, 109; strategies
    of, 109, 147–148, 150, 175, 192. See also
    heterogeneous players; n-player games
policy: on climate, 232; climate change,
    209n2; considerations of, 15; countries
    on, 289–290; on emissions, 39, 295–296;
    on energy price, 46–47; energy tax, 239;
    macroeconomic, 9; responses from, 1; on
    trade, 274, 287. See also environmental
    policy
"polluters pay" principle, 62
polluting good, 276–277
polluting input (fossil fuels), 30
pollution, 1; air, 2–3; Coase theorem on,
    59–60; compensation and, 52n12; in
    countries, 45, 280; debates on, 58–59;
    exporting of, 281–282; flow, 25; from
    fuel types, 30; global levels of, 36–37;
    international, 124n13; international
    authority on, 58; permits for, 239,
    275–276, 294; from production, 45;
    production intensity of, 284;
    reduction of, 119, 124–125; stock, 26;
    upstream, 3n3; utility functions with,
    59–60. See also ambient pollution;
    environmental pollution; global
    pollution
pollution factor, 211–212
"pollution havens," 284, 287
pollution intensity, 211–212
pollution rights, 49, 53, 58–62, 223;
    acceptability of, 63–64; allocation of,
    63; assignment of, 64, 67–68; authority
    on, 124–125; as binding, 228; climate
    change and, 218–223; Coase theorem
    on, 125; competitive markets in,
    126–127, 254; enforcement of, 63, 67–68,
    124–125; exchange of, 225; long-run,
    209; normative criterion of, 208–209;
    self-enforcing, 219; steady state Nash
    equilibrium as, 221–222; subgame-
    perfect cooperative agreements of, 199;
    tradable, 126; well-defined, 67
positive discount rate, 144n13
positive integers, 96n39, 97–98
preferences, 27–29
private goods, 5, 40n22; composite, 165, 167;
    transfer of, 195
production activities, 29–30; in countries,
    46; pollution from, 45
production function, 30, 111–112; aggregate,
    31, 44–45; assumptions of, 165–166, 277;
    Cobb-Douglas, 278; concave, 236–237;
    on damage, 35–36; as identical, 186;
    longevity of, 207–208; Nash equilibrium
    and, 183
production maximization, 44–45
production technology, 207-8, 211–212
profit maximization, 44–45
profit-maximizing firms, 46
property rights, 4; barrier of, 263;
    intellectual, 15
protection, 39–42

Protocol. *See* Kyoto Protocol
Protocol by the United States, 235
public firms, 45
public good, 5, 7, 136n2, 149n15; ambient
    pollution and, 50; dynamic games
    for, 201; efficient provision of, 50n9;
    incentive mechanisms for, 9; models for,
    34, 69; optimal provision of, 214
punishments, 193n16

quasi-concave utility functions, 57n20

reduced-form game, 204
reduced game, 147
reference emissions, 238–242; agreement
    on, 248–249; BAU, 248; equity principles
    for, 249; Nash equilibrium and, 243
regional haze, 103
repeated game, 144–145; equilibrium of,
    146–147; infinitely, 94–102, 292; reduced
    form of, 145; Theorem 4.6, 97–102
resources: for climate change, 232;
    constraints of, 44; forest, 133; land, 133;
    scarcity of, 9n11. *See also* environmental
    resources
rights: allocation of, 124, 223–224;
    assignment of, 125–126; auctioning for,
    62; of countries, 49; efficiency and, 53; of
    emissions, 171n4; to GHGs, 5; objective
    criterion for, 61–62; self-enforcing, 6;
    trade in, 64, 225–227. *See also* pollution
    rights
Rio Convention, 234n2
rules: equal split, 160–161; tie-breaking,
    144n13; time consistent, 249

seasonal haze, 131–134. *See also* Southeast
    Asia
Senate, U.S., 268–269
Shapley value, 92–93
side payments, 158n20

simulation models, 182n12
Singapore, 131–134
singletons, 136, 150; breaking up, 153–154;
    as coalitions, 138–139; non-, 99–100,
    100n44, 110
SO$_2$. *See* sulfur dioxide
social criteria, 43
social interactions, 61
socially optimal steady states, 213–217
social optimality, 232, 265, 295n3
solution concept, 77
solutions, 231–232
Southeast Asia, 103, 131–134
SPNE. *See* subgame-perfect Nash equilibrium
spontaneous equilibria: conditions of,
    54–55; inefficiency of, 54–56; proof for,
    54–55
spontaneous equilibrium, Proposition 3.2,
    54–56
stability concepts, 153n16, 159n24
steady state Nash equilibrium, 219–220;
    emissions of, 223–224; as pollution
    rights, 221–222; stability of, 222–223
stock accumulation equation, 167
stock externality, 3, 6, 21; climate change as,
    64; decay of, 35
stock models, 25
strategic form games, 197
strategic games: of coalitional function,
    78; coalitional game form of, 78–79;
    compact, 77n13; γ-core as, 68, 103; core
    of, 67–68; equilibrium concepts in,
    70–77; general, 296–297; grand coalition
    in, 87; as natural primitive framework,
    69–70; into partition function game, 88,
    90–91; payoff in, 69–70; power structure
    of, 78; Proposition 4.1, 72. *See also*
    environmental game
strategy profile, 73, 159–160
subgame-perfect agreement, 14, 14n14, 164,
    171, 178, 203

subgame-perfect cooperative agreement, 195, 202–205; climate agreements compared to, 200; conditions for, 196; Definition 7.4, 193–194; interpretations of, 199–201; as pollution rights, 199; stability of, 192; Theorem 7.5, 190–193; Theorem 7.6, 196–198

subgame-perfect equilibrium, 157, 188–189

subgame-perfect Nash equilibrium (SPNE), 166, 297; in dynamic games, 183n13; dynamic programming of, 185n14; infinite time horizon and, 186; as noncooperative solution, 185; payoff of, 190; properties of, 185–186; stationary strategies of, 198–199; Theorem 7.3, 183–186

sulfur dioxide (SO_2), 23–24

superadditive, 87; alternative to, 90–91; benefits of, 112; γ-coalitional function as, 90, 111–112; in coalitional games, 87, 205; partial, 91; partition function as, 90–91

supply and demand, 254

supranational authority, 6–8, 125

symmetric games, 91

tatonnement process, 225, 225n16

taxes, 6–7

technology, 201; cleaner, 263–264, 296; green, 202

temperature rise, 218, 268

terminal stock, 171–172

Thailand, 132

thermodynamic theory of energy, 4

time, 165; as continuous, 211–212; profiles of, 164, 216–217. See also consumption time profiles; emissions time profile

time horizon, 164–165; efficiency and, 172–174; infinite, 172–174, 181–182; Nash equilibrium and, 181–182

trade: under CDM, 253–254; in emissions, 7, 200–201, 252–255, 264, 284; forms of,

262–263; free, 276; Kyoto Protocol on, 251; nature of, 252; payoffs of, 126; policy on, 274, 287; in rights, 64, 225–227. See also international trade

trading equilibrium, 249–251

transferable utility, 69, 76, 168–169

transfer function, 2–3, 21–26, 30–33

transfer matrix, 24–25

transfers: appropriately designed, 238; benefits of, 52, 52n13, 188–189; on emissions reduction, 242; "front loading," 200; as individually rational, 242n15; Kyoto Protocol on, 52n12; models and, 52n13; necessity of, 187–190, 241–257, 242n15, 291–292; Paris Agreement on, 52n12; of private goods, 195; profile of, 200; Proposition 7.4, 187–190; schemes of, 242–243, 245; value of, 53n14

transport matrix, 24

two-dimensional commodities, 19–20

two-stage game, 142–143

U.N. See United Nations

UNFCCC. See United Nations Framework Convention on Climate Change

United Nations (U.N.), 234n2, 261

United Nations Framework Convention on Climate Change (UNFCCC), 215, 234n1, 260

United Nations institutions, 233–234

United Nations Sustainable Programme, 273n2

United States (U.S.), 248; on climate change, 156; energy prices in, 240; fossil fuels in, 45; on Kyoto Protocol, 95n35, 255–256, 271; marginal abatement costs in, 240–241; on Paris Agreement, 296; withdrawal of, 271–272, 296

U.S. See United States

usage value, 64–65

utility frontier, *57*, 129–130
utility functions, 36–37; assumptions of, 165–166; with pollution, 59–60; as quasi-linear, 130
utility maximization, 40, 44, 46, 283

variables: for basic flow models, 32; control, 167; current-value adjoint, 220; of linear damage functions, 196–199; state, 167
vector-valued function, 22–23
"victim pays" principle, 63
voluntary pledges, 218

war. *See* attrition war
weights, 112–113, 230–231
welfare economics, 5, 215
welfare weights, 173*n*6, 188–189
willingness to accept, 65–66
willingness to pay, 65–66, *66*
world: countries and, 50; efficiency gain of, 55; income of, 283; temperature of, 22
world efficiency, 49–53
world efficient emissions, 237–238, 237*n*9, 246
World Trade Organization (WTO), 8, 272–273